THE TOTAL SKYWATCHER'S MANUAL

VISTA GRANDE
PUBLIC LIBRARY
T 164466
ASTRONOMICAL SOCIETY OF THE PACIFIC
THE TOTAL
SKYWATCHER'S
MANUAL
275+
SKILLS AND TRICKS FOR EXPLORING
STARS, PLANETS & BEYOND
Linda Shore, David Prosper &
of the Astronomical Society of the
weldonowen

CONTENTS

NAKED-EYE ASTRONOMY TIPS

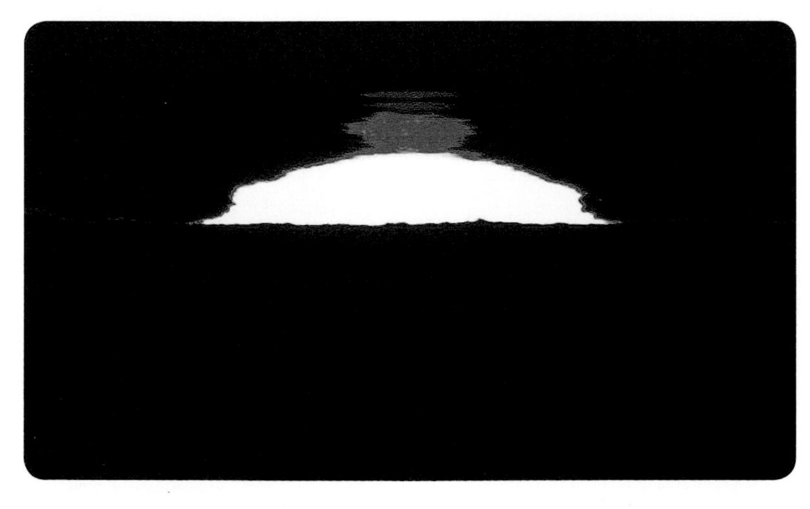

CONTENTS

CONTENTS

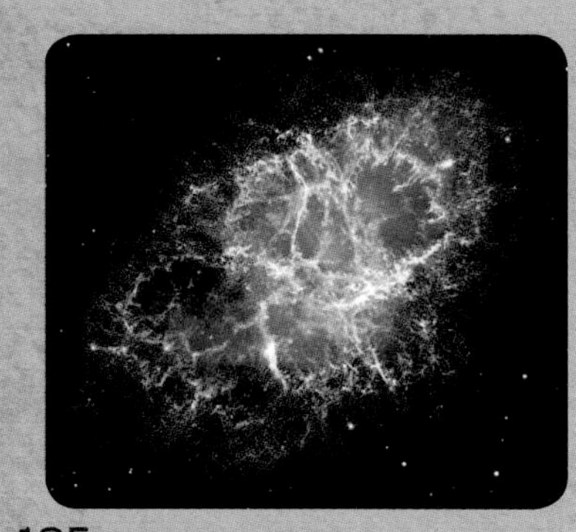

WELCOME TO THE COSMIC LABORATORY

A crystal-clear, pitch-dark, starry night is astonishing. The sight of thousands of stars and the panorama of our own Milky Way Galaxy arching across the sky inspires philosophers, poets, lovers, and scientists. The night sky is also profoundly democratic. It is a science laboratory equally accessible to everyone on Earth who is interested in observing, studying, and exploring. The science of astronomy was developed by virtually every civilization on the planet as a systematic way to decode the very nature of the cosmos. First with unaided eyes, then with simple optical telescopes, and now with sophisticated instruments engineered to look into the very farthest reaches of space and time, we continue to investigate our fascinatingly intricate universe.

The purpose of this book—and the mission of the Astronomical Society of the Pacific—is to help astronomy enthusiasts of all kinds engage with and appreciate the extraordinary laboratory unfolding above your head. Maybe you have a casual interest in astronomy and want to learn more. Perhaps you are interested in buying your first telescope or trying your hand at astrophotography. You might be an experienced amateur astronomer wanting some new ideas. No matter what your prior experience with astronomy, this book will be a resource for you to enjoy at every stage.

You are about to embark on an incredible journey through the universe, and we are thrilled to be your guides. Within the pages of this book, you'll find everything you need to know to study stars, planets, comets, asteroids, nebulae, and galaxies. We start with showing you how to explore the universe with your unaided eyes. Next we introduce you to binoculars and telescopes—tools that give you access to what eyes alone won't reveal. Finally, we open your eyes to some advanced observational techniques that you can try, and we show you the tools that modern astronomers use in their research.

Welcome to your universe. Come on in!

Linda Shore, EdD
Executive Director
Astronomical Society of the Pacific
www.astrosociety.org

NAKED-EYE ASTRONOMY TIPS

01 MEET THE UNIVERSE

The universe is all of space and everything in it—from the smallest speck of subatomic particle to the biggest galaxy, from the planets to the creatures who live (or might live) on them, and all matter and energy. But all this quite literally came from nothing in an event we know as the Big Bang. When the universe exploded into existence, it was infinitely dense, infinitely hot, and infinitely small. In less than a second, the universe had ballooned to the size of a grapefruit and continued to grow rapidly, but at first the temperature was too hot to allow even atoms to exist. Within this first second, the temperature dropped below 18 billion°F (10 billion°C), making it cool enough for protons and neutrons to form. Within a few minutes, these particles combined to make the first hydrogen and helium nuclei. It would take 300,000 years for the universe to cool down enough to allow these nuclei to capture electrons, forming the first atoms.

The first stars and galaxies formed about 1 billion years after the Big Bang. Our solar system is a relative newcomer to the universe, having formed about 8 billion years after the Big Bang. It is here, on the third planet from the Sun, nestled within the Milky Way Galaxy, deep within the Local Group inside the Virgo Cluster, tucked away in the Local Supercluster, that we make our home in the *observable universe*—everything we can see with our eyes, telescopes, and other instruments. Using modern orbiting telescopes like the Hubble (see #262), astronomers have discovered that the universe contains hundreds of billions of galaxies, arranged in large groups and giant clusters, and that within these galaxies are untold scores of stars, planets, asteroids, comets, and other space objects, all ripe for discovery, observation, and study.

LAYERS OF THE UNIVERSE	
①	EARTH
②	SOLAR SYSTEM
③	SOLAR INTERSTELLAR NEIGHBORHOOD
④	MILKY WAY GALAXY
⑤	LOCAL GROUP
⑥	VIRGO SUPERCLUSTER
⑦	LOCAL SUPERCLUSTER
⑧	OBSERVABLE UNIVERSE

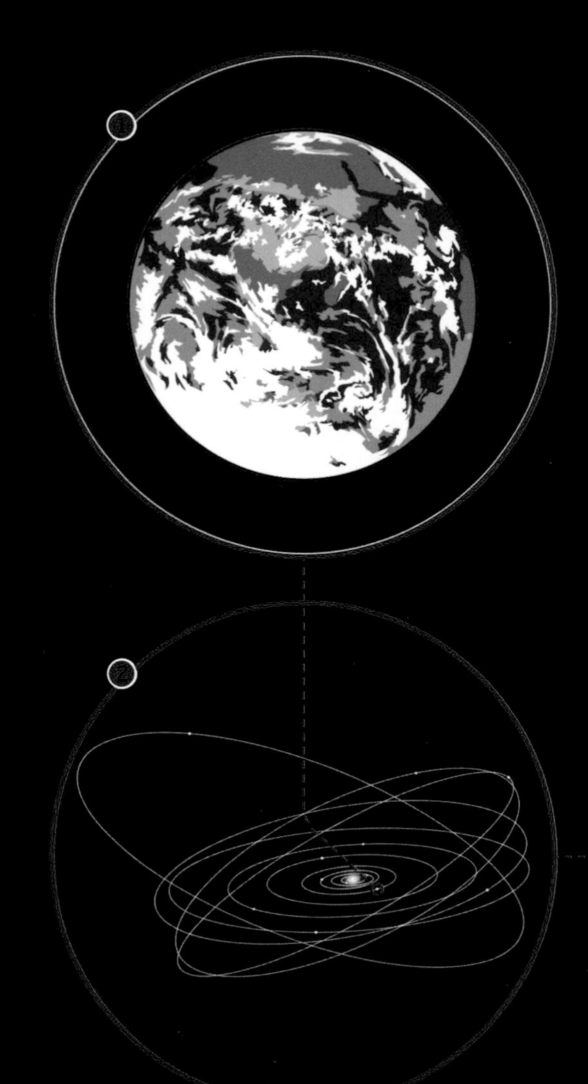

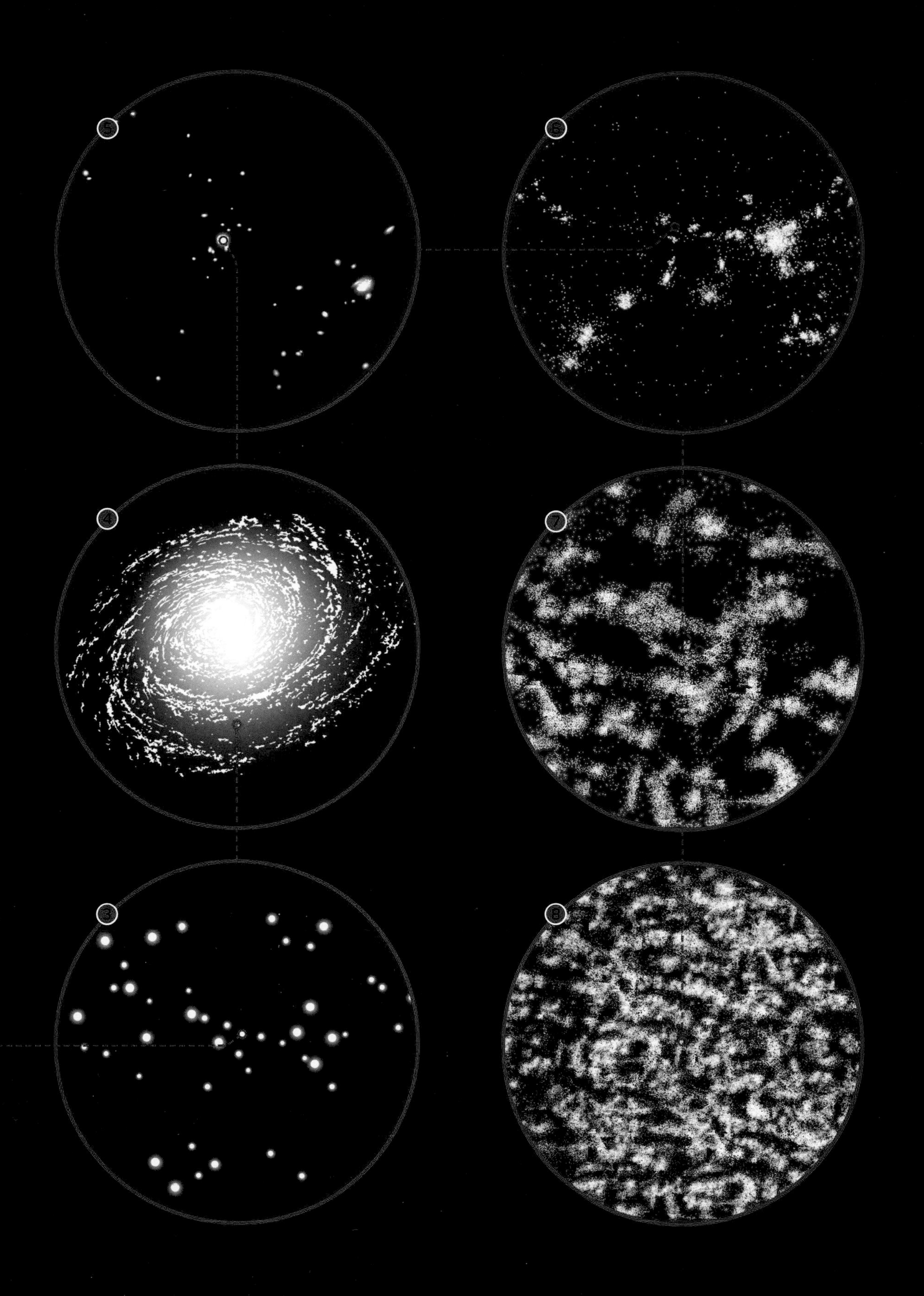
5
6
4
7
3
8

02 DISSECT A STAR

We know the twinkling lights scattered across the night sky as stars. Each of those shimmering beauties is a gigantic ball of mostly hydrogen and helium atoms, all held together by the force of gravity. There is so much mass inside a star that its gravitational forces are powerful enough to raise its core temperature above 27 million°F (15 million°C). At such high temperatures, hydrogen nuclei move incredibly quickly, colliding to create helium nuclei. This process is called *nuclear fusion,* and it releases huge amounts of energy. While a lot of that energy creates the forces that push against gravity, keeping the massive star from collapsing, the rest of it escapes the star in the form of light and heat.

The closest star to Earth is our Sun. Almost every star in the sky is larger than it. But there are also countless smaller stars that are too dim to see. The largest stars, like those of multistar system Eta Carinae (see #13 and #47), are more than 100 times more massive than the Sun and emit more than 5 million times more energy. Large, massive stars have cores that are hot enough to fuse heavier elements into carbon, oxygen, and nitrogen.

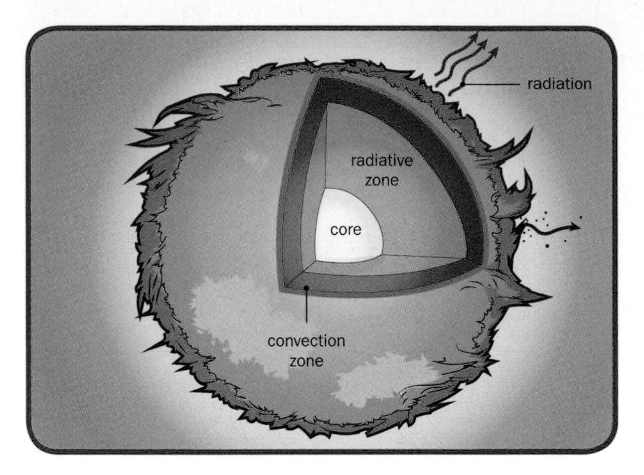

The inside of stars resembles an onion, with different layers responsible for carrying energy from the core into space. Average-size stars have three layers: the *core* (where nuclear reactions make energy), a *radiative zone* (a region of gas where light from the star's core passes randomly between atoms like a game of hot potato), and a *convection layer* (where the star's gases act like water boiling turbulently in a pot). Really big stars don't have a convection zone and transport energy through a huge radiation zone. Tiny stars have only convection layers.

03 LEARN STELLAR CLASSIFICATION

Astronomers have many ways to classify stars. One way is using the Morgan-Keenan (MK) classification, a system based on temperature and *luminosity*—the amount of energy a star emits at a given time. Here's how stars are described using this system.

MAIN-SEQUENCE STARS These are the vast majority of stars in the universe—including our Sun. They have cores hot enough to fuse hydrogen atoms together through nuclear reactions, creating helium atoms and a whole lot of energy. Most of a star's lifespan is spent as a main-sequence star, but these stars come in all sizes and colors: There are yellow main-sequence stars with luminosity and size comparable to our Sun, and there are red, very small, cool, and dim main-sequence stars (called red dwarfs), which are the most abundant in our universe and live longest.

GIANTS A giant star has a large diameter and relatively cool surface compared to a main-sequence star, which results after a star has ceased to burn hydrogen. Instead, the core of a red giant contains helium fusing into carbon. The star Arcturus (Alpha Boötis) in the constellation Boötes is one example (see #30).

SUPERGIANTS Among the largest and most luminous stars are the supergiants, having masses around 10 or more times that of our Sun. There are plenty of supergiants that you can see without a telescope, such as Deneb (Alpha Cygni) in Cygnus or Rigel (Beta Orionis) in Orion. When these massive stars die, they become *supernovas* (immensely bright explosions, followed by star death) and then sometimes *black holes* (invisible areas in outer space with gravity so strong that light cannot escape them) or *neutron stars* (stellar remnants).

VARIABLE STARS Some stars defy categorization. From Earth, *variable stars* seem to fluctuate in brightness. This is due to the shrinking or expanding of a star or to another celestial body eclipsing it. *Cepheid variables* pulsate wildly, giving scientists insight into stars' behavior and their distance from our planet.

04 DISCOVER STAR GROUPS

There's often more than meets the eye when it comes to viewing the heavens—and in some cases, that "more" means more stars than you can make out! Often, stars like to hang out in groups, such as:

A DOUBLE STARS Some stars are not single points of brightness—they're multistar systems that our eyes see as one. In general, there are two types: *optical doubles,* which appear to be one star simply due to our perspective on Earth, and *binary stars,* two stars that are so close they are gravitationally bound and orbit each other. The brightest star in our galaxy, Sirius (Alpha Canis Majoris), belongs to a binary star system. In naming conventions, the brightest star in a system is preceded by "Alpha," the second-brightest by "Beta," and so on.

B OPEN CLUSTERS These families contain thousands of young stars that all came into being at the same time in the same huge molecular cloud. They're bound loosely to each other by weak gravitational forces, but they will eventually drift apart. A majorly famous open cluster that you can see with your naked eyes is the Pleiades (M45) in Taurus (see #58).

C GLOBULAR CLUSTERS Appearing as dense spheres of stars in the sky, globular clusters are kept together longer than open clusters by mighty gravitational forces. Omega Centauri (NGC 5139) is the largest known globular cluster. Hercules's M13 (NGC 6205) is so prominent it was mistaken for a single star back in antiquity.

05 SPOT THE CLOSEST STARS

Besides the Sun, Earth's closest twinkling neighbor is Proxima Centauri, found within the triple-star system Rigel Kentaurus (Alpha Centauri) in Centaurus, a mere 4.24 light years away. Here are more stars in our 'hood—some of which you can see with your naked eyes.

STAR	Distance
SUN	7.5 light seconds
PROXIMA CENTAURI	4.2 light years
ALPHA CENTAURI A & B	4.3 light years
BARNARD'S STAR	6.0 light years
WOLF 359	7.8 light years
LALANDE 21185	8.3 light years
LUYTEN 726-8 A & B	8.4 light years
SIRIUS A & B	8.6 light years
ROSS 154	9.4 light years
ROSS 248	10.3 light years
EPSILON ERIDANI	10.5 light years

06

TOP FIVE
RECOGNIZE IDEAL SKYWATCHING CONDITIONS

Discover the best viewing conditions for skywatching—you'll have more fun the more you can see!

☐ **CHECK THE WEATHER** Cloud cover, haze, fog, dust, and smog all will obscure your view. Even if the night sky is clear, high humidity can affect your visibility and rapidly turn to fog and haze. If you live in a foggy or wet area, monitor the weather closely so you can take advantage of rare nights of good stargazing.

☐ **ENLIST TECHNOLOGY** Available on the web and as mobile apps, Clear Sky Chart is an essential astronomical forecast that alerts you to upcoming clear, dark, and ideal skies. Download it and never doubt! Also, given Earth's orbit around the Sun, you'll see different stars in different seasons. It helps to download a seasonal star chart app so you can plan excursions at optimum times.

☐ **MONITOR THE MOON** It's also a good idea to keep an eye on the lunar phases so you can avoid observing near a full Moon, which will wash out faint objects on an otherwise perfect star-seeking night.

☐ **AVOID LIGHT POLLUTION** Stay away from streetlights, car headlights, and other sources of bright light. Even looking at your phone can ruin your eyes' dark adaption.

☐ **GET UP HIGH** Our atmosphere is full of dust and turbulence, distorting light and limiting telescopes. The less atmospheric dust and turbulence you look though—the higher in altitude you are—the better your seeing will be.

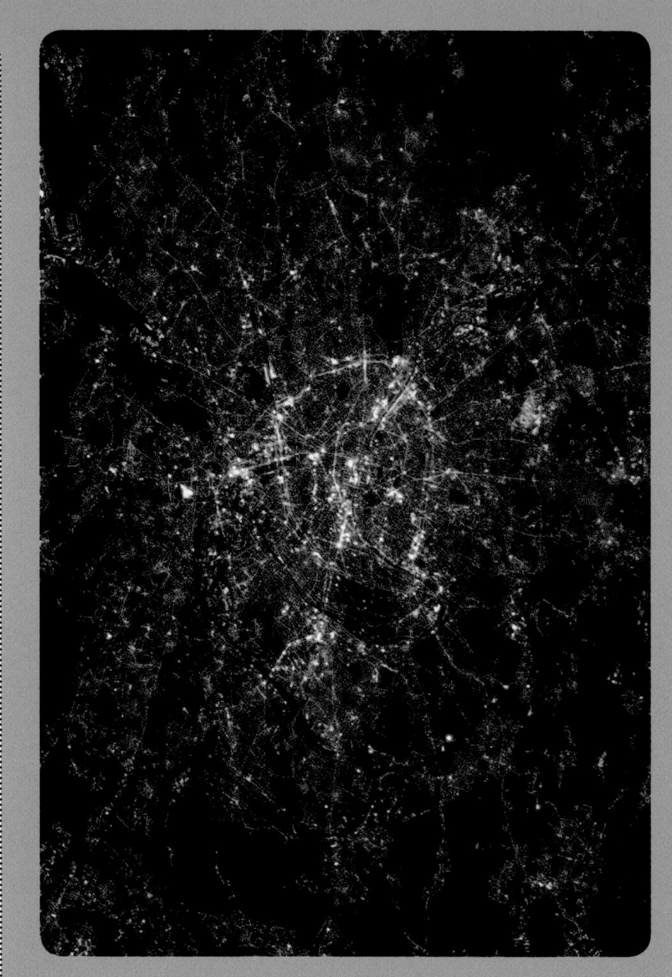

07 GO URBAN OR RURAL

The dark skies of the countryside, jam packed full of stars, can be very confusing for beginners. An odd upside to city light pollution is that the brightest stars are very prominent. Light pollution blocks many stars, the Milky Way, and all but the brightest meteors from city viewers, and, since the brightest stars make up well-known constellations, light pollution makes those constellations easier to find and identify. The lack of background stars also makes it less difficult to find planets.

Meanwhile, viewing the stars in rural areas can be a life-changing experience. Depending on the season, the Milky Way may stretch across the sky, visible as a luminescent cloud. Stars may sprinkle the entire sky, while meteor showers can put on an incredible show. You might even spot distant galaxies and nebulae, such as the Andromeda Galaxy (M31/NGC 224) and the Orion Nebula (M42/NGC 1976).

08 PACK A BEGINNER SKYWATCHING KIT

While you can skywatch with just your eyes, it helps to stock basic supplies. Follow these suggestions and you will be perfectly prepared for observing the night sky.

☐ **BLANKET** As the night progresses and the temperature drops, you'll be glad you have a blanket. It also helps keep you dry when lying on the ground—great for watching meteor showers.

☐ **LAWN CHAIR** Stay comfortable. Adjustable lawn chairs—or even better, reclining ones!—are perfect for observing the sky and keep your neck safe from strain.

☐ **RED LIGHT** Normal flashlights will destroy your ability to see in the dark. As movie ushers learned, red lights are great for helping you see while preserving your night vision. Get one or make your own (see #10).

☐ **DRINKS AND SNACKS** Water will keep you hydrated, while coffee or tea in a thermos can help you stay awake. Also, everyone loves snacks.

☐ **STAR CHARTS** Print out a star chart or bring a planisphere that's specific to your area. (Also see #23–24 for our planisphere templates.) If you are observing in an area with a lot of light pollution, try using a stargazing app on your mobile device.

☐ **EXTRA CLOTHES** Even in warm climates, the night can get chilly, so bring a hat, a scarf, and gloves. Heat packs for your hands and feet are also not a bad idea.

09 OPTIMIZE YOUR VISION

Whether you're embarking on your very first stargazing adventure or have grown frustrated trying to identify constellations within the jumble of stars above, these handy tips will help you have a successful evening of skywatching.

LET YOUR EYES ADAPT Human eyes are sensitive light detectors. Your retina is lined with two types of receptors (*cones* and *rods*) that activate when light enters the eye. While cones respond to color, rods respond to light intensity and help you see the dimmest stars. Rods become more sensitive in the darkness, but they take 20 to 30 minutes to fully adapt. Use the time in the dark to relax or chat with friends.

USE YOUR PERIPHERAL VISION To see faint objects, try *averted vision*: Don't look directly at them but slightly to one side. When you look directly at objects, you use mostly your color receptors (cones), which are terrible for detecting dim objects. Rods, on the other hand, are the receptors located in your peripheral vision that help your brain see faint objects. In using peripheral vision, you are engaging your light-sensitive rods.

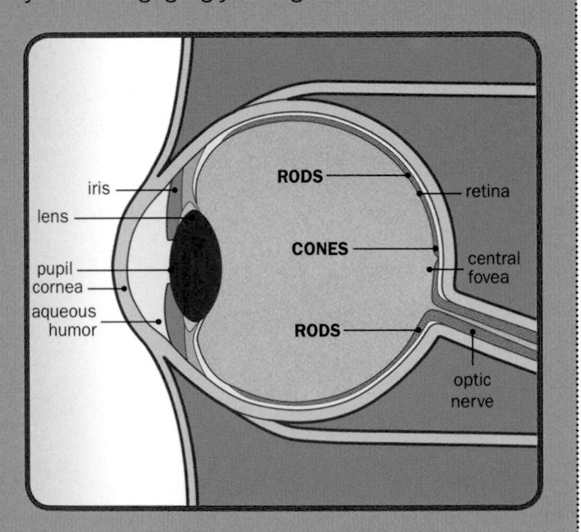

10 IMPROVISE A RED FLASHLIGHT

After you've taken the time to find a prime seeing spot, haul and set up all your gear, and sit in the dark long enough to let your eyes adjust, it'd be a shame to knock your night vision back to square one with a regular flashlight. Thankfully, there is a happy medium between ruining your ability to see in the dark and being unable to read your star chart: red lights.

While you could order a special red flashlight online—or a red filter for a flashlight you already own—you can easily make your own with some basic materials. If you have some red nail polish and are okay with permanently modifying the lens on a flashlight, give both sides of the lens a few coats and let it dry. If you're looking for something more reversible, try these steps.

STEP ONE Get an older flashlight—the kind with a bulb, not some high-powered LED light—and some red plastic wrap. (A red balloon or even tape used to cover brake lights works well, too.)

STEP TWO Pull out the plastic wrap from the roll until its width is enough to cover the whole front of your flashlight, with 1 inch (2.5 cm) or so to spare on either side. Fold this section back until it's about 1/8 inch (3 mm) thick. Tear the whole thing off from the roll. Then fold it in half. This will be your filter.

STEP THREE Hold the filter over the whole lens on the front of your flashlight. Secure it in place with a couple of rubber bands, some string, or good old-fashioned duct tape.

STEP FOUR Test it! In the dark, turn on your newly red light. Make sure there aren't any spots where white light leaks out. Add more film if it's too bright.

11 NOTICE THE BRIGHTNESS AND COLOR OF STARS

Like any lightsource, a star's brightness (or *magnitude*) depends on its size, temperature, and distance from us. There are two types of magnitude: *intrinsic,* which measures how bright stars would appear if they were all the same distance from us, and *apparent* (how bright they look to us at their actual distance from Earth). The brightest stars visible to the naked eye are said to be of magnitude 1, while the faintest fall around 6. Binoculars can show you down to magnitude 10, while a 10-inch (25-cm) telescope can see down to 14 or so. And the scale goes brighter than 0: At its brightest, the planet Venus is of –4.5 magnitude. Knowing the magnitudes of the stars in a constellation will help you gauge which ones to look for first and which may be invisible to you.

But what about color? At first glance, all stars appear white. But once your eyes have adapted, you'll notice some are slightly reddish, orange, yellow, or pale blue. Star colors correspond to their surface temperatures, which we label using the Harvard Spectral Classification system (O, B, A, F, G, K, and M), shown here. The very hottest stars are a pale-blue color. In descending order of surface temperature, stars are white, yellow, orange, and red.

COLOR	SURFACE TEMPERATURE (IN KELVINS)	CLASSIFICATION
BLUE	More than 30, 000	O
BLUE-WHITE	10, 000–30,000	B
BLUE-WHITE TO WHITE	7,500–10,000	A
WHITE	6, 000–7,500	F
YELLOW-WHITE	5,200–6,000	G
YELLOW-ORANGE	3,700–5,200	K
ORANGE-RED	Less than 3,700	M

12 NAVIGATE THE NORTHERN CELESTIAL SPHERE

If you stood at the North Pole, the heavens above would match our chart here, with the constellations shown appearing most visible in the months indicated on the chart's outer ring. Everywhere else, you would see only parts of this sky each evening, and they would change at different times of the year. If you were to travel toward the equator, you'd see constellations near the northern horizon vanish, while new ones became visible on the southern horizon.

But while you're here, let's see the northern hemisphere's most striking sights:

THE MILKY WAY Look for a hazy band of distant starlight arcing across the night sky. This is the edge-on view of our Milky Way Galaxy, as seen from our location in the outskirts of one of the galaxy's spiral arms.

BETELGEUSE Also called Alpha Orionis, this red giant (at Orion's shoulder) could go supernova soon (see #201). When it does, it will look like a bright light in the daytime sky.

VEGA When stars are measured for brightness, it is in comparison to venerable Vega (Alpha Lyrae), whose magnitude is 0. Our whole solar system is moving in the direction of Vega as we make our way around the Milky Way.

BEEHIVE CLUSTER One of the largest clusters of nearby new stars, this compact swarm is a joy to see in dark skies. In ancient times, the Beehive Cluster (M44/NGC 2632) was used to predict weather. If it could not be seen in a clear sky, a storm was brewing.

ANDROMEDA GALAXY Look inside the constellation Andromeda. You'll find a very faint, fuzzy oval near the Milky Way: the Andromeda Galaxy (M31/NGC 224). It's the largest galaxy in the Local Group.

POLARIS It's not the brightest star in the sky or part of any huge constellation, but the North Star (or Alpha Ursae Minoris) gave our northern hemisphere ancestors a leg up in navigating (see #36).

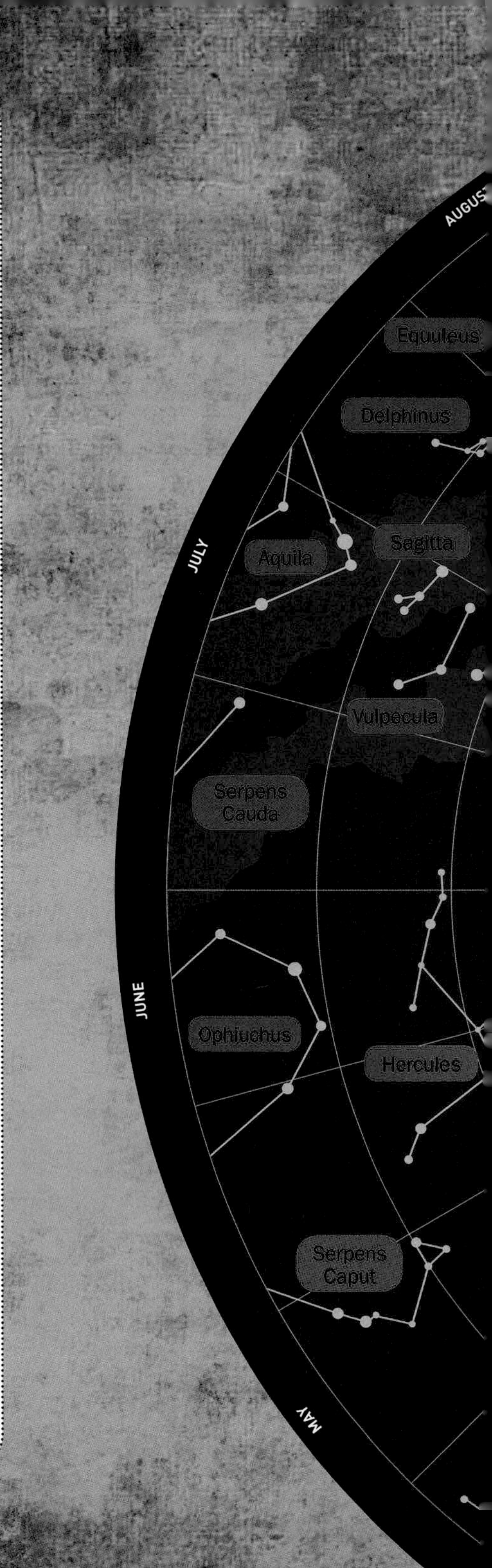

SEPTEMBER
OCTOBER
NOVEMBER
DECEMBER
JANUARY
FEBRUARY
MARCH
APRIL
Pegasus
Lacerta
Andromeda Galaxy
Milky Way Galaxy
Cassiopeia
Andromeda
Perseus
Taurus
Cygnus
Lyra
Vega
Cepheus
Camelopardalis
Auriga
Orion
Polaris
Betelgeuse
Ursa Minor
Gemini
Draco
Canis Minor
Corona Borealis
Lynx
Ursa Major
Beehive Cluster
Leo Minor
Cancer
Canes Venatici
Boötes
Coma Berenices
Hydra
Leo
Virgo

13 SEE THE SIGHTS IN THE SOUTHERN HEMISPHERE

Astronomer Bart Bok, namesake of Bok Globules (the nebulae found in large numbers in the southern hemisphere), once said, "All the good stuff is in the southern hemisphere!" While we're not ones to choose sides, it's a sentiment echoed by most amateur astronomers. Here are highlights, which would appear exactly as they do here if you stood at the South Pole:

CANOPUS A blue supergiant star about 300 light years away, Canopus (Alpha Carinae) is so bright that it is often the default guide star for tracking systems on space probes. It is fairly near Sirius (Alpha Canis Majoris) in our sky. First, find Canis Major (see #59), Sirius's home. Look for a second bright star above it; that's Canopus.

MAGELLANIC CLOUDS On a dark night, you may be able to see what look to be two glowing clouds in the sky. These aren't clouds; they are satellite galaxies to our own Milky Way Galaxy, named after Magellan, the famous global navigator whose crew spotted them on their voyage around the world. To find them, follow a path starting from Sirius down through Canopus. Keep following that line until you arrive near the Large Magellanic Cloud. The Small Magellanic Cloud is nearby.

SOUTHERN CROSS The Southern Cross is the symbol of the southern hemisphere for many. Also known as Crux, it serves as a handy guide to many of the other southern night-sky delights (see #40–43). While Crux should be very noticeable, there is also a False Cross made of two stars from Carina and two from Vela. You can always make sure you have the true cross by following its crossbar to the bright stars Hadar (Beta Centauri) and Rigel Kentaurus (Alpha Centauri).

RIGEL KENTAURUS Check out the nearest star system to our own: Rigel Kentaurus (Alpha Centauri), which is actually three stars in orbit (some of which have planets orbiting them). While this star system's light only takes four years to reach us, travel to it is still out of reach. The fastest craft we've ever sent into space would take almost 20,000 years (more than 600 generations) to reach it.

ETA CARINAE Best seen through a telescope and found in the glowing Carina Nebula (NGC 3372), Eta Carinae are two stars orbiting each other. They flared up very brightly in the 1800s, creating the Homunculus Nebula. The larger of the two is an enormous supergiant and a candidate to go supernova soon—you may see it explode in your lifetime.

CONSTELLATION SEASONS (BEST EVENING VIEWING)

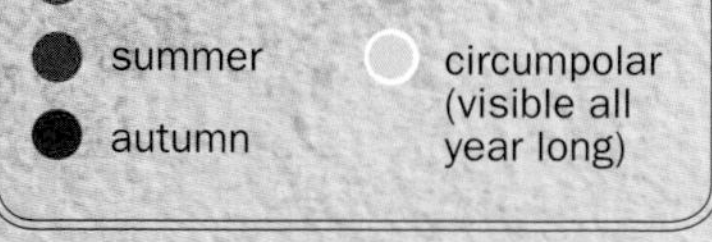

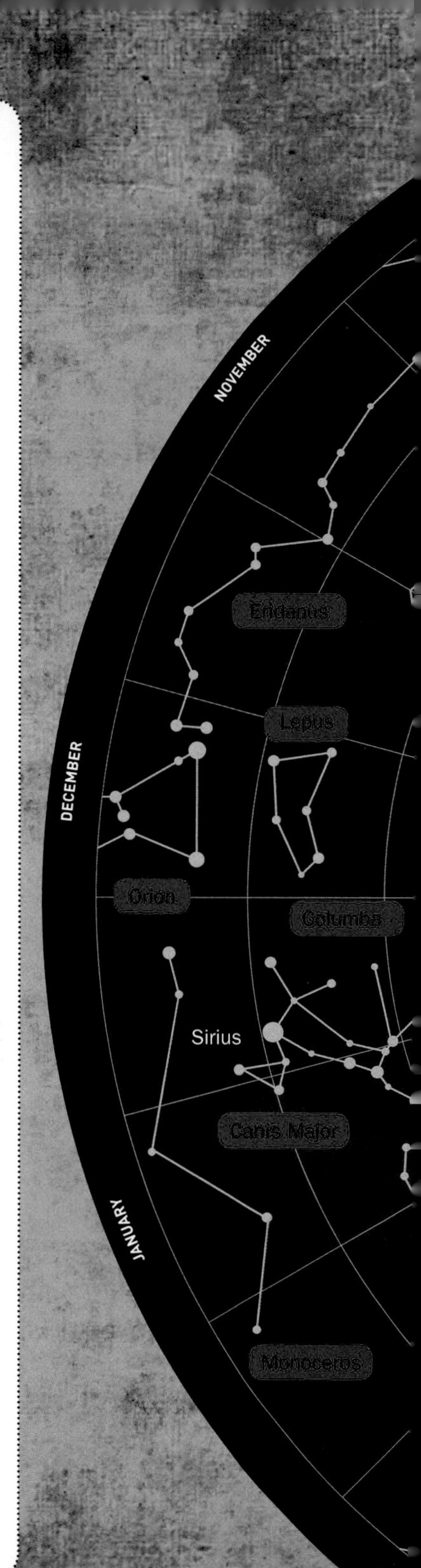

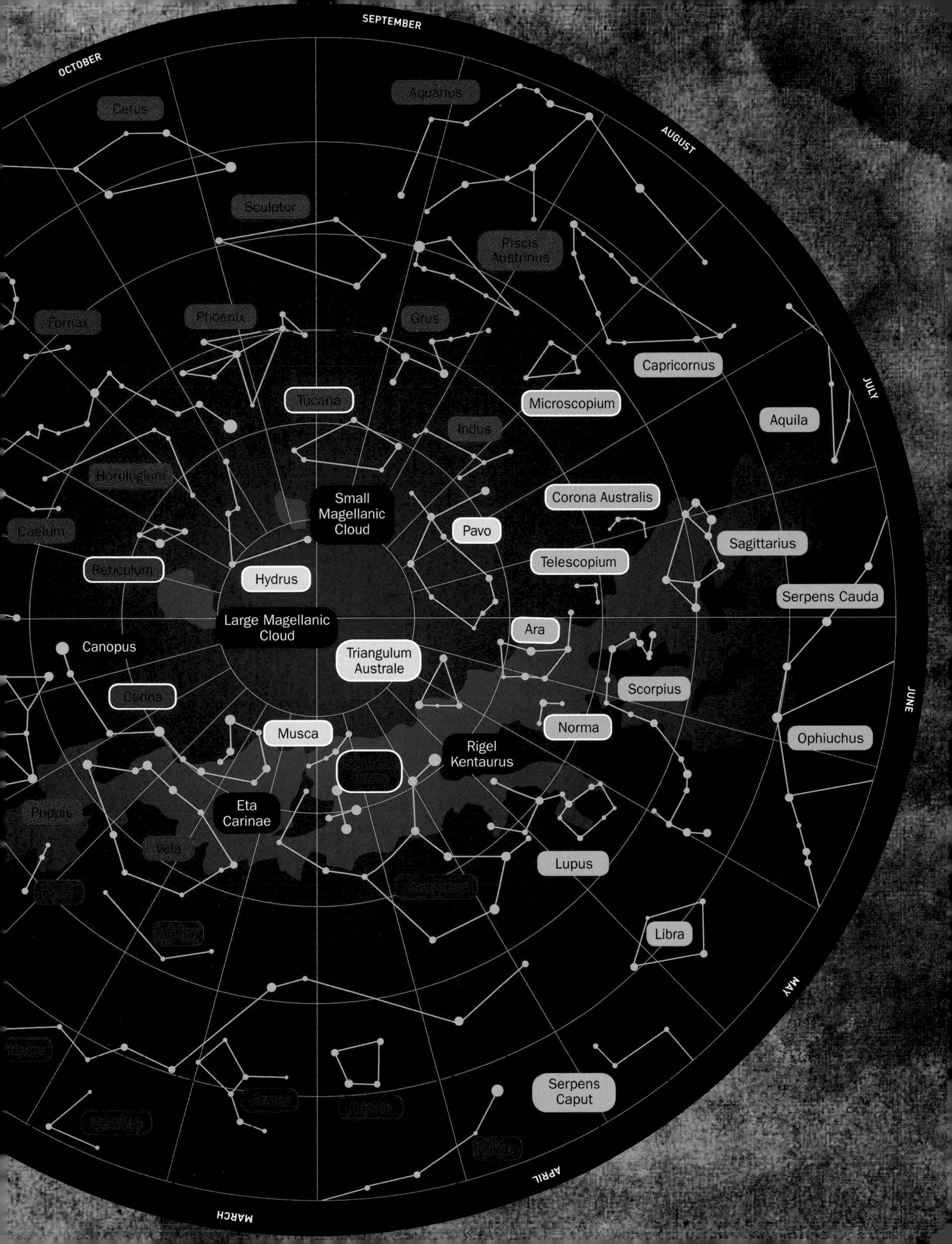
SEPTEMBER
OCTOBER
AUGUST
JULY
JUNE
MAY
APRIL
MARCH
Cetus
Aquarius
Sculptor
Piscis Austrinus
Fornax
Phoenix
Grus
Capricornus
Tucana
Microscopium
Indus
Aquila
Horologium
Small Magellanic Cloud
Corona Australis
Caelum
Pavo
Sagittarius
Reticulum
Hydrus
Telescopium
Serpens Cauda
Large Magellanic Cloud
Ara
Canopus
Triangulum Australe
Scorpius
Carina
Musca
Norma
Ophiuchus
Rigel Kentaurus
Puppis
Eta Carinae
Vela
Lupus
Libra
Serpens Caput

14 WITNESS ZODIACAL LIGHT AT DAWN

Our view of the *ecliptic* (the area of the sky marked by the zodiac constellations) is littered with trace dust that reflects the light from our Sun. While much of these flecks have come from the tails of comets flying into our solar system at its deepest edges, some may remain from the solar system's formation—primordial dust that has never been absorbed into the greater masses of the Sun, planets, and smaller rocky bodies. You can take a peek at the granules of our early solar system by snagging a glimpse of *zodiacal light*.

Fall is the best time to witness the light in the pre-dawn. In the northern hemisphere, this means you can see it illuminating the horizon from September through October; in the southern hemisphere, try catching a glimpse between March and April.

First, find a clear sky free of light pollution on a Moonless night. Look east about half an hour before sunrise. The sky will light up in a large, softly glowing triangle, often called *false dawn*. That is zodiacal light, sunlight reflecting off dust in our solar system.

At certain points in the year you can observe the zodiacal light at sunset, but morning air tends to be clearer, making it much easier to see the angular and radiant flush coupled with sunrise's colors.

15 SPOT MORNING AND EVENING PLANETS

Before telescopes, Venus was known as both the "morning star" and the "evening star," because it is only ever seen near the Sun before sunrise or after sunset. Since Venus and Mercury move against the background constellations, they won't be found on any permanent sky chart. Here's how to find them.

VENUS If you see a very bright "star" low on the horizon just after sunset, there's a good chance you're looking at Venus (see #93). The third-brightest object in the sky after the Sun and the Moon, Venus can sometimes be seen in the daytime if you know where to look. To spot Venus during the day, try this: When Venus is high in the pre-dawn sky, follow its position as the Sun rises. You should be able to see it well after dawn. (Just be careful not to look at the Sun.)

MERCURY To spot speedy Mercury, you will need both an unobstructed horizon and crisp, clear skies. Sometimes Mercury can appear very bright, but the planet often appears dim and can be difficult (yet not impossible) to spot. Be alert: When Mercury is visible just before sunrise or just after sunset, there are only a few days—maybe weeks at best—to spot it before it dips too close to the Sun to be seen with the naked eye. (Check out #78 for more on this planet.)

16 REDISCOVER SUNRISE

You've seen it many times before, sure, but have you really carefully observed the dawn? Before the Sun begins to rise, take a close look at the last moments of night sky. Note how many changes you see are the reverse of the sky you observe at sunset.

Dawn itself begins with light creeping from the east. The dimmest stars fade out first, along with any traces of the Milky Way. Watch all the stars blink out in order of brightness until nothing but the stark outlines of their respective constellations remain; in minutes, those outlines vanish, too. Last, the bright planets fade out from view. You will still be able to see the Moon if it is up, albeit not as brilliantly as at night.

Sunrise starts when the Sun lights up the sky as it climbs over the eastern horizon. Watch as the sky's color changes and brightens up in a sudden brilliant stroke of light. Clouds—especially those high in the atmosphere—are often vividly colored in pinks and purples just before day begins.

17 MEET THE SUN

At the center of our solar system is a near-perfect sphere of pulsing plasma—a yellow dwarf star that we call the Sun. The radiation that it gives off through constant nuclear fusion provides our planet with enough energy to sustain life. While it is dangerous to look at it without tools (see #182 and 218–224), here are some facts about our solar system's star.

DIAMETER 864,900 miles (1.4 million km), or 109 times wider than Earth

MASS 333,000 times Earth's mass

SURFACE TEMPERATURE 9,980°F (5,530°C)

CORE TEMPERATURE 27 million°F (15 million°C)

ROTATION PERIOD 25 Earth days (at the Sun's equator) to 34 Earth days (at its poles)

DISTANCE FROM EARTH An average of 93 million miles (150 million km)

NUMBER OF KNOWN PLANETS 8

AGE 4.6 billion years

SURFACE GRAVITY The Sun's gravity is 28 times Earth's gravity. If you weigh 100 pounds (45 kg) on Earth, you'd weigh 2,800 pounds (1,270 kg) on the surface of the Sun (that is, if it were solid and cool enough that you could actually stand on it).

WORTH THE WAIT Scientists have calculated that it takes tens of thousands of years for the Sun's energy to travel from its core, through its interior, and into space, but only 8 minutes for this energy to reach Earth.

SUCCESSFUL MISSIONS
1990: *Ulysses* orbiter (United States and Europe)
1995: Solar and Heliospheric Observatory orbiter (United States and Europe)
2006: Solar Terrestrial Relations Observatory orbiter (United States)
2010: Solar Dymanics Observatory orbiter (United States)

MANY FACES Our eyes have evolved to detect the Sun's visible light. But the Sun also emits types of "light" our eyes can't see, including radio waves, microwaves, infrared radiation, ultraviolet light, X-rays, and gamma rays. Scientists put telescopes in space to observe the Sun in various kinds of light to better "see" its features.

COMPOSITION The Sun does not have a solid surface. It is a ball of gas—mostly hydrogen (92.1 percent) and helium (7.8 percent)—with various other chemical elements making up the last fraction. Like other stars, it has a core, a radiative zone, and a convection layer (see #02). Its visible surface is called the *photosphere*, and above it is a layer of extremely hot hydrogen called the *chromosphere* (the colors of which you can see through a filter—see #227). The *corona* is a crown of plasma that the Sun radiates out for millions of miles into our solar system.

SOLAR FLARES The Sun regularly releases sudden flashes of light that can be as intense as 1 billion hydrogen bombs. They are often followed by *coronal mass ejections*, huge clouds of electrons and atomic nuclei erupting through the Sun's corona into space. Solar activity fluctuates during an approximate 11-year period between when the magnetic poles switch (the North Pole becomes the South Pole and vice versa). Solar flares are so strong that they can sometimes even interrupt Earth's electrical grid.

LIGHTS OUT In about 5 billion years, our Sun will run out of the hydrogen fuel it has used for billions of years to produce energy. Forced to burn the helium inside its core to fuel reactions, it will swell until it engulfs the orbits of Mercury, Venus, and Earth. At this stage, our Sun will be a red giant. Someday, it will run out of helium and a giant shell of gas (called a *planetary nebula*) will surround the squashed core—now just the size of Earth—which will then have all the characteristics of a white dwarf star.

NUCLEAR FUSION The Sun's enormous gravity pulls gases inward, causing the temperature and pressure at the Sun's center to skyrocket. Inside the core, temperatures and pressures are high enough to support nuclear fusion. Hydrogen nuclei collide at extremely high speeds and fuse together to form helium nuclei. With each collision, a very tiny bit of hydrogen mass is lost and converted to huge amounts of energy. Each second, 700 million tons (635 billion kg) of hydrogen are converted to 695 million tons (630 billion kg) of helium. The 5 million tons (4.5 billion kg) of matter that are lost are converted into an amount of energy equal to the explosion of a whopping 100 billion tons (91 trillion kg) of dynamite.

SOLAR WIND The Sun is continuously streaming charged electrons and protons, sending them into space at 1 million miles per hour (1.6 million km/s). The solar wind shapes the tails of comets and creates auroras on Earth and other planets. Planets with strong magnetic fields, like Earth or Jupiter, can deflect the solar wind, preventing their atmospheres from being stripped away.

SUNSPOTS The Sun's *plasma* (a huge cloud of ions and electrons) creates volatile magnetic fields that wreak beautiful havoc on the surface. Such effects include *sunspots*, which are temporary dark areas that result from a reduction in convection caused by concentrated magnetic fields. This loss in convection makes the area on the surface marginally cooler than the rest of the Sun—hence the dark colors.

18 UNDERSTAND SOLSTICES AND EQUINOXES

Many cultures have long marked the annual *solstices*, the year's shortest and longest days, and the *equinoxes*, the two days of the year on which night and day are equal in length. Both events result from Earth's tilted axis, which is always inclined at 23.5 degrees. As Earth orbits the Sun, the northern hemisphere will be warmest and receive the most direct sunlight in the summer months of June, July, and August, when the North Pole points toward the Sun. Meanwhile, in the southern hemisphere, the warmest months happen in December, January, and February, when the South Pole points toward the Sun and brings direct sunlight to that hemisphere.

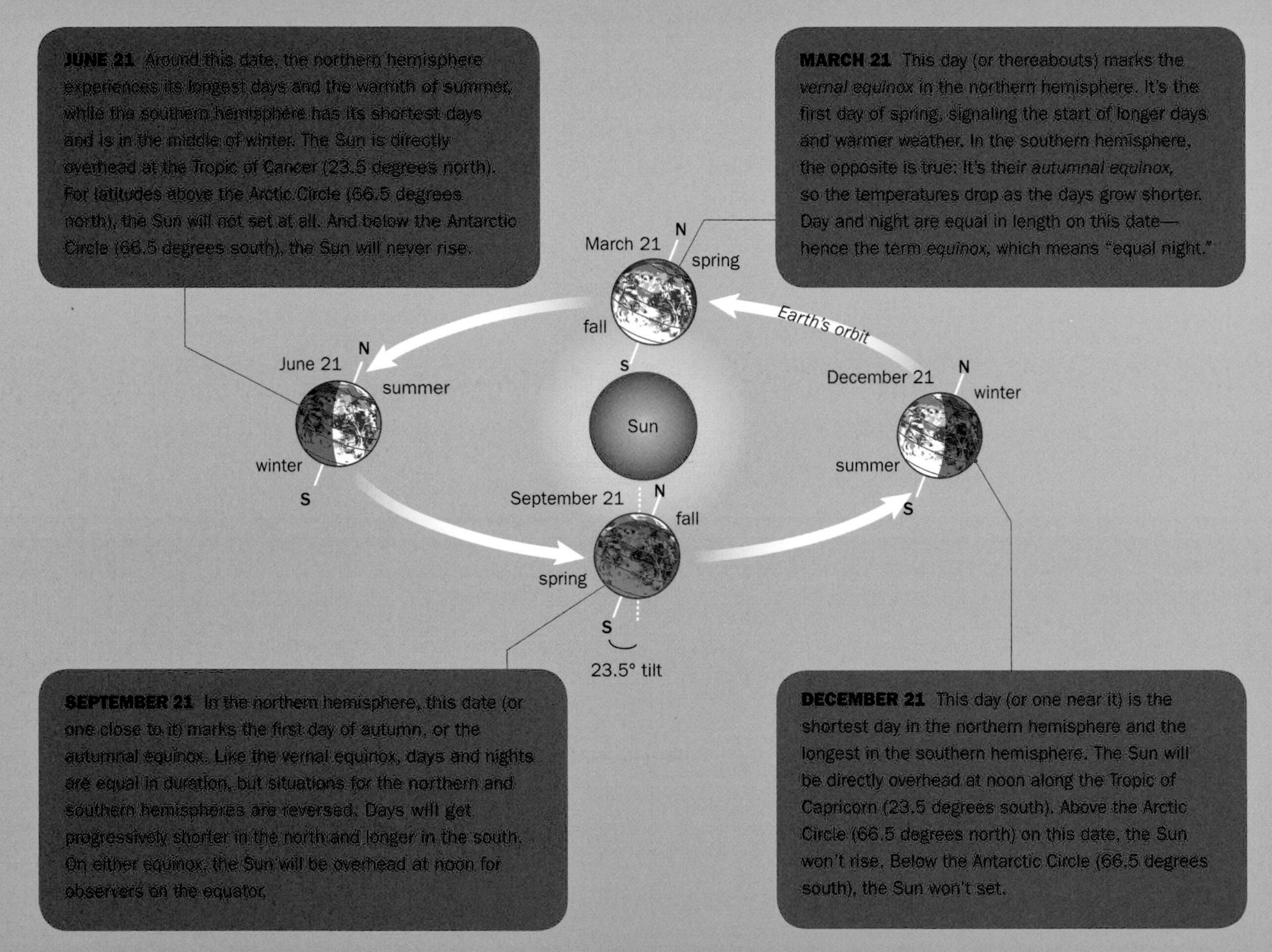

19 GET THE REASONS FOR THE SEASONS

Many people think Earth's changing proximity to the Sun is the reason why seasons change on Earth. While our orbit is not a perfect circle, that small shift in proximity to the Sun doesn't cause the summer and winter. In fact, our orbit is closest to the Sun in January—when much of the northern hemisphere is bundled up in deep freeze. The real reason for the seasons is the tilt of Earth's axis. When one hemisphere or the other is tilted more toward the Sun, it causes that hemisphere to experience longer days and more direct rays from the Sun. Basically, during the summer, sunshine is concentrated in a smaller area—on the part of Earth facing the Sun. So goes the saying, "It's the length of the days and the angle of the rays."

20 EXPLORE ANCIENT ASTRONOMICAL MARKERS

When humans first began cultivating the land thousands of years ago, they needed a way to figure out when to plant and harvest crops each season. Careful observation of the changes in daylight throughout the year helped them discover the importance of the solstices and equinoxes, and many of the ancient markers that they built to track the seasons still exist in the world. Here are just a few.

Ⓐ STONEHENGE A 5,000-year-old observatory located in Wiltshire, England, Stonehenge appears to be aligned with the winter solstice sunset and summer solstice sunrise. Unlikely to have been making detailed astronomical observations, humans probably used Stonehenge for ritual around the seasons.

Ⓑ CHACO CANYON SUN DAGGER Built by the ancient Anasazi people of the Chaco Canyon in modern-day New Mexico in the United States, this observatory was designed so that when sunlight passed through slabs of rocks during the equinoxes and solstices, a dagger of light appeared to pierce the spiral carved into the rocks. Sadly, these rocks shifted in 1989 and the Sun dagger no longer works.

Ⓒ NEWGRANGE On the winter solstice, the rising Sun shines on one of the chambers inside this observatory in Newgrange, Ireland. Built around 3200 BCE, this observatory is older than the pyramids. Each year, thousands enter a lottery to see the Sun light up this chamber for 17 minutes.

21 TRACK THE SUN WITH AN ANALEMMA

If you mark the Sun's position on a wall, working at the same time once a week for a year, you'll end up sketching a figure eight called an *analemma*. This neat shape is caused by Earth's tilted rotational axis and elliptical orbit around the Sun.

Earth's tilted axis makes the Sun appear to be higher or lower in the sky at various times of year—thus creating marks of varying heights along your analemma's vertical axis. For example, in the summer, when the northern hemisphere is angled toward the Sun, our star seems higher; in the winter, when the northern hemisphere slants away from the Sun, it looks lower. Meanwhile, the Sun's elliptical orbit makes it rise a little to the east or west (depending on the season), creating the two loops in the analemma shape. You can make your own analemma using paper pieces, a mirror, and a bright spot at home.

STEP ONE Choose a day and time when you can Sun-chase weekly, and then find a south-facing window. Make sure the Sun will shine through that window at your chosen time throughout the year.

STEP TWO Set a small mirror on the windowsill so it will reflect sunlight onto a blank wall. Tape the mirror to the sill, and ask that no one mess with it for a full year.

STEP THREE On the first day, stick a piece of paper to the wall where the reflection hits and note the date. Each week at the same time, put another dated sticky on the reflection. (Remember to adjust for daylight saving time!)

STEP FOUR At year's end, see which shape your notes make. To make your analemma for keeps, lightly trace the pattern in pencil.

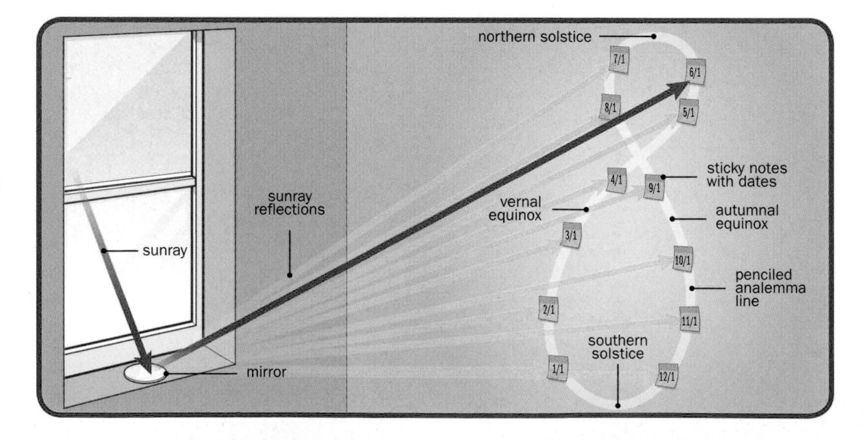

22 CREATE YOUR OWN SOLARGRAPHS

There's more than one way to catch the Sun's annual journey across the sky—and this project results in an amazing image, too. A *solargraph* is the product of a six-month time-lapse exposure made with a pinhole camera, revealing an arc of colored bands created by the Sun as it rises and sets every day in a slightly different pattern. You can make ethereal images of the Sun's path with this low-tech, low-maintenance procedure.

STEP ONE Buy some light sensitive, black and white, semimatte photographic paper.

STEP TWO Make a simple, sturdy pinhole camera by poking a small hole in one side of an opaque container. (Aluminum cans and 35mm film canisters will work.) Make it light tight (besides the pinhole) with duct tape.

STEP THREE Cover the pinhole with a "shutter"—a tiny piece of electrical tape. You'll remove this when it's time to start your exposure.

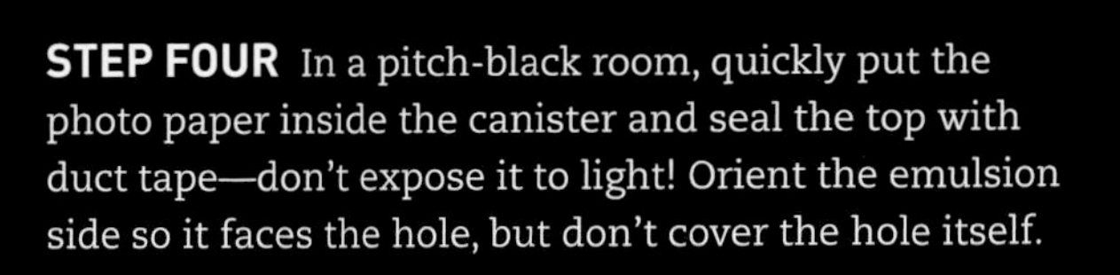

STEP FOUR In a pitch-black room, quickly put the photo paper inside the canister and seal the top with duct tape—don't expose it to light! Orient the emulsion side so it faces the hole, but don't cover the hole itself.

STEP FIVE Find a safe, protected spot facing south in the northern hemisphere or north in the southern hemisphere. Snap a test shot to get a composition you like first. Try to include a man-made object for context. Then, tie down or tape your pinhole rig in place.

STEP SIX Remove the shutter. Leave the camera in place for one to six months.

STEP SEVEN In a darkened room, quickly place the unprocessed paper on a decent scanner. Scan the image with 500+ dpi (don't preview) for a negative. Use any photo software to get the inverse of the image, then adjust contrast and brightness. Marvel at the variation in the Sun's path over time.

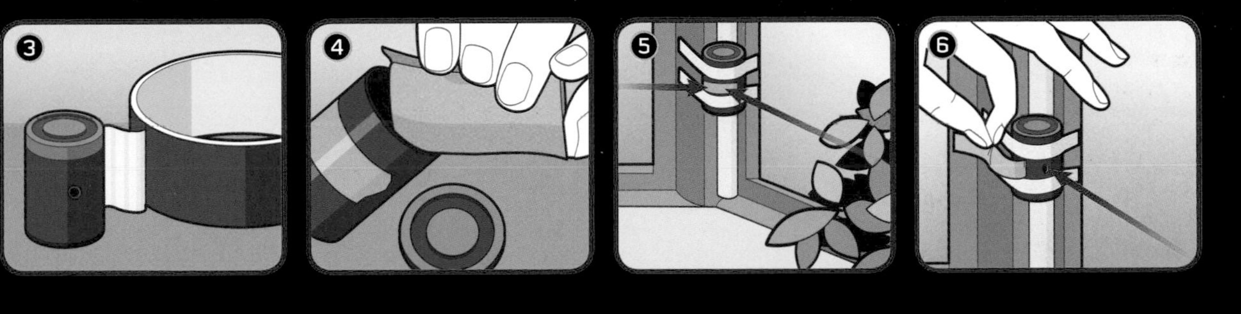

23 MAKE YOUR OWN PLANISPHERE

There are many constellation websites and apps out there to help you find what you're after in the night sky. But you can also go low-tech and make a planisphere (aka a star finder) out of paper. Of course, planispheres are not one-size-fits-all. In fact, they only work well for a specific latitude range (usually of about 10 degrees). When making yours, cut the lines that correspond closest to your own latitude.

STEP ONE Using a photocopier (or scanner and printer), make two copies of the yellow template below. Glue each copy to a piece of cardstock or thin cardboard.

STEP TWO On the template that will become the front side, cut along the light-blue line corresponding to your latitude. For example, if you're at 30 degrees latitude (north or south), cut along the light-blue line marked 30 below the horizontal line (marked 0), making a bowl shape. If you're in the northern hemisphere, write "N" in the center, "W" on the left, and "E" on the right. This will represent your orientation relative to the northern horizon. If you're in the southern hemisphere, write "S" in the center, "E" on the left, and "W" on the right to represent your orientation relative to the southern horizon. Cut out the four shaded arcs under the numbers.

STEP THREE For the second template that will be the backside, cut along the dark-blue line at the same latitude but above the horizontal line, making a bump. If you're at 30 degrees, cut along the 30-degree line above the horizontal line (marked 0). In the northern hemisphere, write "S" in the center under the top of the hump, "E" on the left, and "W" on the right to represent your southern horizon. If you are in the southern hemisphere, write "N" in the center, "W" on the left, and "E" on the right. This represents your northern horizon. You now have two covers—one is a bowl, one is a hump (unless you're right on the equator). Staple them together along the outer edge, making a sleeve.

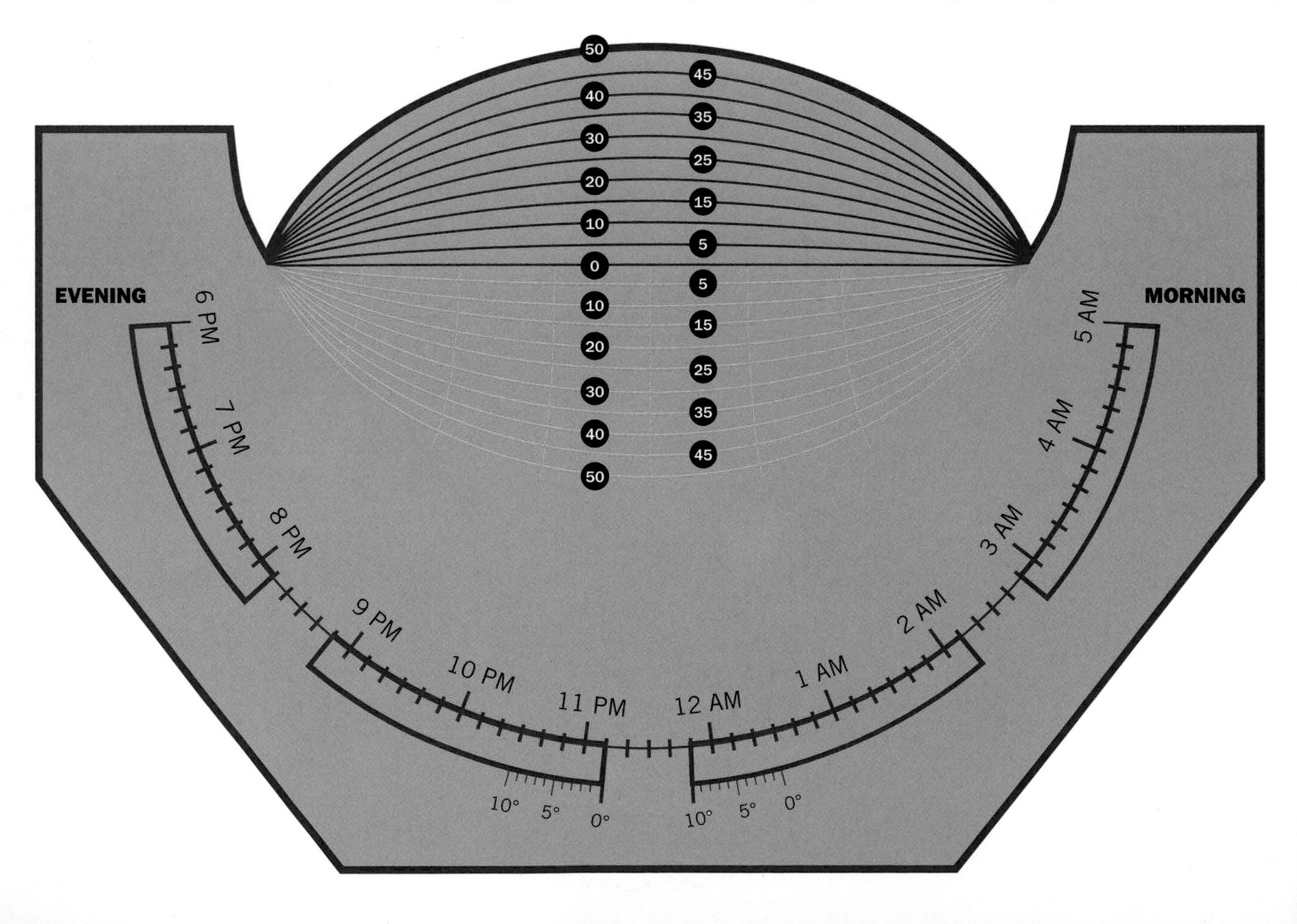

STEP FOUR Copy the northern hemisphere star wheel on this page, as well as the southern hemisphere star wheel on the next page (#24). Glue the northern star wheel to a piece of manila folder or cardstock, then cut it out. Cut out the southern star wheel and glue it to the opposite side of the cardstock, making sure the months line up. You should have one double-sided star wheel.

STEP FIVE To use your planisphere, insert the wheel so that your hemisphere's star field is visible through the bowl side of the opening. Rotate the wheel so today's date (displayed in the window) lines up with the time displayed on the star finder's round edge.

STEP SIX Go outside. Hold the planisphere up toward the sky, orienting the "N" on the cover so it is facing the direction of the northern horizon. What you will see in the sky are the constellations just above the N on the planisphere. To view the constellations in the southern sky, flip your planisphere over and use the same method, keeping the star finder above your head and aligning the "S" with the southern horizon.

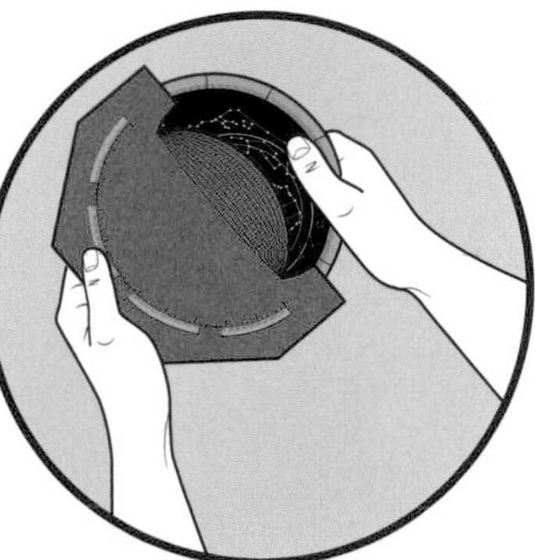

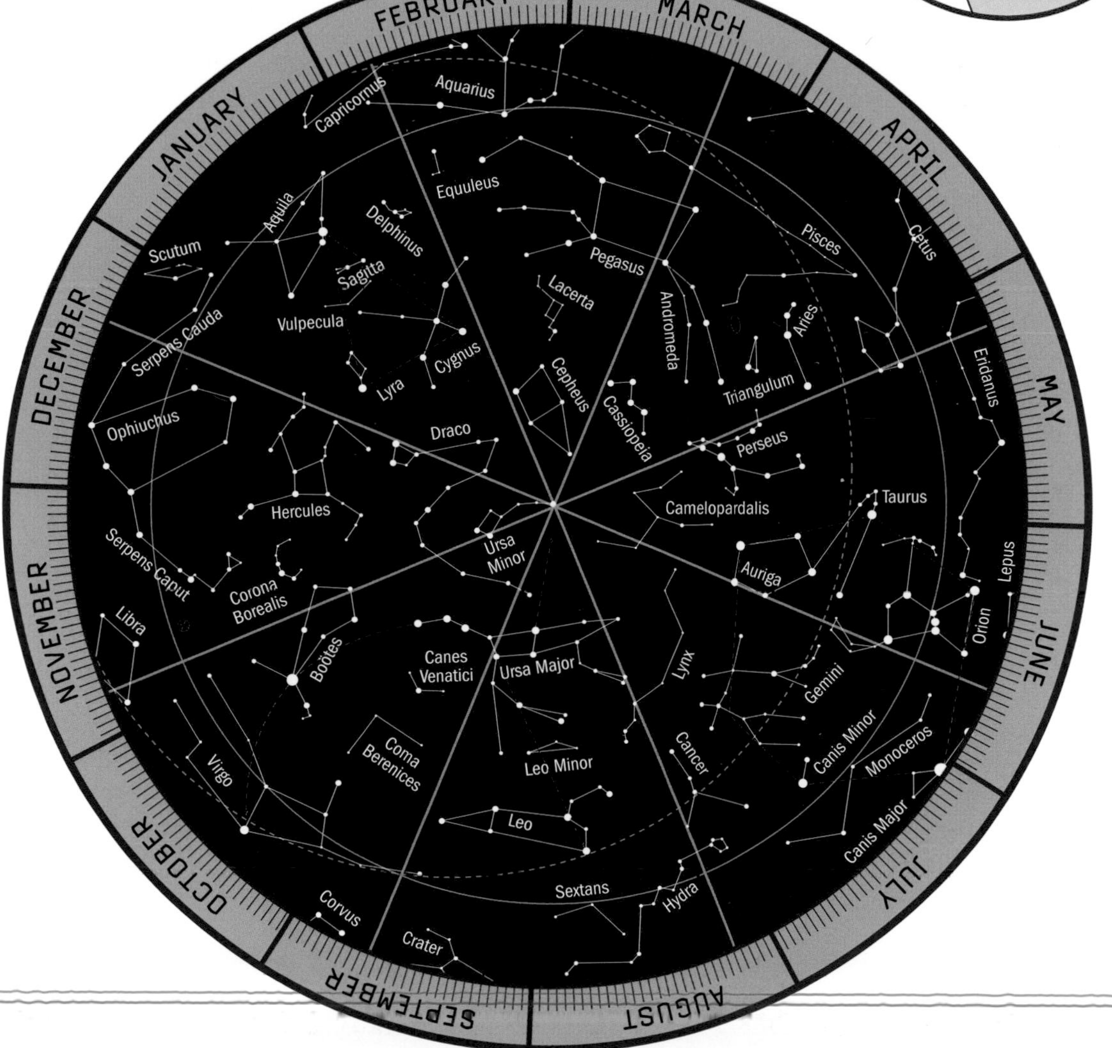

24 LOOK SOUTH WITH YOUR PLANISPHERE

Planispheres make navigating the night sky easier in a lot of the same ways that maps help us navigate roadways and wilderness: They show us direction, major landmarks, and relative distance between objects and areas. Unless you're observing at the north pole, you will certainly see part of the southern hemisphere's sky when you look up, so you'll need this chart, too. The same goes for those in the southern hemisphere: You will also need the northern hemisphere star template on the preceding page (#23).

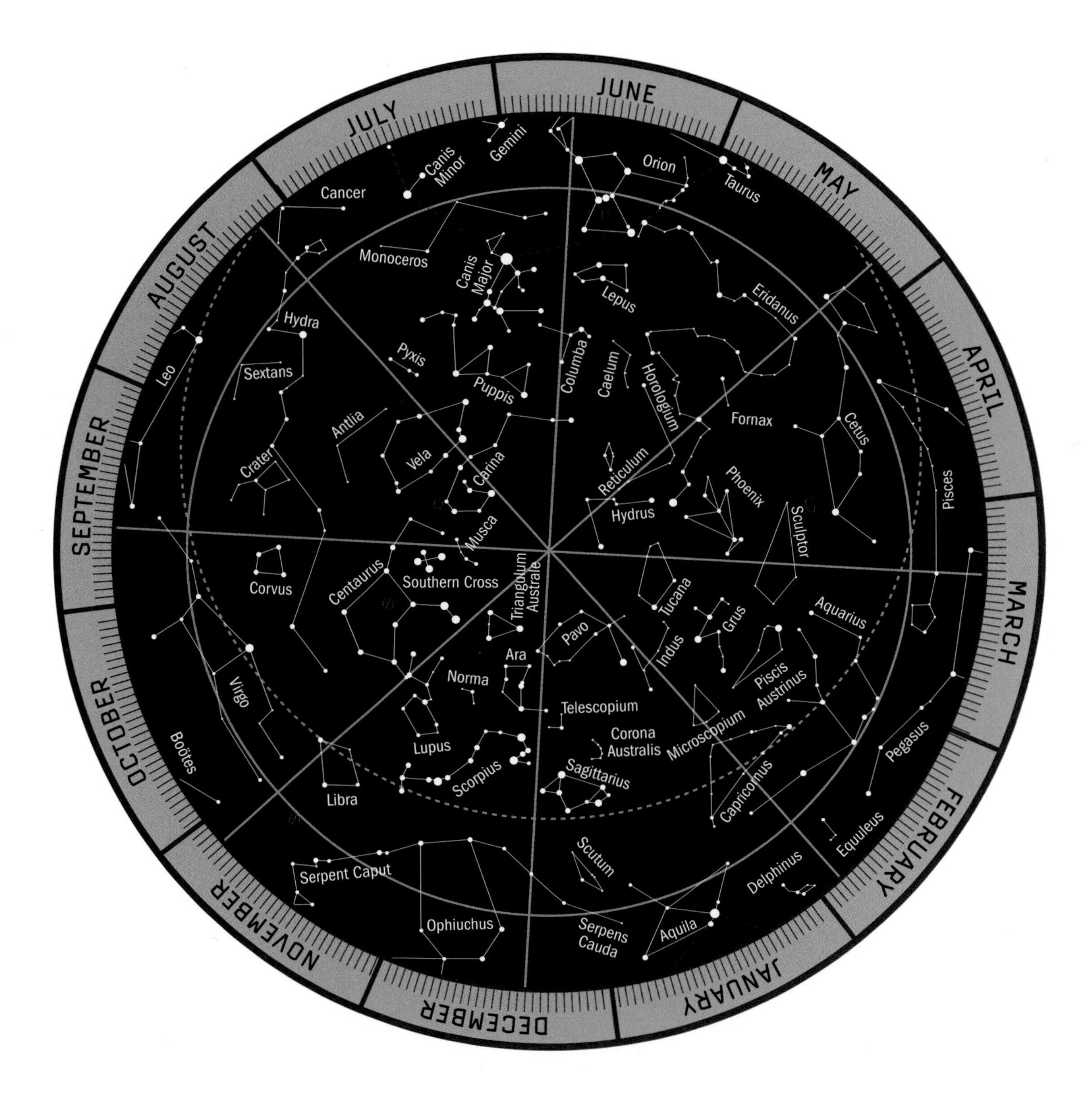

25 MEASURE THE HEAVENS

You would never use a yardstick to measure a mountain range. In much the same way, the cosmos requires its own units of measurement.

ASTRONOMICAL UNITS The tape measure of the solar system, *astronomical units* (au) are used to gauge the distance between objects in relatively close proximity. One au is 93 million miles (150 million km), or approximately the distance from Earth to the Sun (which changes throughout the year).

MINUTES AND SECONDS OF ARC When astronomers describe the size of a celestial object or the distance between two celestial bodies, they often speak in terms of *angular size*—how large an object or distance appears from our vantage point in our sky, measured in degrees (°). For example, the Large Magellanic Cloud in the southern hemisphere takes up 5 degrees of the sky, while the Andromeda Galaxy (M31/NGC 224) in the northern hemisphere takes up 3 degrees. But in the heavens, even 1 degree is often too large a unit of measure. That's where minutes and seconds of arc come in. Each arc minute (') is 1⁄60 of a degree; each arc second (") is 1⁄60 of an arc minute. For a sense of how big (or small) these units are, consider that the Moon as viewed from Earth is about 0.5 degrees, or 30 arc seconds.

LIGHT YEARS Nothing travels faster than light, so astronomers use the enormous distances it can travel as a convenient ruler for measuring space. About 6 trillion miles (10 trillion km), one light year is the inconceivable distance that light travels in a vacuum in one year. But don't be fooled by its name: Some mistake light years for a unit of time instead of distance.

PARSECS Hold your thumb up at arm's length and close one eye. Note where your thumb appears against the background. Switch eyes—see how your thumb "moved"? That's called *parallax*. Scientists use it to gauge a star's apparent distance using a *parsec* (pc), or the parallax of 1 arc second. If you imagine the Sun and Earth making up two corners on the short side of a triangle, the distance between them is 1 au. When the angle opposite the short side (the angle created by the star whose distance you're measuring) measures 1 arc second, that star is 1 parsec—or 3.26 light years—away.

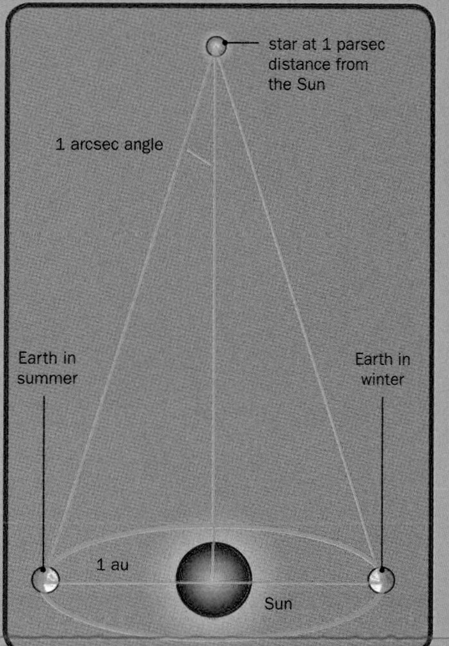

26

TOP FIVE
ZONE THE SKY

Get acquainted with these basic signposts.

☐ **HORIZON** The *horizon* is just where the line of Earth meets the sky. It's best to find a place with low horizons so you can see as much of the sky as possible. This varies dramatically depending on your location.

☐ **ZENITH** The very center of the sky (aka the point directly above your head) is known as the *zenith*. If you're using the right sky chart at the right time, you should see whatever is at the center of your chart when you look up.

☐ **POLES** These are the points directly above the northern and southern poles on Earth. All other stars appear to circle around these points.

☐ **CELESTIAL EQUATOR** This circle on the celestial sphere is in the same plane as Earth's equator. Stars near the equator (like Orion's belt) are visible from any spot on the globe. It's also 0 degrees declination (see #49).

☐ **ECLIPTIC** This is the path that the Sun and the Moon appear to take along the stars. All the planets will travel the sky along it, too.

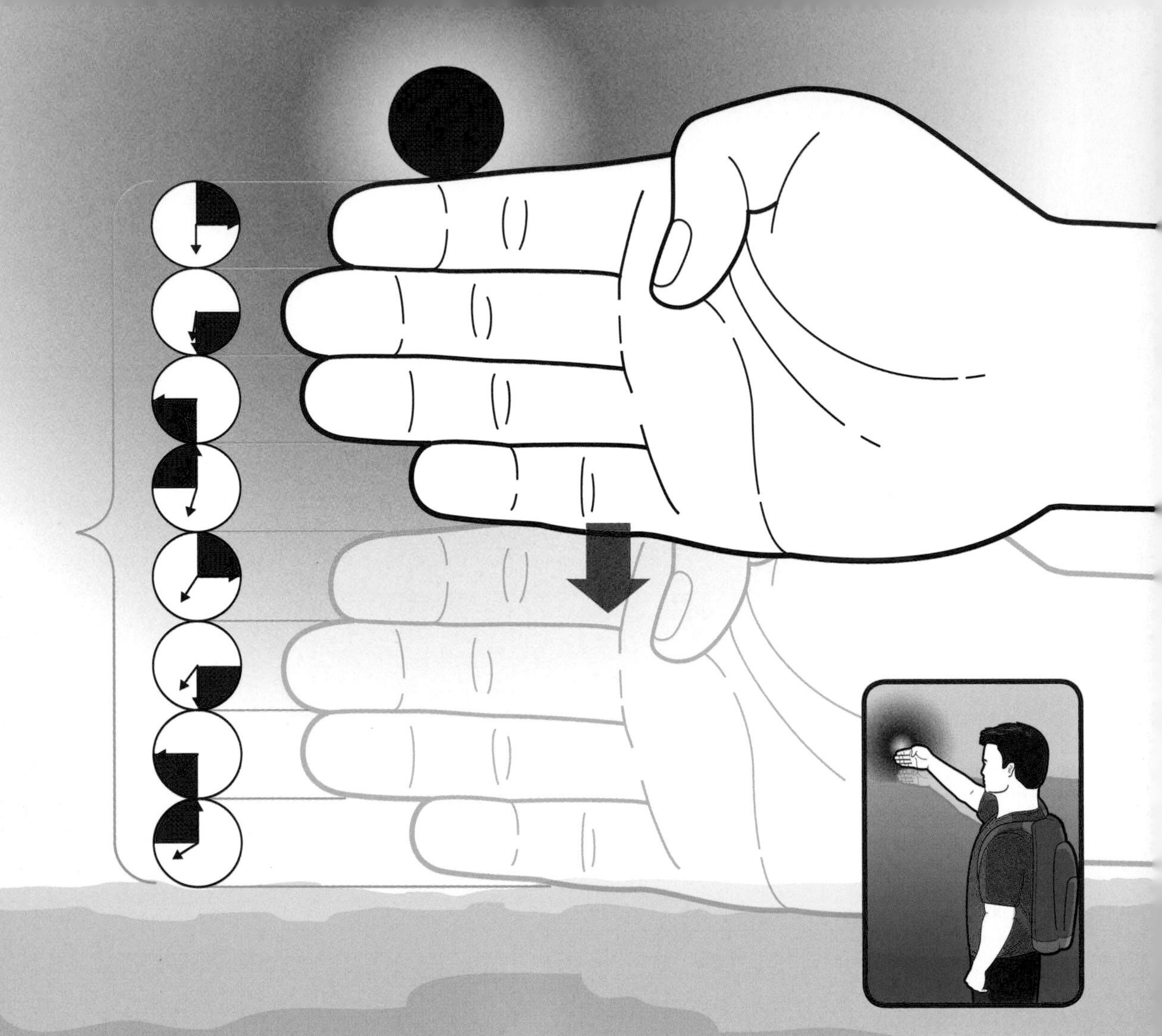

27 TIME THE SUNSET WITH YOUR FINGERS

When you're counting down the minutes to your next evening star gaze, it's crucial to know what time the Sun will dip over the horizon and when the skies should be thoroughly dark enough for viewing. Of course, not everyone carries around a sunset calendar or is constantly in range of the all-knowing Internet. This trick can come in handy when you need to measure how much light remains in a day.

STEP ONE Hold either hand out at arm's length with your elbow straight. Turn your hand sideways at the wrist, so that the palm faces you. Tuck your thumb into your palm (or just ignore it).

STEP TWO Line up the top of your index finger with the bottom border of the Sun. Count the number of finger widths between the bottom edge of the Sun and the horizon, moving your hand as necessary. (As always, don't look directly at the Sun!)

STEP THREE Multiply the number of finger widths between the Sun and the horizon by 15. Each finger width represents approximately 15 minutes until sunset, so this calculation should give you a ballpark estimate of how much time you have until the Sun goes down. After sunset, it takes at least half an hour to begin seeing stars.

All space above you makes up 180 degrees.

The angle of separation between the horizon and the point directly above you is 90 degrees.

When measuring the height of an object in the sky, you are measuring the angle from the horizon.

Imagine the universe is a 360-degree sphere with you at its center.

Everything under your feet—the ground, the center of Earth, the other side of Earth, and all space below Earth—makes up 180 degrees.

28 MEASURE DEGREES OF SEPARATION IN THE SKY

When astronomers want to describe the distance between celestial bodies, they speak in terms of *angular separation:* the angle of distance between objects as we perceive them, which is measured in degrees, arc minutes, and arc seconds (see #25). Using units smaller than degrees generally requires a telescope, but you can get a rough idea of the angular separation between stars with your hands and some basic math.

It helps to imagine all of space as a sphere, with Earth at its center—not as a reversion to some pre-Copernican worldview but because, for the sake of this exercise, you actually are the center of the universe. There are 360 degrees in a circle, and the horizon splits that circle in two, meaning the sky above you and the ground beneath your feet both measure 180 degrees. The angular separation between the point directly above you and the horizon is 90 degrees. To measure smaller than 90-degree increments, try these hand positions, with your hand at arm's length, elbows straight, palm facing outward. A good starter object to measure? The Sun or the Moon, which each take up 0.5 degrees and can be covered by your pinky!

USING YOUR HAND

A 1° The width of your pinky finger represents roughly 1 degree of separation.

B 5° Together, your three middle fingers cover about 5 degrees.

C 10° A clenched fist is 10 degrees.

D 15° To measure 15 degrees, separate your pinky and your index finger.

E 25° The distance between the tip of your pinky and thumb is around 25 degrees.

A 1°

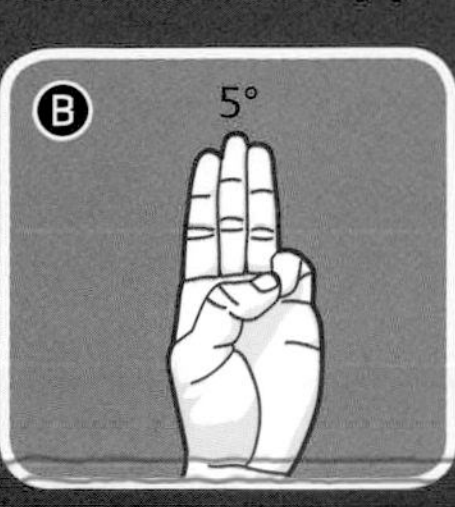

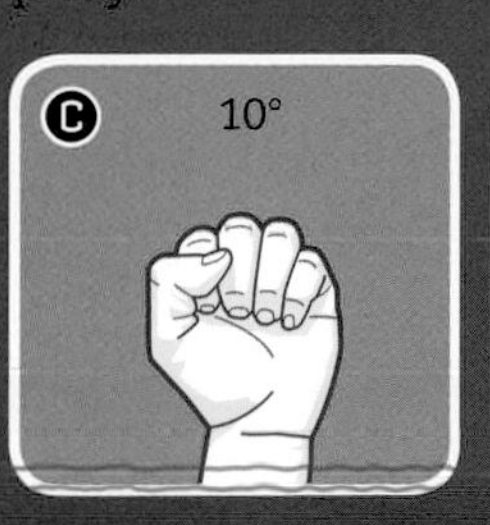

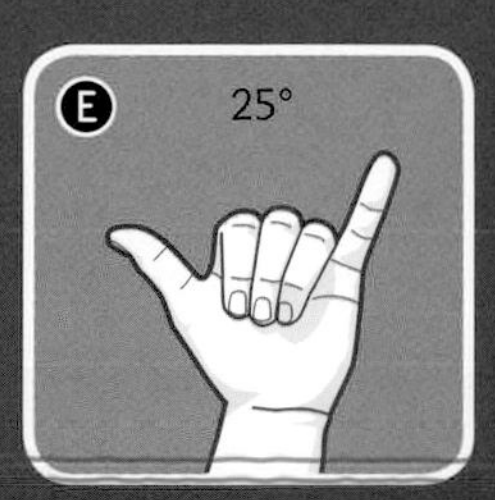

29 START AT THE BIG DIPPER

Making up the tail and rump of the large constellation Ursa Major (aka the Great Bear; see #53), the Big Dipper is one of the most familiar patterns in the northern hemisphere. It's long been important to navigators because it is visible most of the year and can always be seen near Polaris (the North Star, or Alpha Ursae Minoris).

To find it, use a magnetic compass to face north, then look up from the horizon. You should spot the Big Dipper about halfway up the sky. Since its orientation depends on the time of night and the season of the year, be prepared to see the Big Dipper with its bowl facing up, sideways, or down.

Just as people use familiar landmarks to navigate unfamiliar areas, the Big Dipper can help you find many other constellations. Here are a few celestial tours you can take with the Big Dipper.

30 PINPOINT ARCTURUS

Arcturus (Alpha Boötis) is the fourth-brightest star in the sky and the second-brightest in the northern hemisphere. It's in the kite-shape constellation Boötes the Herdsman, which ancient skywatchers saw as a farmer tilling the land with a plow (the Big Dipper). About 25 times larger than our Sun, Arcturus is an orange star whose light takes more than 36 years to reach our solar system. To find it, start at the three stars that form the Big Dipper's curved handle. Follow this curve away from the Dipper's bowl until you reach a bright orange star. You can remember this by the mnemonic "arc to Arcturus."

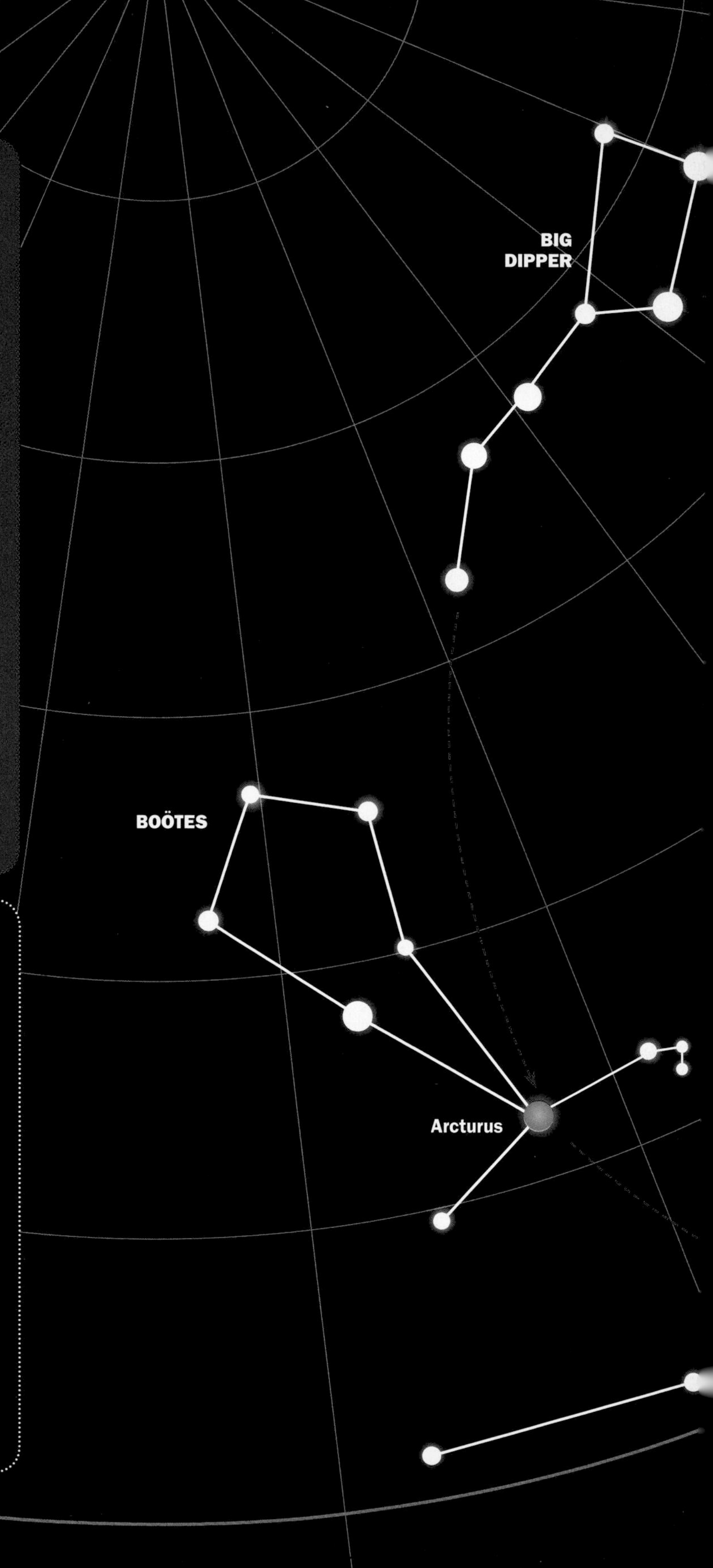

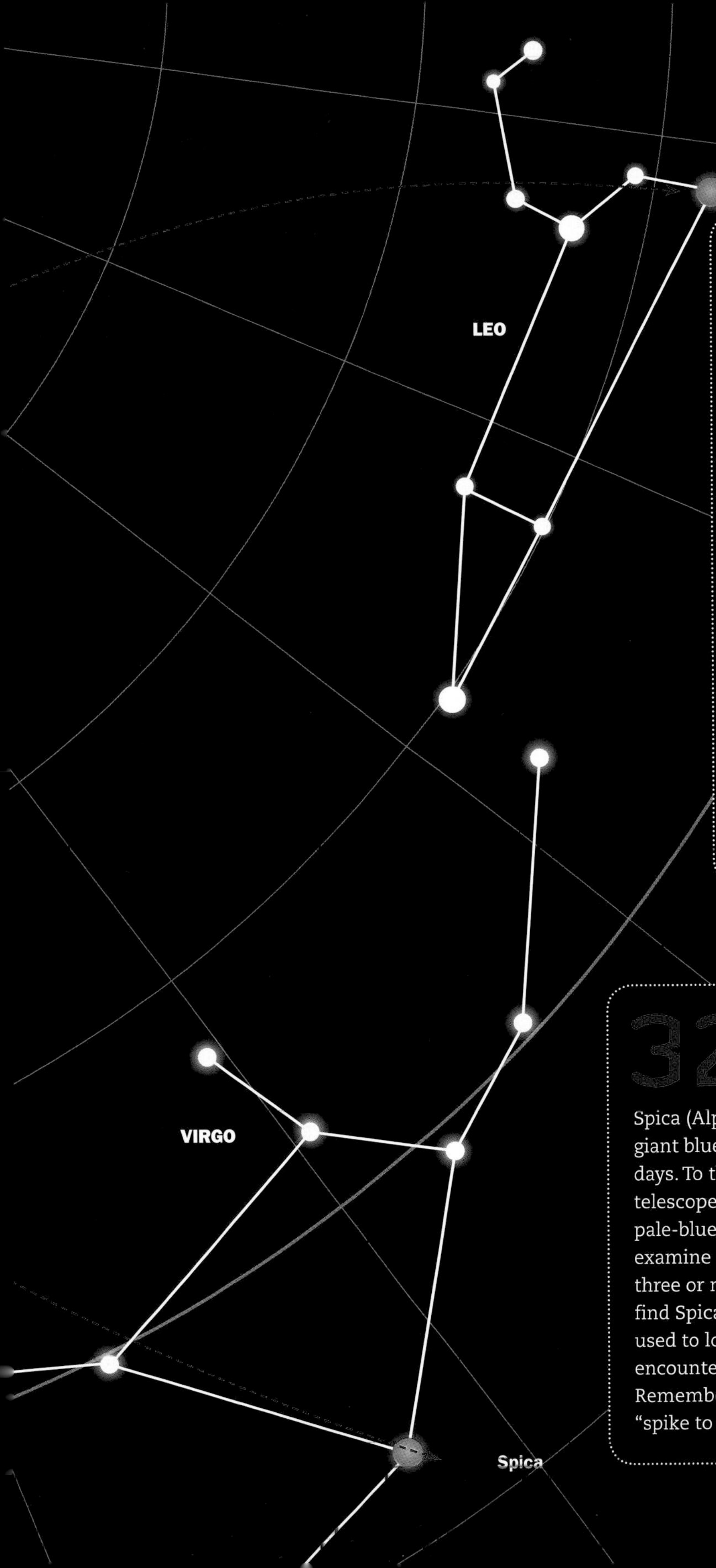

31 LOCATE REGULUS

Regulus (Alpha Leonis), the brightest star in the constellation Leo the Lion, is actually four stars. The system includes the very large Regulus A (three times the size of our Sun) and its tiny, as-yet-unseen orbiting companion—which astronomers think is a *stellar remnant*, also called a white dwarf star. Farther away, Regulus B and Regulus C (each half the size of our Sun) orbit each other. Light from this quartet takes 79 light years to get to Earth. To find Regulus, start with the two stars that make the handle side of the bowl of the Big Dipper and connect them with a line. Extend that line outward from the bottom of the bowl. The first bright star you encounter will be Regulus

32 SUSS OUT SPICA

Spica (Alpha Virginis) is not one but (at least) two giant blue stars orbiting each other once every four days. To the unaided eye—and even through a telescope—Spica looks like a single, very bright, pale-blue star. Still, some instruments used to examine Spica reveal that there may actually be three or more stars in orbit around each other! To find Spica, continue following the curved path used to locate Arcturus. The next bright star you encounter will be Spica, located in Virgo the Virgin. Remember these directions using the mnemonics "spike to Spica" or "speed on to Spica."

33 TEST VISION WITH THE BIG DIPPER

Next time you spot the Big Dipper, see if you can make out the double star, the second star from the end of its handle. If you can see it, congratulations—in ancient times, you could have officially worked as a soldier or a hunter, or held some other job that required perfect eyesight. Romans who could distinguish Mizar (Zeta Ursae Majoris) and Alcor (80 Ursae Majoris), the two stars that make up this double star, were considered to have excellent vision.

Modern astronomers have taken a closer look at Mizar and Alcor and discovered that both stars are much more complex than the ancients originally knew. Mizar is actually a system of four stars orbiting each other, and Alcor is a system of two stars, making six gravitationally connected stars within a couple of light years from each other. This star system is a distance of 83 light years from Earth. But no matter how good your eyesight is, you won't be able to see these six stars with the unaided eye.

34 TRACK TIME WITH THE LITTLE DIPPER

Need to know how much time has passed since you stepped outside to stargaze? If you're in the northern hemisphere, you can use the Little Dipper as a timer.

The Little Dipper completes a counterclockwise circle around Polaris, or the North Star (Alpha Ursae Minoris), every 24 hours. This means that every hour it moves 15 degrees (divide a full 360-degree rotation by 24 and you get 15). If you track the Little Dipper and estimate how many degrees it's traveled in the circle, you should be able to guess how much time has passed.

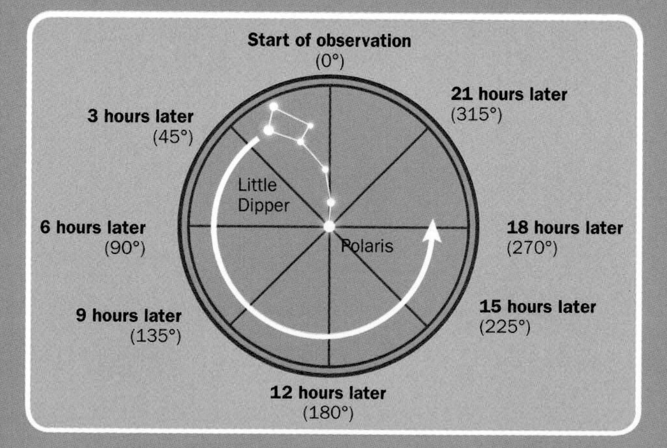

35

ASSEMBLE A STAR CLOCK

Before the invention of mechanical clocks, people used the motion of the stars to measure time at night. Turns out that the Big Dipper is a handy clock if you forget your watch.

Think of the Big Dipper as the hand of a clock with Polaris in the dial's center—only the hours of the clock are printed backward. That's because the Big Dipper moves counterclockwise, so this clock will need to rotate in that direction, too. Second, the hours aren't stationary but change depending on what month it is. In other words, midnight will not always be at the top of the dial.

TO MAKE Photocopy and cut out the circles, then attach the centers with a paper fastener.

TO USE Make sure that the current month is at the top of the outer circle. Rotate the dark circle until the diagram of the Big Dipper matches the real Big Dipper's position in the sky. The current time will appear in the cutout. If daylight saving time is in effect (in summer), add one hour to the time.

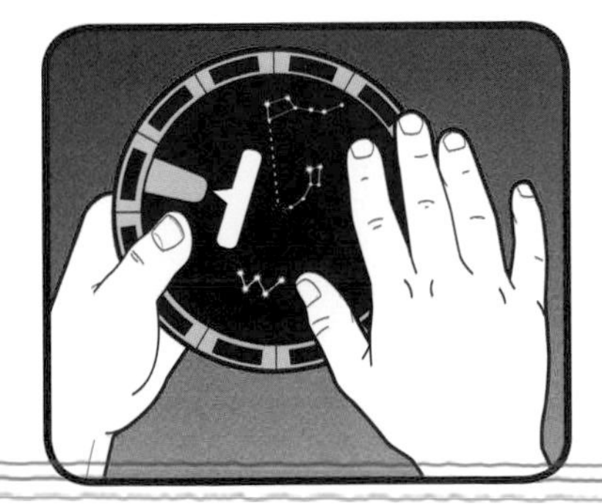

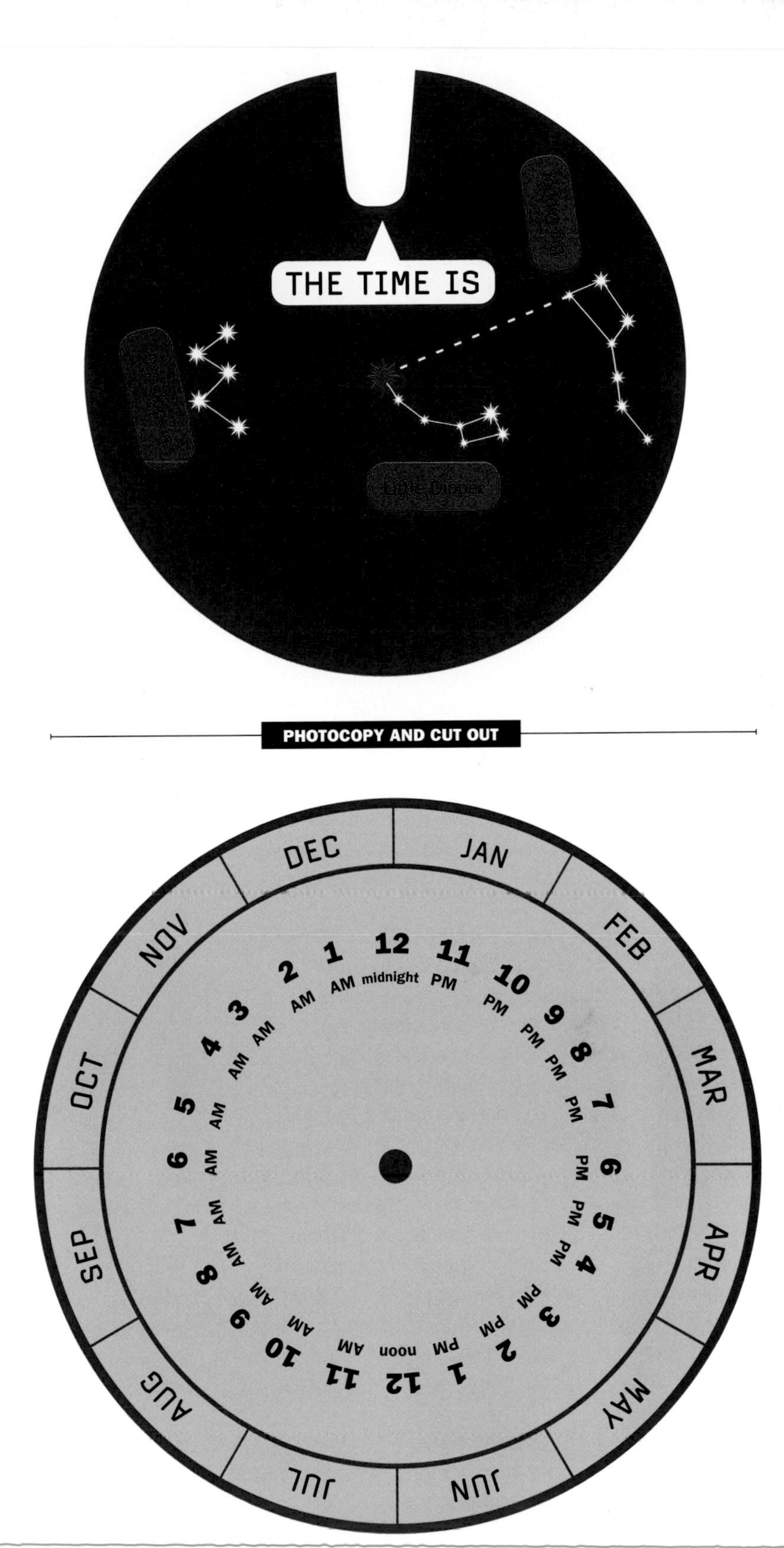

36

FIND YOUR WAY BY THE WIND, STARS, AND WAVES

Unless you're a shark, getting lost in the ocean is pretty much a death sentence. Without the tools of modern astronomy, the people of the South Pacific managed to turn an otherwise incredibly dangerous and impossible adventure into a routine journey between islands.

Long before the oceans and skies were scientifically mapped and charted—and even longer before satellites and GPS allowed seafaring adventurers to determine their location with pinpoint accuracy—Polynesian navigators relied on the winds, stars, and waves to find their way among the many tiny islands of the Pacific. Wayfinding was much less a science of getting from point A to point B than it was a way of being in the world; it involved becoming closely attuned to one's surroundings and developing an almost felt facility for traversing the seas. With no map, no compass, and no sextant to speak of, Polynesian navigators used their own powers of observation, coupled with knowledge passed down through generations in sometimes decades-long apprenticeships, to guide their outrigger canoes around the Pacific with incredible precision. But how did it work?

Wayfinding was a sophisticated and complicated system, which, at its most basic, involved the memorization of a "star chart"—more a mental map than anything written down—and almost constant observation of one's surroundings, speed, direction, and time passed. But to call this mental navigational tool a "star chart" is not really doing it justice. Beyond mapping the rising and setting locations of a number of important celestial bodies at different times during the year (the "houses" of the stars, or where they appeared to "go into" and "come out of" the ocean), this star chart also accounted for the directions of the waves and winds, as well as the flight patterns of birds. The system allowed navigators to determine the direction in which they were traveling—and more or less where they were—in relation to both the stars above and their starting location.

While many of the secrets of ancient Polynesian celestial navigation have been lost to time, something of a resurgence in non-instrument navigation has resulted in a mix of Western and traditional methods being practiced today. Here are just a few of the many signposts Pacific Islanders were using to find their way, long before anyone else dared leave the relative safety of land.

THE SUN Early in the day, the Sun creates a narrow reflection on the water, which widens as the Sun rises in the sky, then narrows again as the Sun sets. During the morning and evenings, when our star is not so high overhead, navigators would rely on signals from the water, as well as the reflections made by the Sun on the waves, to determine overall direction.

WAVES When the Sun was too high in the sky to be helpful for navigation—or the night skies too cloudy to

make use of the stars—navigators would shift their attention to the sorts of clues given by the waves below. In the Pacific Ocean, winds and currents tend to follow more predictable patterns than those of some other oceans, resulting in a relatively steady stream of waves. The one exception? Swells specifically kicked up by ocean storms.

STAR PATHS More than just memorizing tips like "The North Star points north," wayfinders learned the varied paths that numerous stars take across the sky on a nightly basis—to the point where many navigators could identify the direction in which they were headed with only a few visible stars in the sky. Navigators also learned cues like pointer stars—where certain stars always line up to point a certain direction—to help find the way.

BIRDS AND OTHER ANIMALS While the behavior of certain animals, like dolphins, offered some clues to ocean navigators of the Pacific, birds were a voyager's best friend. Encountering a flock of winged fishermen was just one telltale sign that land was near, and the flight paths of birds could give away the direction of the nearest island. Of course, a skilled navigator would follow the right birds. Some birds head out to sea soon after hatching and only return to land to nest. Following those birds could have you lost at sea.

37 USE THE BIG DIPPER TO LOCATE POLARIS

At night, stars appear to drift from east to west, but they aren't actually moving. Instead, Earth spins once every 24 hours, giving us the illusion that the sky is turning the other way.

But there are two points in the sky that do not drift at all—in fact, the stars appear to revolve around them. One of these stationary spots (the North Celestial Pole) is visible in the northern hemisphere, while the other (the South Celestial Pole) is visible in the southern hemisphere.

In the north, there is a cosmic coincidence that is useful if you get lost at night without a compass. The bright star Polaris is located almost exactly at the North Celestial Pole. The star is often called the North Star since you can use it to find north—and get yourself out of any navigatory pickles. Here's how.

STEP ONE Find the Big Dipper (see #29). Note the two stars that make up its bowl's outer edge: Merak (Beta Ursae Minoris) and Dubhe (Alpha Ursae Majoris).

STEP TWO Connect Merak and Dubhe with an imaginary line, and extend that line 30 degrees until you reach the first bright star—Polaris.

STEP THREE Since Polaris is due north, use it to orient yourself and navigate back on course.

Readers scanning the skies thousands of years from now: Polaris will not be your North Star. As Earth spins, its axis wiggles—an effect called *precession*—which slowly shifts the direction of the North Celestial Pole. In 14,000 years, Vega (Alpha Lyrae) will be our North Star!

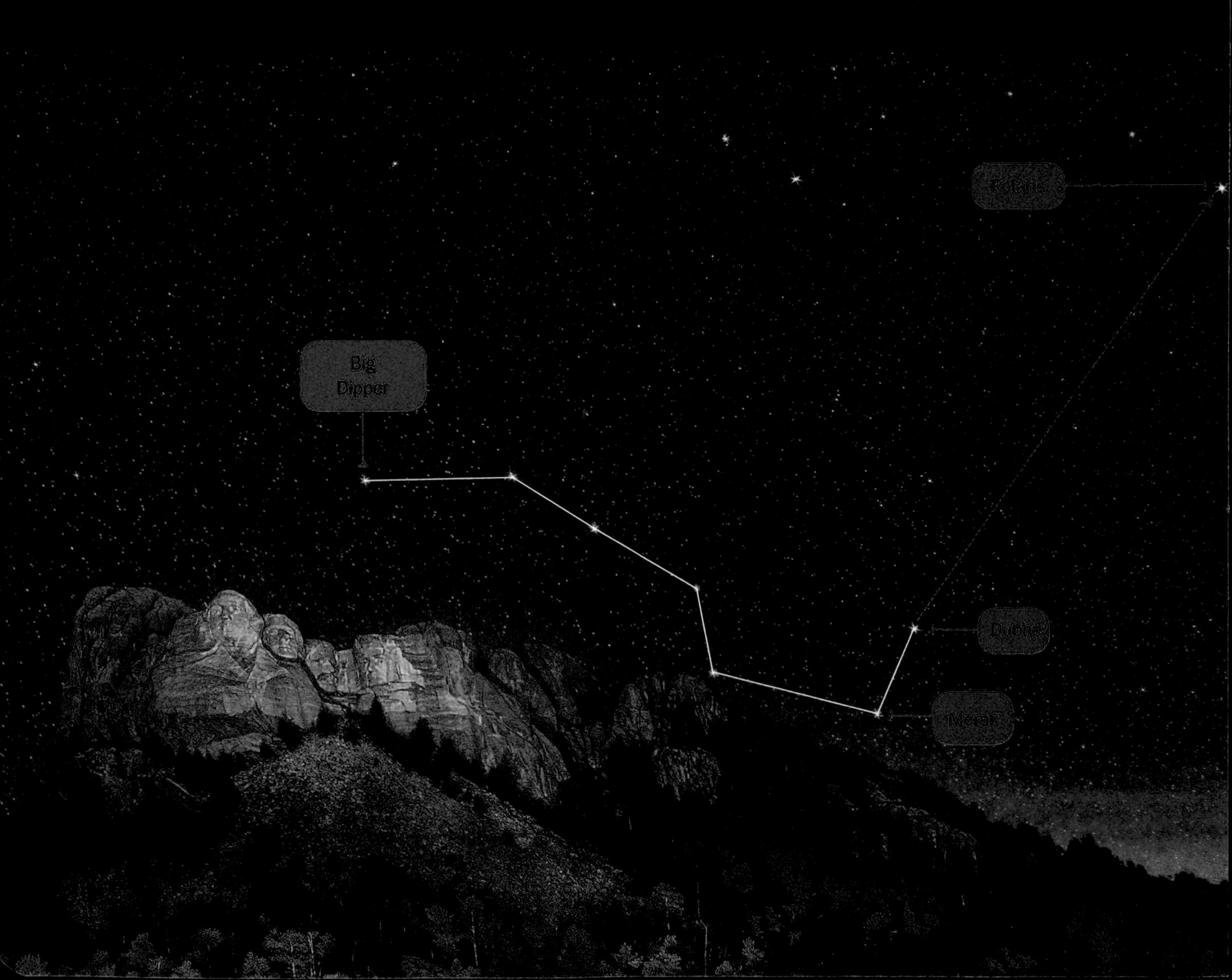

38 ORIENT YOURSELF WITH A GNOMON

If you've never tracked the Sun during the course of a day, it can be, well, illuminating! The next time it's sunny out, use this method to find north (or south in the southern hemisphere).

STEP ONE Start early and place a vertical stick on a large flat spot that will stay in sunlight all day. This is called a *gnomon*. A yardstick in a can of dirt, or a (clean!) plunger, will work. Mark its spot in case it moves.

STEP TWO Mark the line of the plunger's shadow in chalk or place a stone at the tip of the shadow. You can also write the time.

STEP THREE Make this same observation as often as possible, especially in the middle of the day. Every 30 minutes is ideal.

STEP FOUR Find the shortest shadow marking or the stone closest to the gnomon. That line represents when the Sun was highest in the sky—midday (it can be as late as 1:30 PM during daylight saving time). It is also the point when the Sun was southernmost in the northern sky, or northernmost in the southern sky.

STEP FIVE Draw a line between the plunger and the tip of the shortest shadow. This line points directly north in the northern hemisphere or south in the southern hemisphere. With a compass (or compass app), see for yourself the difference between true north, which you just found, and magnetic north.

39 FIND SOUTH IN THE SOUTHERN HEMISPHERE

Sadly, there's no Polaris (aka the North Star, or Alpha Ursae Minoris) of the southern hemisphere; you won't be able to use one star as a stand-in for due south. But the Southern Cross is still handy in finding your way.

Start by locating the Cross (see #40 for tips), which is 5 degrees in length along its long arm (about the width of three middle fingers when measured with your arm extended). Then draw a line from the Cross's top to its bottom and stack two fists (again, with your arms extended) to elongate the line about 20 degrees. Add a couple more finger widths and you will be very close to due south.

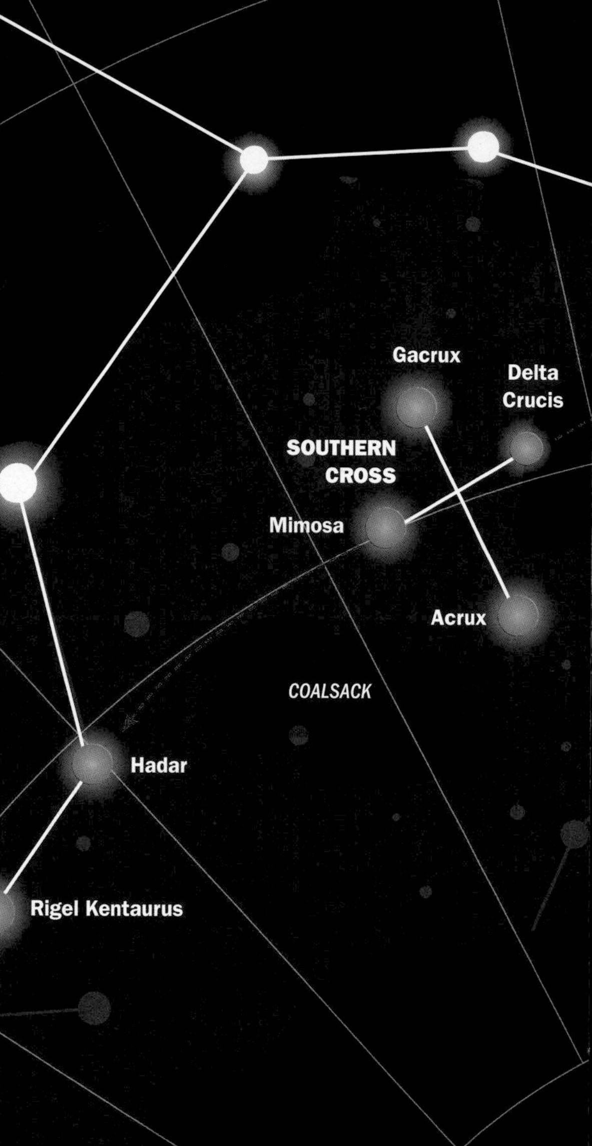

40 FIND THE SOUTHERN CROSS

The five stars of the Southern Cross, also known as the Crux, serve the same role to astronomers in the southern hemisphere as the Big Dipper does for folks in the northern hemisphere: They're the guideposts of the southern skies. Just as the Big Dipper can only be seen from areas north of about 25 degrees south latitude, to observe the Southern Cross you must be somewhere south of 25 degrees north latitude. Learn both constellations to skyhop from anywhere on Earth.

The four main stars of Crux, in order of brightest to least bright, are: Acrux (Alpha Crucis), Mimosa (Beta Crucis), Gacrux (Gamma Crucis), and Delta Crucis. Gacrux is a giant star with orange and red colors, contrasting with the bright blue of the other stars in this constellation.

VELA

Kappa Velorum

Caldwell 85

Delta Velorum

41 BAG THE COALSACK

The Coalsack is a massive cloud of dust in the Milky Way and one of the best-known examples of a dark nebula. It is a very distinctive night-sky object because it lacks stars! The cloud of interstellar dust obscures many stars that would otherwise be visible in the bright lanes of the Milky Way; humans have noted it since prehistoric times. To find the Coalsack, spot Mimosa and Acrux in the Southern Cross. Now, look for the void in the stars just beyond those two points, almost as large as the cross itself.

42 LOCATE CENTAURUS

Centaurus is one of the most famous constellations in the southern hemisphere. It contains Rigel Kentaurus (Alpha Centauri), the star system nearest to us. Centaurus surrounds the Southern Cross on three sides, resembling a centaur—a mythical beast with a man's torso and a horse's body. To find Centaurus, make a line from Delta Crucis through Mimosa until you hit the two brightest stars in Centaurus: Hadar and Rigel Kentaurus (aka Beta and Alpha Centauri). Trace the arc of bright stars around Crux and you have the horse-like body of Centaurus. Complete Centaurus by drawing a line from Acrux and Mimosa past the outline of Centaurus's body to find the stars making up the Centaur's head, arms, and torso.

43 SAIL OVER TO VELA

The constellation Vela is one part of a set of three constellations that, when combined, make up the mythical ship Argo Navis—named after the famed ship of Jason and the Argonauts. "Vela" means "the sail." To find it, use the shorter bar of the Southern Cross. Start with Mimosa and draw a line over and through the star Delta Crucis. Continue until you hit a bright group of stars in a roughly boxy shape nearby. You have found Vela! Don't be confused by the False Cross—made up of two stars in Vela (Kappa and Delta Velorum) and two stars in Carina. It is sometimes mistaken for the Southern Cross because it is nearby and considerably larger. There is also a small open cluster called Caldwell 85 just off the arm of the False Cross.

44 CHECK OUT CAPRICORNUS

Capricornus, the zodiac's least visible constellation, can be found in the southern hemisphere by joining Aquila's three brightest stars in a line southward. Just higher than Cancer, Capricornus the Sea Goat wades with the water constellations Pisces, Aquarius, and Eridanus.

Capricornus has been represented by some form of goat since the time of the Babylonians. It is commonly depicted as a goat with the tail of a fish. This might relate to a story about the god Pan, who leaped into the Nile when fleeing the monster Typhon. The part of him that was underwater turned into a fish tail, while his top half remained that of a goat.

The brightest star in Capricornus is variable star Deneb Algedi (Delta Capricorni). In Arabic, its name means "the tail of the goat." Other notable stars include Dabih (Beta Capricorni), a double star made of a yellow, 3.1-magnitude giant star and a blue-white, 6.1-magnitude star. You need binoculars to resolve them into view.

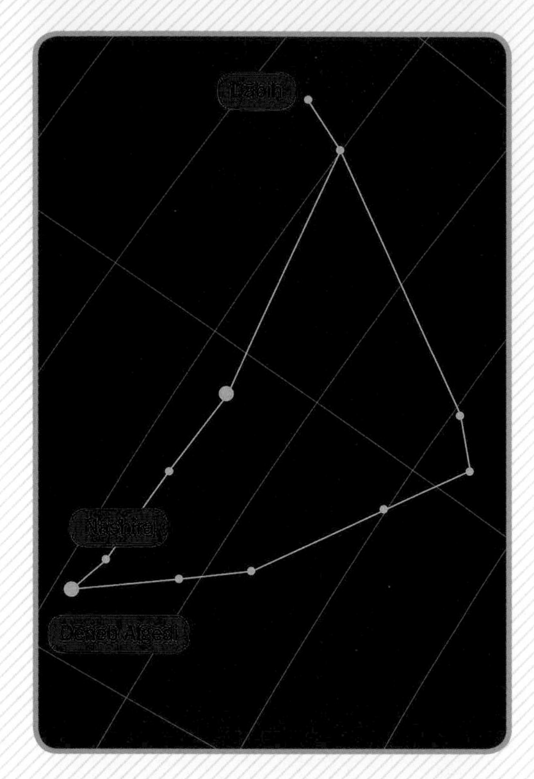

45 HUNT DOWN HYDRUS

Unlike many constellations, Hydrus isn't linked to any myth; as a southern constellation, it was out of sight for the Greeks and the Romans. Flemish astronomer Peter Plancius came up with the Water Snake, but it was German astronomer Johann Bayer who first published it in his star atlas of 1603. He placed it near Achernar (Alpha Eridani), the star at the mouth of the river Eridanus, nestled between the Large and Small Magellanic Clouds. It is sometimes called the Male Water Snake to avoid confusion with Hydra, the female water snake.

As far as notable sites within Hydrus go, VW Hydri is the most popular cataclysmic variable with southern hemisphere observers. When in its usual state, it shines at a faint magnitude 13, but when it goes into outburst (an event which occurs about once a month), it can get brighter than magnitude 8 in just a few hours. Another star of interest is Beta Hydri, the constellation's brightest star. It's about 24 light years from Earth and is the closest bright star to the Southern Celestial Pole.

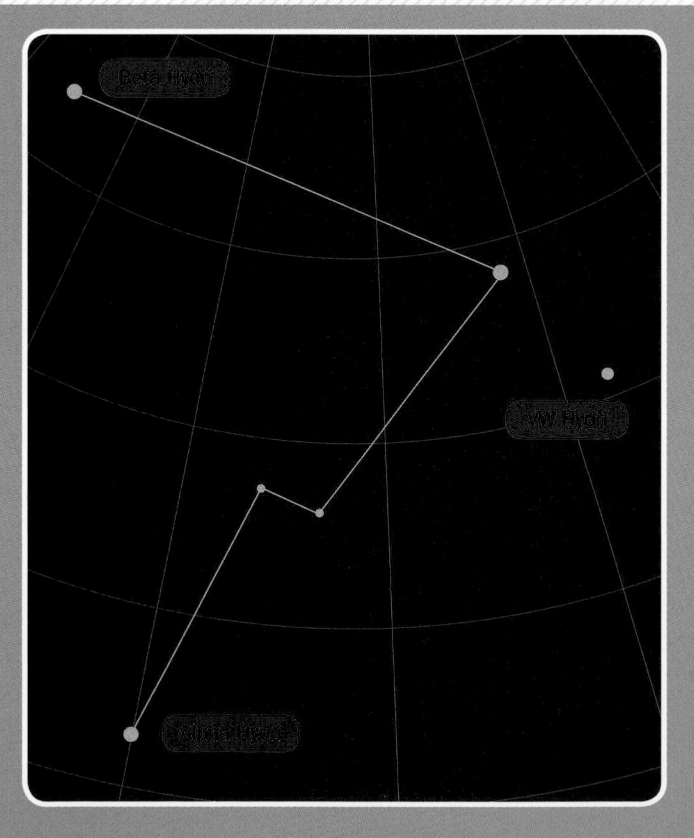

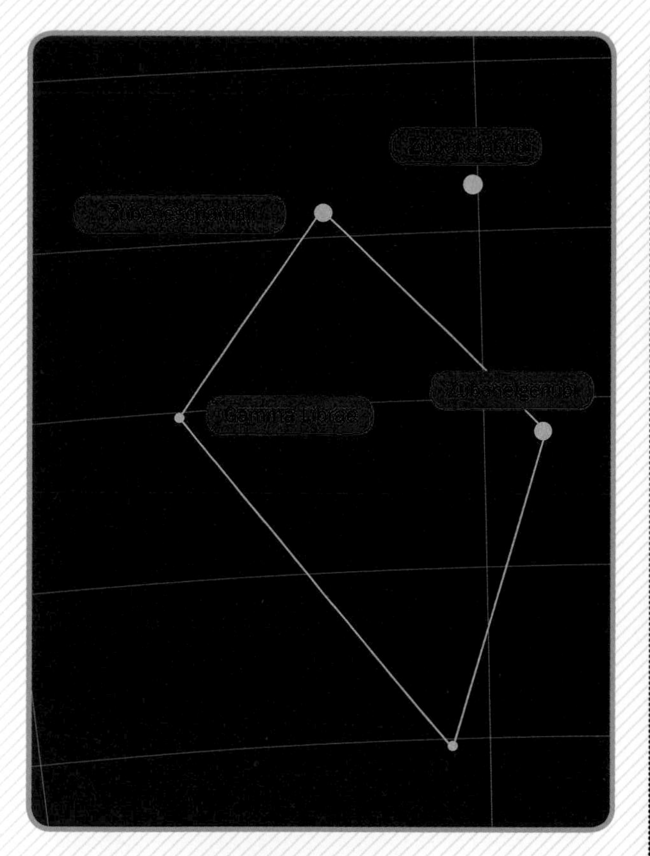

46 LOOK FOR LIBRA

Appearing as a high-flying kite, Libra is situated in the southern sky, bordered by Virgo, Scorpius, Hydra, Centaurus, Ophiuchus, and Lupus. It's easy to find it by extending a line westward from Antares (Alpha Scorpii) and its two bright neighbors in Scorpius. The line reaches a point between Zubenelgenubi and Zubeneschamali (Alpha and Beta Librae).

Libra is one of the constellations of the zodiac and was associated with Themis, the Greek goddess of justice, whose attribute was a pair of scales. Most sources say that Libra became a separate constellation during the time of the ancient Romans, but prior to that these stars were considered part of Scorpius. In fact, the alpha and beta stars still carry the original Arabic names from Scorpius: Zubenelgenubi (Southern Claw) and Zubeneschamali (Northern Claw).

Similar to Algol (Beta Persei)—the so-called demon star said to depict the head of Medusa—Zubenelakribi (Delta Librae) is an eclipsing variable star that fades by approximately 1 magnitude every 2.3 days, from 4.9 to 5.9. The entire cycle is visible to the naked eye.

47 GAZE UPON CARINA

Carina the Keel is a southern hemisphere constellation in the middle of one of the Milky Way's richest parts. Under a dark sky, it is breathtaking. Carina is part of what was once a huge constellation known as Argo Navis (Argo the Ship—named after the vessel that Jason and his Argonauts sailed in on their search for the Golden Fleece). Argo Navis covered such a big area that it was divided into four separate constellations: Pyxis (meaning compass), Puppis (poop deck), Vela (sail), and Carina (keel).

Carina boasts the second-brightest star in the sky: Canopus (Alpha Carinae), after the mythological navigator for Menelaus, King of Sparta. A yellow supergiant, it is 74 light years away. Carina is also home to bright stars Miaplacidus (Beta Carinae) and Avior (Epsilon Carinae). While you can't see much of the Carina Nebula (NGC 3372) with the naked eye, its central region may shine bright enough to make out without a telescope.

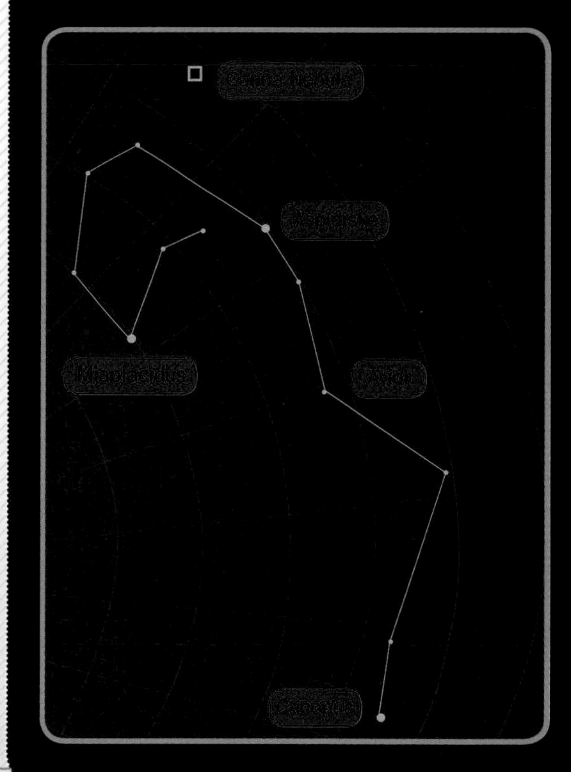

48

TOP FIVE

TRACK THINGS THAT MOVE ABOVE

Now that your eyes are adapted to the dark, you are ready to spot some fast movers in the night sky. Many of these are visible on any clear night.

☐ **THE REVOLVING SKY** Due to Earth's rotation, the sky appears to move slowly, rising in the east and setting in the west. Because the 360-degree Earth rotates once in 24 hours, the sky will appear to move 15 degrees per hour. Settle in and look to the eastern horizon—you'll see the stars slowly come into view.

☐ **METEORS** Shooting stars are not stars at all but pieces of space rock (meteoroid) often no bigger than a grain of sand. Meteoroids enter our atmosphere at such high speeds that heat produced by pressure and friction vaporizes them, leaving only a glowing trail—called a *meteor*—behind. On most nights under dark skies, you'll see a few each hour (see #89 and #92).

☐ **SATELLITES** Several hundred orbiting satellites can be seen with the unaided eye, appearing as either a steady or flashing point of light drifting across the sky. The best viewing time is 45 minutes after sunset or before sunrise, when the sky is dark but the satellite still reflects light back to you.

☐ **IRIDIUM FLARES** Often reported as a UFO, one type of communications satellite (called *iridium satellites*) produces flashes of light 30 times brighter than Venus. These satellites have three large, reflective antennae that can sometimes bounce sunlight back to Earth, and when positioned just right, these flares can be visible for up to 20 seconds—even in the daytime.

☑ **INTERNATIONAL SPACE STATION (ISS)** The ISS (shown here) outshines most stars, appearing as a steady point of light traveling about as fast as an airplane. There are many websites that describe where and when to see ISS from your location.

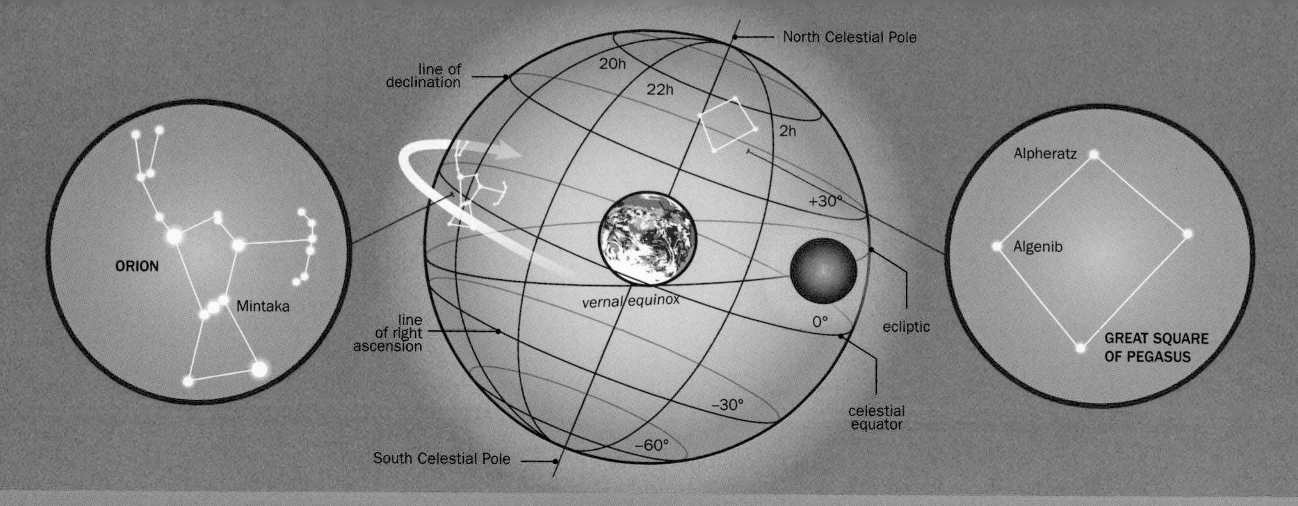

49 LOCATE THE CELESTIAL EQUATOR

To locate something precisely on Earth, you need to know its *latitude* and *longitude*. While latitude measures how far north or south you are from the equator (0 degrees latitude), longitude measures how far to the east or west you are from the prime meridian—an imaginary line at 0 degrees longitude, running between the North and South Poles.

We map the sky by projecting Earth's latitude and longitude system into space. Instead of plotting location in terms of latitude and longitude, astronomers use the terms *declination* (dec) and *right ascension* (RA), respectively. They consider the projection of Earth's equator—the *celestial equator*—to be the 0-degree mark for declination (dec). Objects north of the celestial equator have declinations between 0 and 90 degrees north, while objects south have declinations between 0 and 90 degrees south. To locate the celestial equator:

STEP ONE Since the celestial equator passes directly through Orion's belt, start by finding these three stars. The star almost exactly at 0 degrees is Mintaka (Delta Orionis), the star closest to Orion's left shoulder.

STEP TWO Watch Orion as it rises in the east and sets in the west. The path that Mintaka takes as the evening goes on will trace the celestial equator (0 degrees declination). The celestial equator passes through many other constellations, such as Pisces, Eridanus, Orion, Monoceros, Canis Minor, Hydra, Sextans, Leo, Virgo, Aquila, and Aquarius.

50 FIND 0HR RIGHT ASCENSION

Astronomers call the measurement of astronomical east and west *right ascension* (RA)—basically the projection of Earth's longitude lines into the sky. While longitude is measured in degrees, we measure right ascension in hours, minutes, and seconds, starting at 0hr and ending at 24hr. As the prime meridian is the imaginary line used to connect north and south on Earth, the location of 0hr (and also 24hr) is an imaginary line that connects the North and South Celestial Poles. It passes through the Sun's location on the first day of spring (March 21, the vernal equinox).

Knowing how this coordinate system works will help you navigate star charts and even judge the amount of time that has passed during long observing nights. Find 0hr right ascension like so:

STEP ONE In the northern hemisphere, find Polaris (the location of the North Celestial Pole—check out #37). In the southern hemisphere, find the South Celestial Pole (see #39).

STEP TWO Find the Great Square of Pegasus (see #79) and within it the star Algenib (Gamma Pegasi) and the corner that's lost to Andromeda: Alpheratz (Alpha Andromedae). The line of 0hr runs parallel to the line between them, just inside the square.

51 USE ORION TO MEASURE SPHERICAL COORDINATES

While right ascension (RA) and declination (dec) are regularly used by professional astronomers, observatories, and planetariums, most new amateur astronomers don't use them to find things in the night sky—at least not until they graduate to looking for more obscure objects with the aid of a detailed star atlas. But it's still a handy system to learn!

STEP ONE See #49 at left to find Mintaka (Delta Orionis), the star in Orion's belt closest to his left shoulder. This gets you to the celestial equator (0-degree declination).

STEP TWO Extend your arm straight out and use three or four fingers to find the bright red star Betelgeuse (Alpha Orionis) at 7.5 degrees north and the bright star Rigel (Beta Orionis) at 8 degrees south.

STEP THREE To see a whole 90-degree declination, just find the North Star (from the northern hemisphere) or the Southern Cross (from the southern hemisphere). From Orion's belt to the pole stars is 90-degree declination.

To measure right ascension is no harder, but you'll need a little time:

STEP ONE Find a fixed point—such as a flagpole or tree—that reaches Orion's belt from your view. Stand so that one edge of the belt of Orion is lined up with the top of the tree.

STEP TWO Wait to see which way the sky appears to move. You'll want to watch each of the belt stars move past the fixed point. Start timing as the first star passes.

STEP THREE Keep track of the time and you'll see that it takes just more than 4 minutes for Earth to rotate between each of the stars in Orion's belt, which means from end to end the belt is almost 9 arc minutes across!

ADDRESSES OF THE 10 BRIGHTEST STARS

STAR	Constellation	Right Ascension (RA)	Declination (dec)
1 SIRIUS	**Canis Major**	6 hr 45 min 8.9 sec	–16° 42 arcmin 58 arcsec
2 CANOPUS	**Carina**	6 hr 23 min 57.1 sec	–52° 41 arcmin 45 arcsec
3 ARCTURUS	**Boötes**	14 hr 15 min 39.7 sec	+19° 10 arcmin 57 arcsec
4 RIGEL KENTAURUS	**Centaurus**	14 hr 39 min 35.9 sec	–60° 50 arcmin 07 arcsec
5 VEGA	**Lyra**	18 hr 36 min 56.3 sec	+38° 47 arcmin 01 arcsec
6 CAPELLA	**Auriga**	5 hr 16 min 41.4 sec	+45° 59 arcmin 53 arcsec
7 RIGEL	**Orion**	5 hr 14 min 32.3 sec	–8° 12 arcmin 06 arcsec
8 PROCYON	**Canis Minor**	7 hr 39 min 18.1 sec	+5° 13 arcmin 30 arcsec
9 ACHERNAR	**Eridanus**	1 hr 37 min 42.9 sec	–57° 14 arcmin 12 arcsec
10 BETELGEUSE	**Orion**	5 hr 55 min 10.3 sec	+7° 24 arcmin 25 arcsec

52 TRACE DRACO THE DRAGON

When viewed from much of the northern hemisphere, the large and faint constellation Draco the Dragon is *circumpolar,* meaning it never drops below the horizon. Best seen during the warmer months, Draco is visible all year long for northern viewers—though it may be hard to trace from head to tail as it winds between Ursa Major, Boötes, Hercules, Lyra, Cygnus, and Cepheus.

The Chaldeans, Greeks, and Romans all saw a dragon here, while Hindu mythology claims that the creature is an alligator, and the Persians saw a man-eating serpent. Draco has been linked to dragons in a number of ancient Greek stories, like the dragon that Hercules killed in the Garden of the Hesperides, and the dragon that attacked Athena as she was fighting the Titans and threw her into the sky.

Thuban (Alpha Draconis), the brightest star in the constellation, was the pole star in ancient times, but Earth's precession—the slow and continuous change of its axis caused by gravity—has since moved the pole to Polaris (the North Star, or Alpha Ursae Minoris).

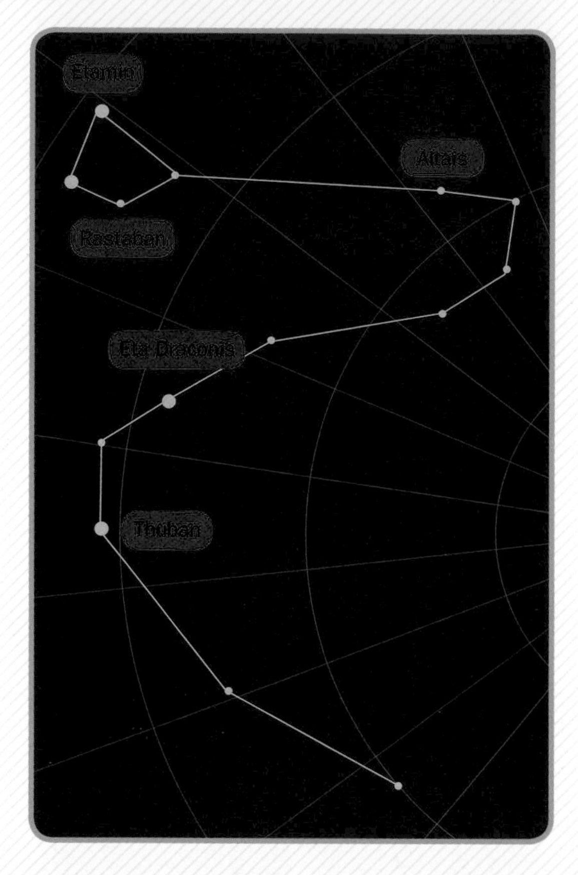

53 HUNT DOWN URSA MAJOR

One of the oldest of the constellations, Ursa Major (the Great Bear) is also perhaps the best known, and numerous legends have been associated with it. The Cherokee and Iroquois see different versions of a bear hunt, while the Sioux don't see a bear—they see a long-tailed skunk.

Particularly famous among this group are the seven stars that make up an asterism commonly known as the Big Dipper, or the Plough. According to one Chinese legend, the stars of the Big Dipper form a bushel measure to portion food in times of famine. The ancient Hebrews saw a similar bushel measure. Beyond bears and bushels, early Britons saw King Arthur's chariot, and the Romans saw a team of seven oxen, driven by Arcturus (Alpha Boötis).

No matter what you see, look for the famous double star Mizar (Zeta Ursae Majoris) and Alcor (80 Ursae Majoris). Separated by 12 arc minutes, this pair in the middle of the Dipper's handle (or the bear's tail, or the line of oxen) is viewable with the naked eye. (See #33 for more info.)

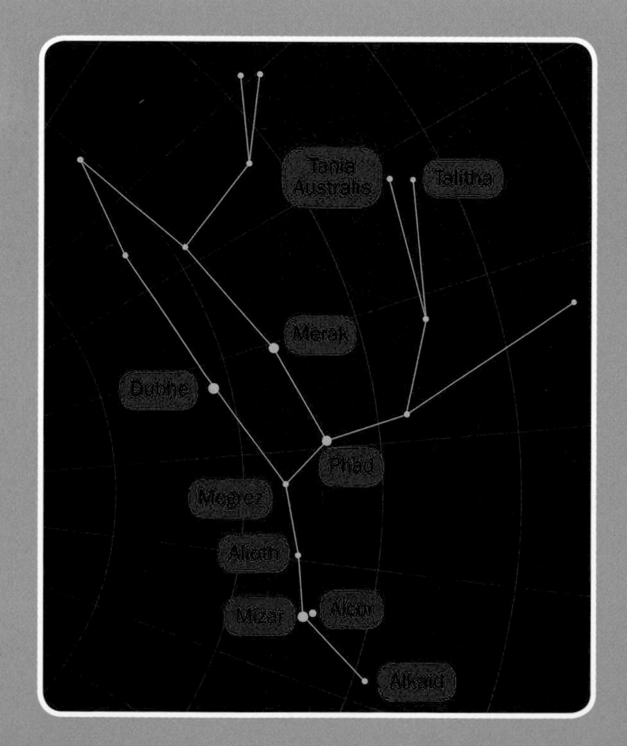

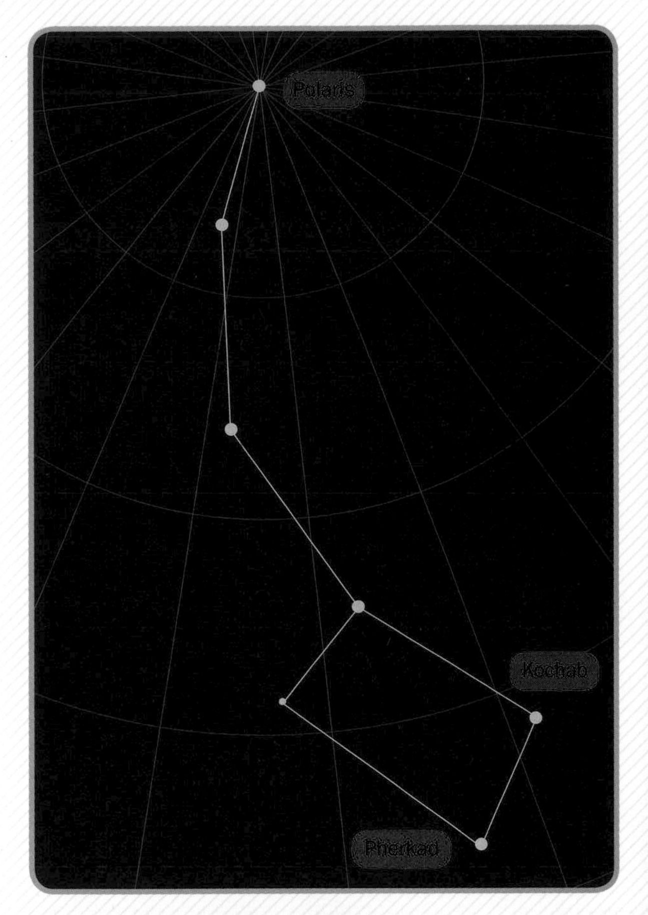

54 FOLLOW URSA MINOR

Also known as the Little Dipper, Ursa Minor (the Little Bear) looks a bit like a spoon with a bent-back handle. Greek astronomer Thales recognized this group of stars as a constellation in 600 BCE. According to Greek legend, the Little Bear (Arcas) and his mother, Ursa Major (Callisto), were both placed in the heavens by Zeus. They follow each other endlessly around the North Celestial Pole. It would be remiss not to mention that Polaris (the North Star, or Alpha Ursae Minoris), the pole star for the northern hemisphere and the brightest star in Ursa Minor, is at the end of the Little Dipper's handle.

Polaris is a Cepheid variable and is currently almost 1 degree from the exact pole. The precession of Earth's axis will carry the pole to within about 27 arc minutes of Polaris around the year 2100, and then it will start to move away again. Eventually, another star will take its turn atop our world. But don't worry; Polaris won't be lonely. It also has a 9th magnitude companion some 18.5 arc seconds away.

GET A GLIMPSE OF THE GIRAFFE

So what exactly is a giraffe doing next to two bears and a dragon, in the frigid sky, near the North Star? The constellation Camelopardalis was thrown into its harsh surroundings in 1624 when German astronomer Jakob Bartsch published star charts depicting new constellations earlier dreamed up by Petrus Plancius. The new constellation represented the camel that brought Rebecca to Isaac in the Bible. ("Camel-leopard" was the name the Greeks gave to the giraffe, seeing the head of a camel and a leopard's spots.) The constellation lies in the large space between Auriga and the bears.

Z Camelopardalis is a cataclysmic variable star, erupting every two or three weeks from its minimum of magnitude 13 to a maximum of 9.6—which is still pretty faint compared to many others. (You'll need a telescope to view this one.) Z Cam's resemblance to other variables ceases when, while fading, it hovers at a middle magnitude. This "standstill" can last for months before the decline resumes. In the late 1970s, Z Cam stayed at magnitude 11.7 for several years.

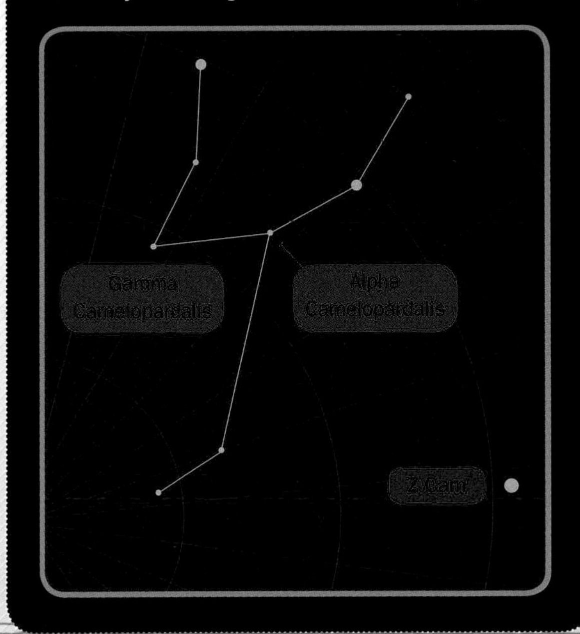

56 BEGIN AT ORION

Orion is one of the more recognizable constellations. Most people can see the three stars that form his belt: Alnitak, Alnilam, and Mintaka (Zeta, Epsilon, and Delta Orionis). Two very bright stars—Betelgeuse and Bellatrix (Alpha and Gamma Orionis)—mark his shoulders, and another pair—Saiph and Rigel (Kappa and Beta Orionis)—mark his knees. You might even glimpse a shield in his left hand, a club held overhead in his right hand, and a sword hanging from his belt. In the northern hemisphere, Orion is easiest to spot between November and February. In the southern hemisphere, May through August is best, although Orion will be upside down.

To find Orion, look south in the northern hemisphere and north in the southern hemisphere. His height above the horizon depends on your latitude: The closer you are to the equator, the higher he will appear.

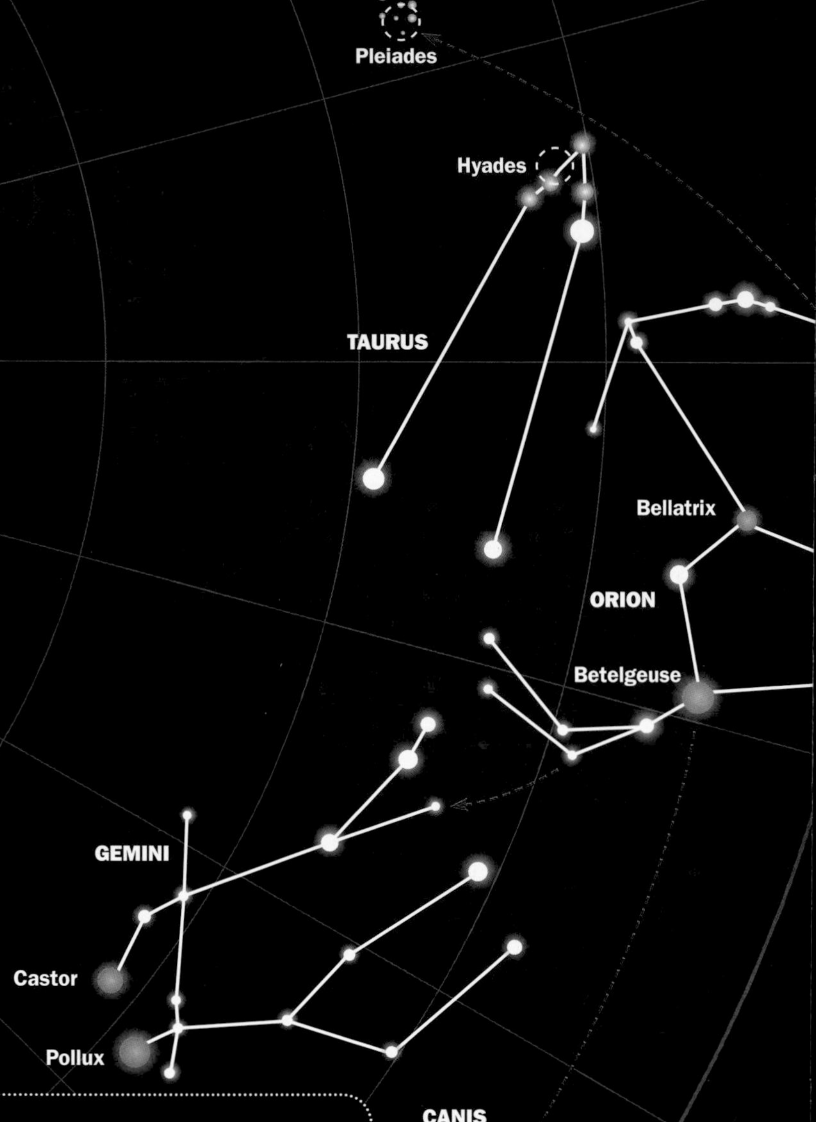

57 GET YOURSELF TO GEMINI

Gemini, or the Twins, is often visible in the winter sky. Two bright stars—Castor (Alpha Geminorum) and Pollux (Beta Geminorum)—form the heads of the stick-figure twins, while the remaining stars make up the duo's arms, legs, and torsos. (The twins look like they have their arms over each other's shoulders.) To find them, start with Betelgeuse (Alpha Orionis) in Orion's right shoulder. Draw a line connecting Betelgeuse and the stars that make up Orion's raised club. The group of stars you reach form one twin's stick-figure legs. In the northern hemisphere, Gemini will be higher in the sky than Orion. In the southern hemisphere, it will be lower, maybe even out of sight below the horizon if you live at the equator.

58 TRACK DOWN TAURUS THE BULL

Orion is engaged in eternal battle with another constellation: Taurus the Bull. Smaller star patterns that make up parts of Taurus include the Pleiades (a cluster of seven visible stars on the back of the bull) and the Hyades (the bull's V-shape face). Pleiades (M45) and Hyades are separate open clusters that formed hundreds of millions of years ago inside huge clouds of interstellar gas. Gravity holds them together, but they will eventually drift apart (likely in a few hundred million years) as they are pulled by the gravity of other objects in the galaxy.

To locate Taurus, imagine drawing a line from the three stars of Orion's belt through the middle of his shield. Extend this line and you will pass just below the Hyades. Extend the line farther and you will reach the Pleiades.

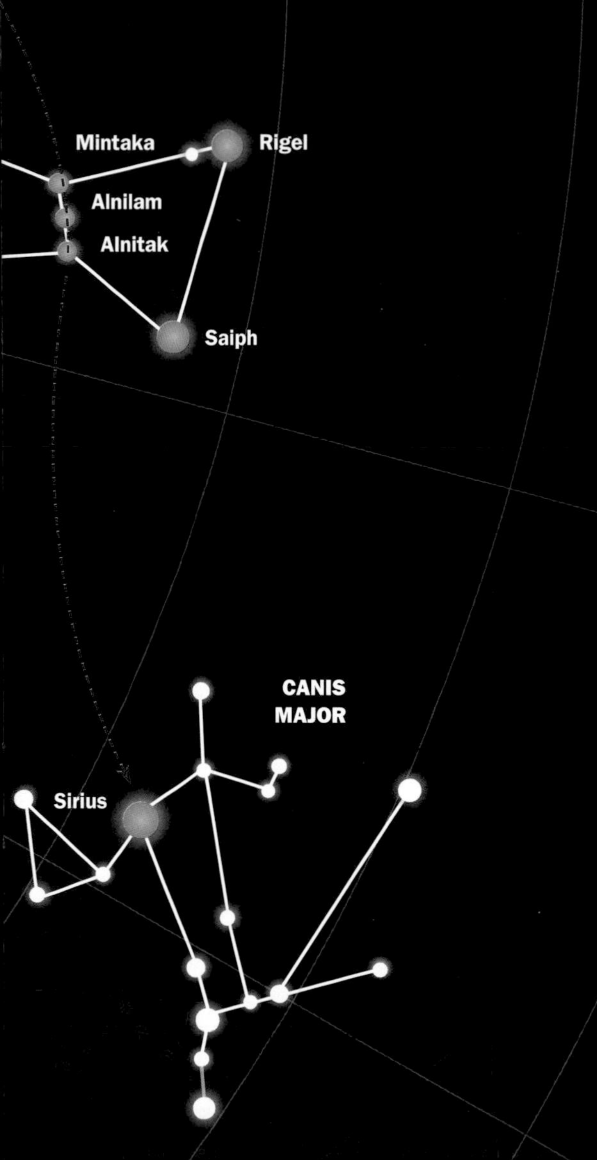

59 FIND CANIS MAJOR AND CANIS MINOR

Orion pursues Taurus across the sky accompanied by two hunting dogs: Canis Major (the Big Dog) and Canis Minor (the Little Dog). The stars that make up Canis Major form the pattern of a stick-figure dog with a triangular face, four straight legs, and a long skinny tail. The very bright star at the heart of the dog, Sirius (Alpha Canis Majoris), is sometimes called the Dog Star. While Canis Major looks a lot like a dog, you really have to use your imagination to see Canis Minor as much of anything at all. Frankly, it resembles a hot dog more than a real dog.

To locate Canis Major, draw an imaginary line through Orion's belt away from his shield until you reach Sirius at the dog's neck. To find Canis Minor, draw a line connecting Betelgeuse (Alpha Orionis) and Bellatrix (Gamma Orionis)—the stars located on Orion's right and left shoulders—away from Orion's shield, until you reach the hot dog.

60 GAZE AT THE MAJESTIC MILKY WAY

Look up at night. Every star you see is in the Milky Way Galaxy—a thin, spiral disk of stars, gas, and dust spread wide like an enormous record. When we observe this luminous belt, we are actually looking inward on our galaxy from the outer part of the disk. The Milky Way contains more than 200 billion stars and perhaps 100 billion planets or more, but we only see the brightest individual stars, all of which are located within a span of 2,000 light years away.

Since before recorded history, humans have gazed upon this mottled streak of light and created explanations for its existence. (Many call it the Milky Way because of an ancient Greek myth in which the goddess Hera was made to feed Zeus's half-mortal son Hercules in her sleep and spilled milk in the sky when she awoke.) Then, about 400 years ago, Galileo first pointed a telescope at this dim, glowing arch and discovered that each patch of diffuse light is actually hundreds of stars, smaller than the eye can see.

On most evenings, if you can find a dark place away from bright city lights, look up and you will see the hazy band of starlight running from horizon to horizon. But be aware that light pollution has dimmed the beautiful sight for many places on Earth; in fact, most babies born today will never experience observing our inspiring galaxy. (To tour it with binoculars, see #136.)

61 GLIMPSE SPILLED GRAIN IN THE CHEROKEE MILKY WAY STORY

The Cherokee of North America have their own story explaining the origins of the Milky Way. An old man and his wife had stored grain for the winter in a basket. One morning, they discovered it was missing and saw paw prints of a giant dog near the basket. The man and his wife hid near the basket with noisemakers and waited. When a giant spirit dog swooped down from the sky and started eating their cornmeal, the man and wife scared the dog back into the heavens. Grain spilled from the dog's mouth, forming the pathway of stars we now know as the Milky Way.

62 SEE MORE MILKY WAY MYTHS

It seems every culture found a narrative among the Milky Way's glimmering streak.

EGYPTIAN The Egyptians had a few myths explaining the origins of the Milky Way. In one, the goddess Nut stretches across the sky, the shape of her body forming the Milky Way.

HINDU The Hindu compared all the stars and planets to a giant swimming dolphin. The Milky Way itself was called Akasaganga—the Ganges River of the Sky. It is no wonder that Hindus associated the Milky Way with the Ganges—named after their holiest waterway.

CHINESE The Chinese called the Milky Way Tianhe, or the Silver River. Nine bright stars in Cygnus the Swan represented Tianjin, the path across a shallow part of this heavenly river. Once a year, on the seventh day of the seventh month of the Chinese calendar, a weaver girl and a cowherd (whose love for each other had been forbidden) cross this bridge to meet. This day is celebrated as the Chinese version of Valentine's day.

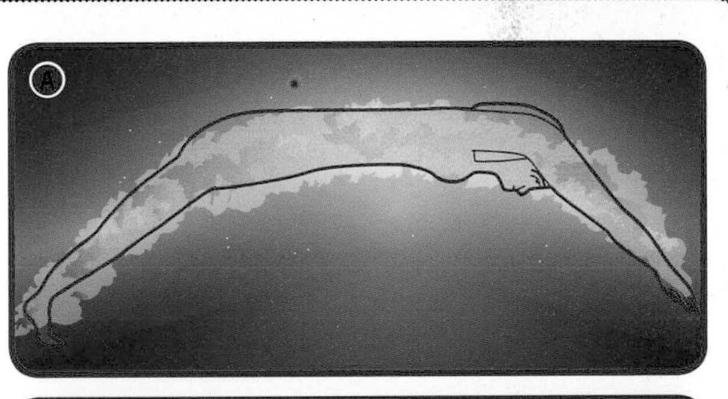

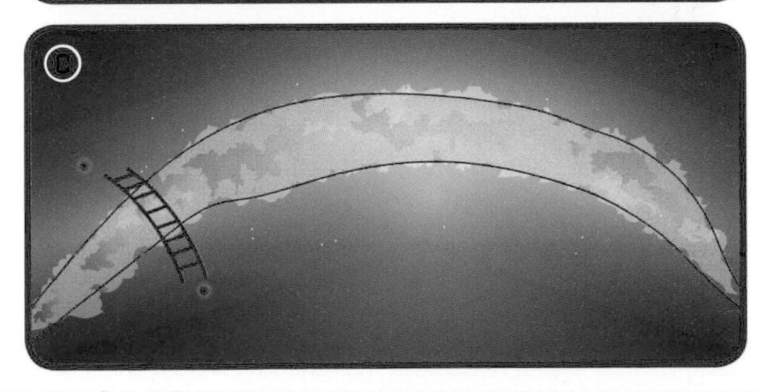

63 SPOT A LLAMA IN THE INCAN MILKY WAY

In the Andes, the Inca saw shapes in both the stars and the darkness. They believed that the dark patches inside the Milky Way were various animals drinking water from what they called the "sky river." In their mythology, a mother llama—named Urcuchillay—walks through the sky river with her baby. The farther the mother llama walks, the darker and more sacred her fur becomes. Her eyes are the bright stars Rigel Kentaurus and Hadar (Alpha and Beta Centauri). The Inca also saw a fox, partridge, toad, and serpent in the Milky Way's darkness.

LEARN HOW THE ABORIGINALS SEE THE MILKY WAY

The indigenous of Australia have observed the sky for 40,000 years, making extremely careful observations of the night sky and passing them down through generations. They are experts in the southern hemisphere skies, where the Milky Way is very easy to see and utterly spectacular. In fact, skies are so clear that you can easily see dark clouds of gas and dust that obscure distant stars.

Some aboriginal groups imagine the Milky Way as a river containing fish (bright stars) and water lily bulbs (dim stars). A legend from the Yirrkala community says the two dark patches of the Milky Way near Sagittarius (often called the Coalsack today) represent the bodies of brothers who drowned while fishing. To several other clans, the Coalsack is part of a much larger dark feature of the Milky Way that represents a gigantic emu in the sky.

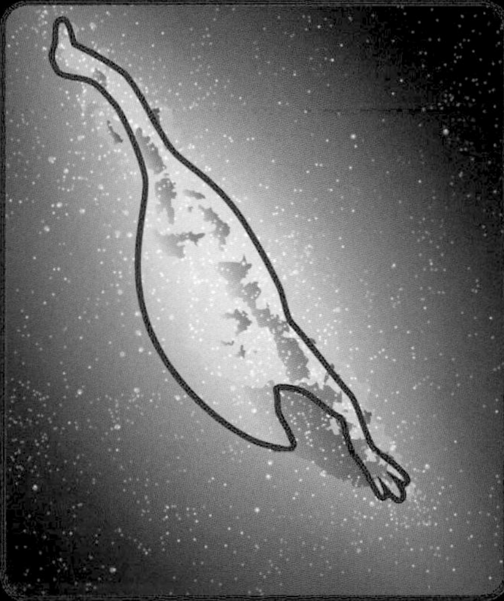

65 LUCK OUT WITH NOCTILUCENT CLOUDS

Have you ever witnessed wispy, rippling trails of electric blue glowing in the sky just after sunset? Then you're a very fortunate skygazer. Called *noctilucent clouds* (which is Latin for "night shining"), this rare and spooky effect occurs when light reflects back from the recently set Sun. That's right: While the Sun may have set from our point of view on the ground, its light still shines on the clouds right above us, making for a beautiful display. The highest in our skies, these clouds of tiny ice crystals and dust particles cling to an atmospheric layer known as the *mesosphere,* right at the edge of space. (See #98 for more about the mesosphere.) They are viewable in the summer months from both the northern and southern hemispheres, but only if you are watching between 50 and 70 degrees north and south of the equator.

66 FEAST YOUR EYES ON AFTERGLOW

Ever wonder what causes those rosy-hued skies after sunset? Even though the Sun is below the horizon, its light can be reflected in high clouds or fine dust particles, resulting in an effect called *afterglow.* The colors often appear as individual bands due to *refraction,* or the bending and separating of white light as it passes through a substance (in this case, large air molecules or dust), which is the same process that gives us rainbows. Some of the most stunning afterglows occur following events that throw enormous amounts of dust into the upper atmosphere—like massive volcanic eruptions or major forest fires. Dust particles from these events then catch and scatter the Sun's light long after it has set for observers on the ground.

67 GLIMPSE THE GREEN FLASH

Next time you have an unobstructed view of the sunset on an especially clear evening, keep your eyes peeled for the green flash—a fleeting mirage caused by the atmosphere being hotter than usual at higher altitudes. Since catching a green flash is easiest if you have an unobstructed view of the western horizon, a mountain or beach location works best. If you're ever on a plane traveling west, look out the window at sunset.

What causes the green flash? The answer has to do with prisms and rainbows. If you hold a prism in sunlight at just the right angle, you can break its white light into a spectrum of colors. Our atmosphere acts like a prism: When the Sun sets low on the horizon, the angle separates sunlight into its component colors: red, orange, yellow, green, blue, indigo, and violet. The red, orange, and yellow hues of the Sun's disc disappear over the horizon first, leaving the green, blue, and violet parts of the light to linger behind. So when you see a red-orange Sun dip below the horizon, look for a very brief flash of green light over the top of the setting Sun.

68 MEET THE MOON

Ah, the Moon—our natural satellite and the brightest thing we see in the sky besides the Sun. But did you know that the Moon was likely once part of Earth? When our planet was young, a large body struck it and knocked molten material into space, where it cooled into our Moon. Here are more neat facts about our orbiting companion.

DIAMETER 2,100 miles (3,400 km), or one-fourth the diameter of Earth

MASS 73.5 sextillion kg, or one-hundredth the mass of Earth

SURFACE GRAVITY 16 percent of Earth's gravity. If you weighed 100 pounds (45 kg) on Earth, you'd weigh 16 pounds (7 kg) on the surface of the Moon.

LENGTH OF DAY 29.5 Earth days. Since the Moon is tidally locked to Earth (and so always faces the same side of Earth), its day lasts one whole orbit of Earth. That means that lunar nights last two weeks!

TIME TO ORBIT EARTH
It takes about 29.5 days for the Moon to orbit Earth.

DISTANCE FROM EARTH
239,000 miles (385,000 km)

AGE 4.5 billion years

SURFACE TEMPERATURE 253°F to −387°F (123°C to −233°C). Because the Moon does not have an atmosphere to moderate its surface temperature, you would find it unbearably hot during the daytime and dangerously frigid at night.

SUCCESSFUL MISSIONS
1959: *Luna 2* lander and *Luna 3* orbiter (Soviet Union)
1968–1969: *Apollo 8* and *11* orbiters (United States)
1994: NASA *Clementine* orbiter (United States)
2009: NASA *Lunar Reconnaissance Orbiter* (United States)

ROBOTIC TELESCOPES Some astronomers fancy the Moon as a potential base for robotic telescopes. It is relatively nearby, doesn't have an atmosphere to disrupt viewing, and is permanently dark and cold at the bottom of some polar craters. Plus, radio telescopes placed on the far side would be unaffected by Earth's radio noise. Some even think that large telescope mirrors, up to 164 feet (50 m) across, could be manufactured on the Moon using lunar soil mixed with other ingredients.

INSIDE THE MOON We know that our Moon has a *differentiated body*, meaning it possesses a crust, a mantle, and a core. Based on experiments conducted by the *Apollo* astronauts, scientists believe that our satellite has a small core, about 217½ miles (350 km) in diameter, made up primarily of iron, sulfur, and nickel. This core's outer part might be liquid or molten. The Moon's mantle is 62 miles (100 km) thick and consists mainly of minerals, while the crust—another 31 miles (50 km)—contains mainly oxygen, iron, and silicon.

BOMBS AWAY! The Moon likely owes its numerous craters to a cataclysmic event called the Late Heavy Bombardment, a heavy rain of asteroids and comets that occurred 4.1 billion years ago. The largest and oldest impact crater on the Moon, the South Pole–Aitken Basin, stretches 1,600 miles (2,600 km) wide and nearly 4 miles (6 km) deep in some areas. We can't see this crater from Earth because it is on the Moon's far side.

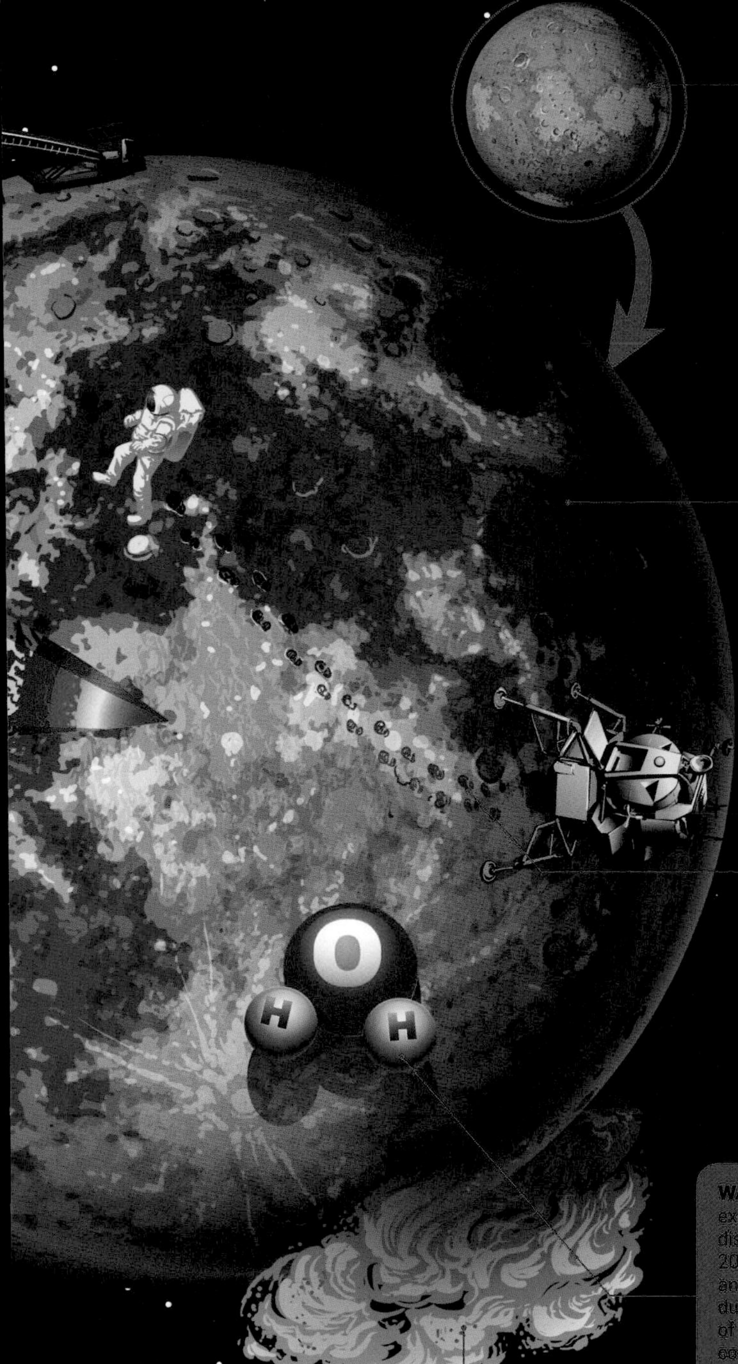

THE FAR SIDE The Moon only shows us Earthlings the side that is nearest to us. Its far side (often erroneously called the "dark side") is never visible to our planet. That's because the time it takes the Moon to rotate once on its axis is equal to the time it takes to orbit once around Earth. We call this phenomenon a *tidal lock* or *gravitational lock*. Most planets and their moons are tidally locked—some double stars are, too!

SURFACE FEATURES The first astronomers to use telescopes to study the Moon in the late 1600s believed its large dark patches were oceans, dubbing them *mares*, Latin for "seas." These are actually pools of cooled basaltic lava that squeezed out of the Moon's mantle; they cover about one-third of the lunar surface. The largest is Oceanus Procellarum, on the near side of the Moon. The *terrae* are the white areas of the Moon, which stargazers first believed were land on water. The highest is Mons Huygens, but it's not even half the height of Mount Everest. (See #153 for more lunar features.)

FOOTPRINTS ARE FOREVER Because the Moon has no wind, water, volcanoes, moving continental plates, or other mechanisms that change the geological features on Earth, the Moon's surface has essentially remained unchanged, except for meteorite strikes. This is why you can see so many craters on the Moon but hardly any on Earth. The footprints left on the Moon by the *Apollo* astronauts will last for millions of years. Eventually they will be erased by the constant "rain" of meteorite dust.

WATER ON THE MOON While liquid water can't exist on the Moon (radiation would cause it to disassociate quickly), the LCROSS mission of 2009 confirmed water ice on the lunar surface and in craters. Tests of volcanic rocks gathered during the *Apollo* mission showed the presence of water molecules, too. The Moon is also constantly bathed in solar wind, which contains hydrogen ions that could react with the oxygen in lunar rocks to create water—someday.

MOONDUST STORMS During the lunar landing of 1969, astronauts struggled with plumes of small particles dancing just above the Moon's surface. Their feet sunk deep into the dust and it covered their suits.

69 DECODE THE PHASES OF THE MOON

The Moon completes a phase cycle from new to full every 29.5 days. As it circles our planet, it goes through phases in which parts of it are visible to us and others are in darkness. These changes are caused by the alignment of Earth, the Moon, and the Sun. We always gaze upon the same side of our natural satellite (there is no "dark side of the Moon"), but that side gets a varying amount of sunlight as it revolves around us. Months were first based on the cycles of the Moon, with one full cycle equaling one month. Consult a lunar calendar for your area to see which Moon you're in store for.

NEW MOON A new Moon always occurs when the Moon is between the Sun and Earth. Since the Moon only shines via sunlight, we can't see it in this position. This is also the only time when solar eclipses can happen.

CRESCENT OR GIBBOUS MOON A sliver of Moon grows, or waxes, from new Moon to full as it swings around Earth. Once it passes full Moon, it wanes, returning to a tiny silver arch.

QUARTER MOON We call the phase in which half of the Moon is visible the quarter Moon—because the Moon is one-quarter or three-quarters around Earth.

FULL MOON A full Moon always happens when the Moon is on the side of Earth opposite the Sun and is fully illuminated. This is also the only time when a lunar eclipse can happen (see #75).

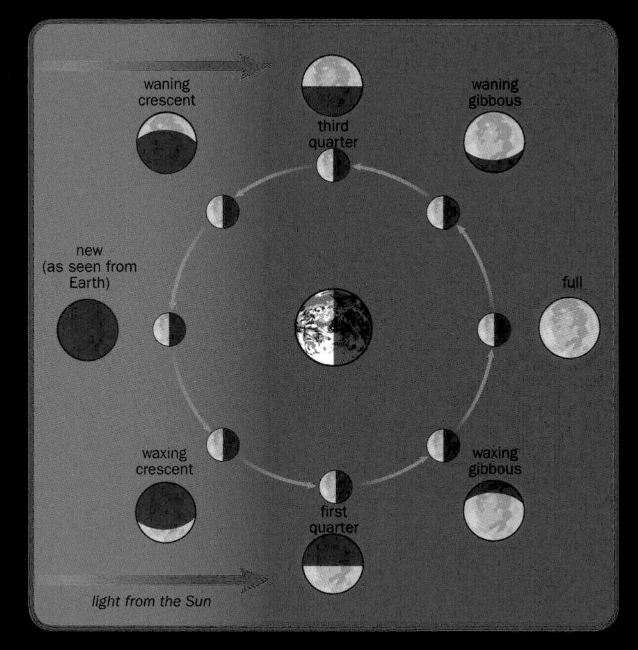

70 UNDERSTAND THE MOON ILLUSION

Have you ever seen the full Moon looking gigantic as it rises above a hillside or behind buildings? Some think this megamoon occurs because the Moon is closer to Earth when it is on the horizon. But no, the full Moon measures 0.5 degrees in the sky (or about half your pinky's width), no matter where it is. Try it: Close one eye and compare the size of the Moon to the tip of your pinky held out at arm's length with your elbow straight. Do the experiment twice: once when the full Moon is near the horizon and a few hours later when it's higher.

While no one doubts that the Moon really does appear bigger, no one has come up with a satisfactory explanation. One popular theory—shown by the Ponzo illusion illustrated at right—posits that when the Moon is just above the horizon, your brain uses perspective to compare the Moon's size to other objects and concludes that the Moon is huge. But once the Moon moves away from the horizon, your brain can't use perspective to gauge the Moon's size, so it doesn't seem as big.

71 WATCH THE TIDES COME AND GO

As Earth rotates, its gravity keeps our oceans at a relatively even level all around the globe. But the Moon's gravity tugs on the water, creating two bulges in the seas: one on the side of the planet facing the Moon and one on the opposite side. These bulges move around as Earth rotates and the Moon exerts its force all over the planet. (The Sun exerts gravity on it, too, but it's less strong due to the Sun's distance.)

Where the Moon pulls at a section of water to create a bulge, this is *high tide*; simultaneously, *low tide* occurs where water has shifted away into areas of high tide.

A SPRING TIDES During each month's new Moon and full Moon, the Sun and Moon are directly in line with Earth. In this scenario, the Sun's gravitational pull adds to that of the Moon, creating a greater tidal swell. These are called *spring tides*, but they occur in all seasons. When eclipses happen, the direct alignment can cause *supertides*.

B NEAP TIDES In the first- and third-quarter Moons, the Sun and Moon pull in different directions and the oceans undergo *neap tides*: high tides that are lower than usual and low tides that are higher.

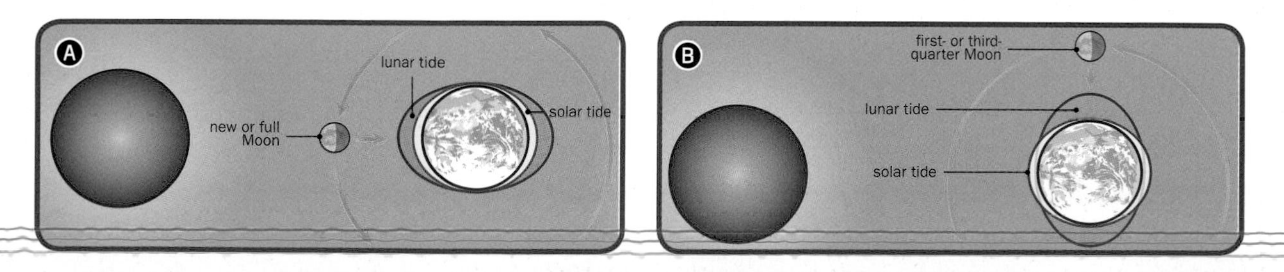

72 SPOT SOME STRANGE MOONS

For something that we see nearly every night in the sky, our plain old Moon can do some surprising things. Here are a few favorites from the trusty *Farmer's Almanac*.

SEE THE SUPERMOON Ever heard the local news talking about an upcoming "supermoon"? Since the Moon orbits Earth in a slight ellipse (not a circle), it varies in its distance from Earth by 26,600 miles (42,800 km). When the Moon is at its closest, that's when we get the supermoon, which can appear up to 14 percent larger and 30 percent brighter in the sky. While "supermoon" is what you'll hear on TV, astronomers call the Moon at its closest to us a *perigee Moon* (*perigee* means "closest to Earth"). Whatever the opposite of a supermoon might be in popular terms, the Moon at its farthest from us is an *apogee Moon* in astronomer speak.

SEE BLUE MOONS You may have heard the expression "once in a blue Moon," but have you ever wondered just how often that actually is? The original meaning from the *Farmer's Almanac* was the third full Moon in a rare season with four full Moons. The 12 usual Moons of a season are already named, so this extra one was called a blue Moon. Starting in the 1930s, we now count the second full Moon in any calendar month as a blue Moon. How rare is it? We're likely to get a blue Moon every two or three years. Two blue Moons in a year (by the new definition) happen about every 19 years—when there is no full Moon in February.

As far as color goes, blue Moons are rarely actually blue. Still, smoke and dust particles floating in the atmosphere can in fact make the Moon appear in varying shades—from red to blue—yearlong.

73 MEET THE MAN ON THE MOON, ALL OVER THE WORLD

While the "man in the Moon" is a European tradition, people around the world see the Moon's many seas, highlands, and craters differently. Flip to #153 to learn about the lunar features that create these illusions.

RABBIT An Aztec story tells of two gods who lit themselves on fire and became suns in the sky. A third god, finding it too bright, threw a rabbit at one, making the Moon.

FOX In Peru, some see a fox who braided a rope to get up to the Moon. He wasn't alone, according to the myth, but had a little bit of help from some bird friends.

FROG One Chinese myth talks of a woman flying to the Moon after finding her husband's immortality tonic. When the husband found out, he turned her into a frog.

74 SEE THE CRESCENT MOON AROUND THE GLOBE

Check out the Moon as you travel. Depending on your location on Earth, you should be able to notice differences in our celestial companion's appearance. The farther you travel, the more noticeable these lunar changes become.

Here is a fun experiment to try: Look at the crescent Moon in your sky and call or email a friend in another hemisphere (either northern or southern). Ask your friend how the Moon's crescent is pointed. From your different points of view, you will discover the Moon's crescent is pointing in opposite directions.

Why does the Moon look different to each of you? Since you are looking at the Moon from very different angles as it orbits around the equator, its appearance in the sky varies. Look at any object across the room from you; now stand on your head (or just imagine standing on your head) and look at that same object again. From your upside-down vantage point, the object will of course appear to be reversed. The Moon appears "flipped" between hemispheres for the same reason. The angle of the Moon's phases will gradually change as you travel from the North Pole to the South Pole, rotating 180 degrees. At the equator, the crescent Moon will actually point up from or down at the horizon as it rises or sets.

75 WATCH THE MOON TURN RED IN A LUNAR ECLIPSE

Lunar eclipses occur when the Moon passes through Earth's dark central shadow (called the *umbra*). For this to happen, the Moon must be in its full phase and lined up with the Sun and Earth. Since the Moon's orbit is inclined 5 degrees from Earth's orbit, this alignment only happens about twice a year. Luckily, it is safe to watch a *total lunar eclipse* with your naked eyes. (Not so much for solar eclipses; see #222–224.) During a total lunar eclipse, the full Moon will take on a spectacular red color, as molecules in Earth's atmosphere remove blue from sunlight and let reddish-colored light pass through to illuminate the Moon.

Partial lunar eclipses occur when only part of the Moon passes through Earth's umbra, creating a semicircular shadow. The three bodies can also align imperfectly so the Moon passes through only the outer portion of Earth's shadow (a *penumbral eclipse*). Most of the Moon gets full sunlight, so these can be hard to see.

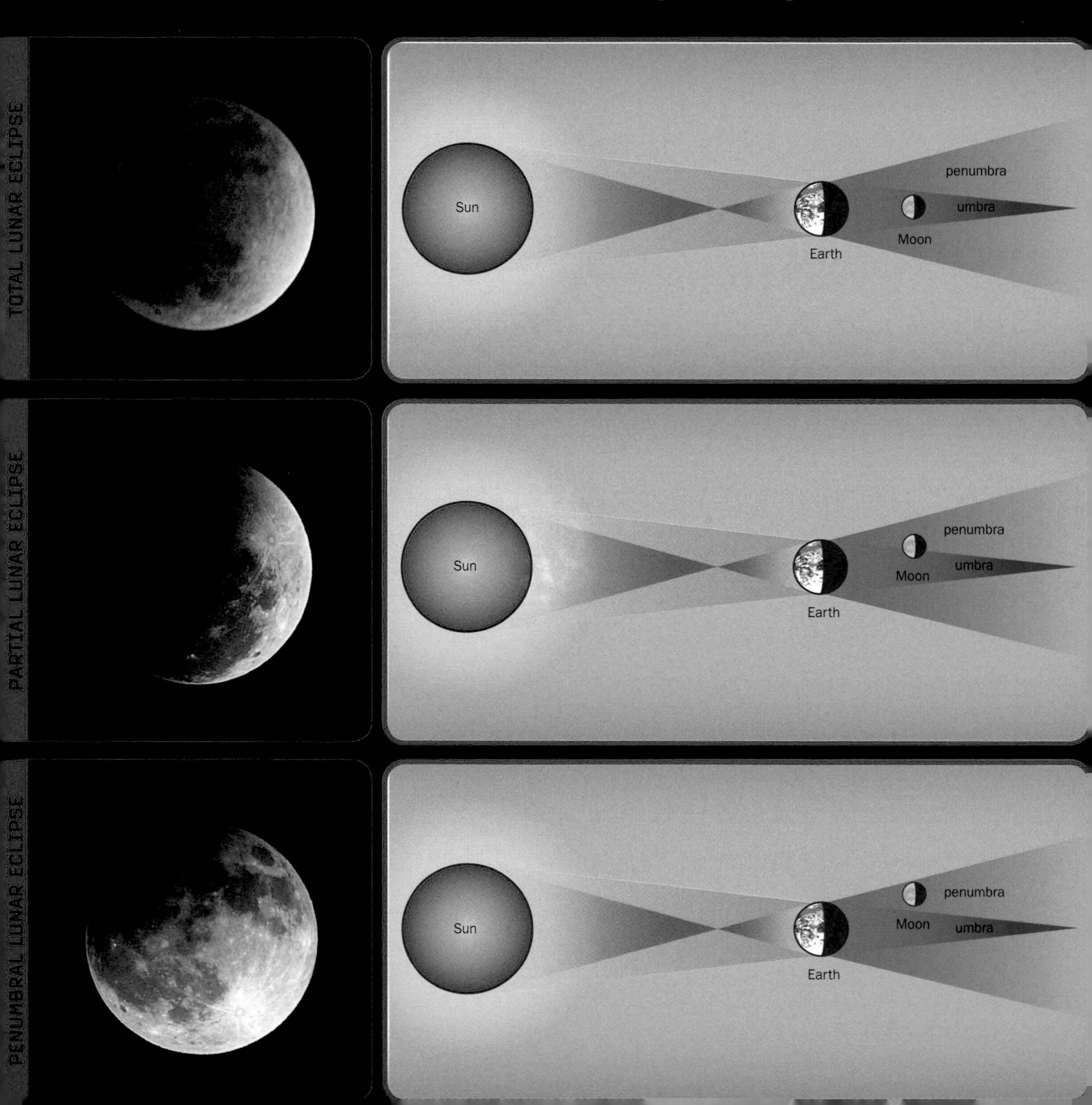

76 READ A LUNAR ECLIPSE MAP

You can use an eclipse map to find out how much of an upcoming lunar eclipse you can view from your location. An eclipse map shows how deeply the Moon will fall into Earth's shadow. These maps often use universal time to indicate the major events in the eclipse, including its beginning, ending, and the time of totality. Everyone around the globe who can see the eclipse will see it simultaneously.

The contour shading on the map shows how deep the Moon will be in eclipse. The lighter the shade, the less noticeable the eclipse, with the darkest shade indicating where a total eclipse will be visible. Shades in between indicate a partial eclipse. You may see bonus info in your chosen eclipse map, like the Moon's apparent size, the magnitude of Earth's penumbral and umbral shadows on the Moon, or the Saros cycle the eclipse is a part of. What's the Saros cycle? The cycle by which every 18 years, 11 days, and 8 hours the pattern of eclipses repeats itself.

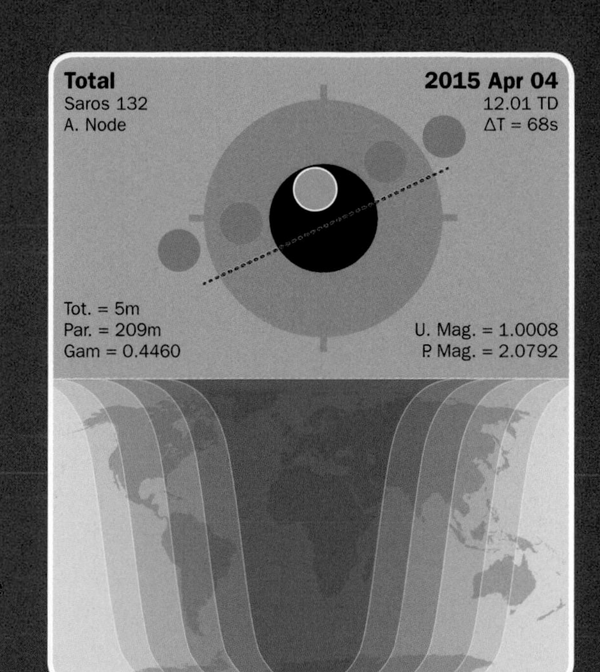

77 PUT TOGETHER A LUNAR ECLIPSE MODEL

What causes a lunar eclipse? Try this activity and you'll remember it for sure.

STEP ONE Working with clay or dough, create an Earth that's 1 inch (2.5 cm) in diameter, then make yourself a Moon ¼ inch (6.4 mm) in diameter. A ratio tip: Four Moons should fit across Earth's face.

STEP TWO To attach Earth and the Moon to scale on a yardstick, just stick a toothpick in each and clip them 2½ feet (76 cm) apart. About 30 Earths should be able to fit between Earth and the Moon.

STEP THREE Outside on a sunny day or in a dark room with one really bright lightbulb posing as the Sun, start your investigation. Holding the yardstick horizontally with Earth closest to your lightsource, turn it until Earth's shadow covers the Moon. Think about what side of Earth you would have to be on to see this play out. During a real lunar eclipse, everyone with clear skies on the night side of Earth sees a red or orange full Moon.

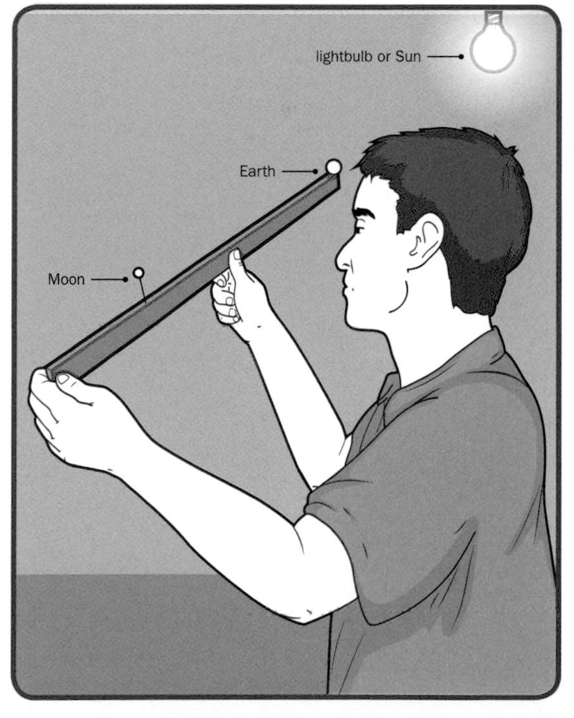

78 MEET MERCURY

Tiny, dense, and the closest to the Sun of any of our solar system's planets, Mercury is a rocky planet that greatly resembles our own Moon—largely gray and covered in craters. The similarities, however, end there.

DIAMETER 3,030 miles (4,880 km)

DISTANCE FROM THE SUN On average, 36 million miles (58 million km), or about two-fifths the distance between Earth and the Sun

MASS 328.5 sextillion kg

MOONLESS Together with Venus, Mercury is one of two planets in our solar system lacking known moons.

LENGTH OF YEAR 88 Earth days

LENGTH OF DAY Mercury revolves rapidly around the Sun but rotates on its axis relatively slowly. While a year on Mercury is just 88 Earth days long, it takes almost 60 Earth days for Mercury to complete one rotation on its axis. The combination of a very fast orbit and slow spin results in a very long Mercurian day. A day on Mercury (sunrise to sunrise) is a surprising 176 Earth days long.

SURFACE TEMPERATURE Much like our Moon, Mercury lacks an atmosphere sufficient to regulate its temperature. The sunlit side heats up to 800°F (425°C), while nighttime temperatures can dip down to −280°F (−175°C). In fact, in the deepest craters never exposed to sunlight, temperatures don't get above −276°F (−171°C).

GRAVITY 0.38 times Earth's gravity. In other words, if you weighed 100 pounds (45 kg) on Earth, you would weight 38 pounds (17 kg) on Mercury.

WHAT'S IN A NAME? In the 14th century BCE, Assyrians nicknamed Mercury "the jumping planet" because it alternates between being on the left and right of the Sun in our sky. Today, we name its craters after artists, musicians, and authors; its ridges after scientists who made significant discoveries about Mercury; its escarpments after scientific expeditions; and the valleys after radio telescopes.

SUCCESSFUL MISSIONS
1973: *Mariner 10* orbiter (United States)
2004: *MESSENGER* orbiter (United States)

ELONGATED ORBIT Like the other planets in the solar system, Mercury orbits the Sun in an ellipse. But curiously, its elliptical orbit is the most elongated. Its distance from the Sun ranges from 28 to 44 million miles (45–70 million km).

THIN ATMOSPHERE Mercury's gravity is too weak to hold on to an atmosphere for very long, making the atmosphere it does have very thin and fleeting. What gases it can capture come mostly from the solar wind—charged particles streaming from the Sun at 560 miles per second (900 km/sec). As the atmosphere floats into space, it is replenished again by the solar wind.

COMPOSITION Mercury's core makes up a huge 42 percent of its volume. In comparison, Earth's core makes up only about 17 percent. Scientists used to think that the cores of very small planets had to be solid because of how rapidly small planets cool. But Mercury has a weak magnetic field—strong evidence of molten iron under the surface.

CRYPTIC SUN On Mercury, you would occasionally see the Sun rise halfway above the horizon, reverse direction, set, and then rise again. This happens when Mercury is closest to the Sun and traveling at its highest speed along its elliptical orbit. During these times of year, Mercury's speed along its orbit is so fast compared to its sluggish rotation that the Sun will temporarily appear to move backward in the Mercurian sky.

SODIUM TAIL As the closest planet to our Sun (and because of its slow rotation), Mercury is continuously blasted with solar wind for long periods. This wind blows sodium ions off the planet's surface, creating a comet-like gas tail that has reportedly reached 1½ million miles (241,400 km) in length.

THE SPIDER Mercury's largest basin, the Caloris Basin, contains a creepy, spider-shape impact crater called the Pantheon Fossae. Its radiating troughs were once disturbed by a meteor impact, which created the Apollodorus crater—aka the spider's body.

MAGNETIC TWISTERS Mercury's magnetic field is known to emit vortexes of magnetic energy into space. These form when the nearby Sun blasts the planet with solar wind, which carries strong magnetic fields. These twist and whirl with Mercury's magnetic field, creating magnetic twisters that can grow as large as half the planet.

CRATERS Mercury has more craters than any other planet in the solar system. Unlike other planets, where craters are often worn away by winds, water, earthquakes, volcanoes, and other geological processes, there are no comparable processes on Mercury to erase craters from the surface.

WRINKLES As the iron core of Mercury cooled and shrunk billions of years ago, its surface wrinkled. Called *lobate scarps*, these wrinkles can be up to 1 mile (1.6 km) high and hundreds of miles long.

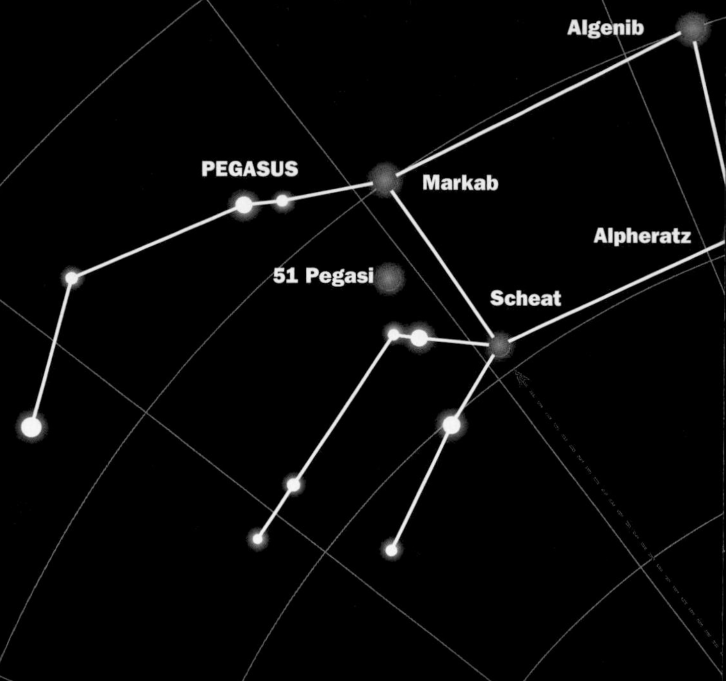

Algenib

PEGASUS

Markab

Alpheratz

51 Pegasi

Scheat

79 BEGIN WITH PEGASUS THE FLYING HORSE

Four bright stars mark the corners of an asterism known as the Great Square of Pegasus—the body of the magnificent flying horse of Greek mythology, and a prominent feature of the fall sky. Often depicted as if Pegasus has just pierced the heavens and is flying down to Earth, the constellation includes a body, head, front legs, and wings. Pegasus can be seen by observers in the northern hemisphere and as far south as 60 degrees latitude. The best time to view it is in October from the northern hemisphere, or in early spring from the southern hemisphere.

To find Pegasus, start by locating the bowl of the Big Dipper. Once there, pay particular attention to the two stars forming the front of the bowl. Then connect these two outside stars with an imaginary line, extending it away from the bowl to Polaris, the North Star (Alpha Ursae Minoris). Continue extending the line until you touch the belly of the flying horse, or the bottom of the Great Square of Pegasus.

80 PINPOINT 51 PEGASUS

Want to see where the first planet ever discovered orbiting another Sun-like star was found? In 1995, we detected Bellerophon orbiting 51 Pegasi, a dim star located between the neck and front legs of the flying horse, barely bright enough to be seen. The planet was named Bellerophon after the Greek hero who tamed and rode Pegasus to slay a terrible monster.

81 FIND ANDROMEDA THE PRINCESS

Fall sky constellations tell the story of Princess Andromeda, chained to a rock as a sacrifice to the sea monster Cetus. To find Andromeda, first locate the Great Square of Pegasus. Pinpoint the star where the horse's belly meets its back legs (Alpheratz, or Alpha Andromedae). Notice that this star marks the vertex of a group of seven stars that also look like they could be

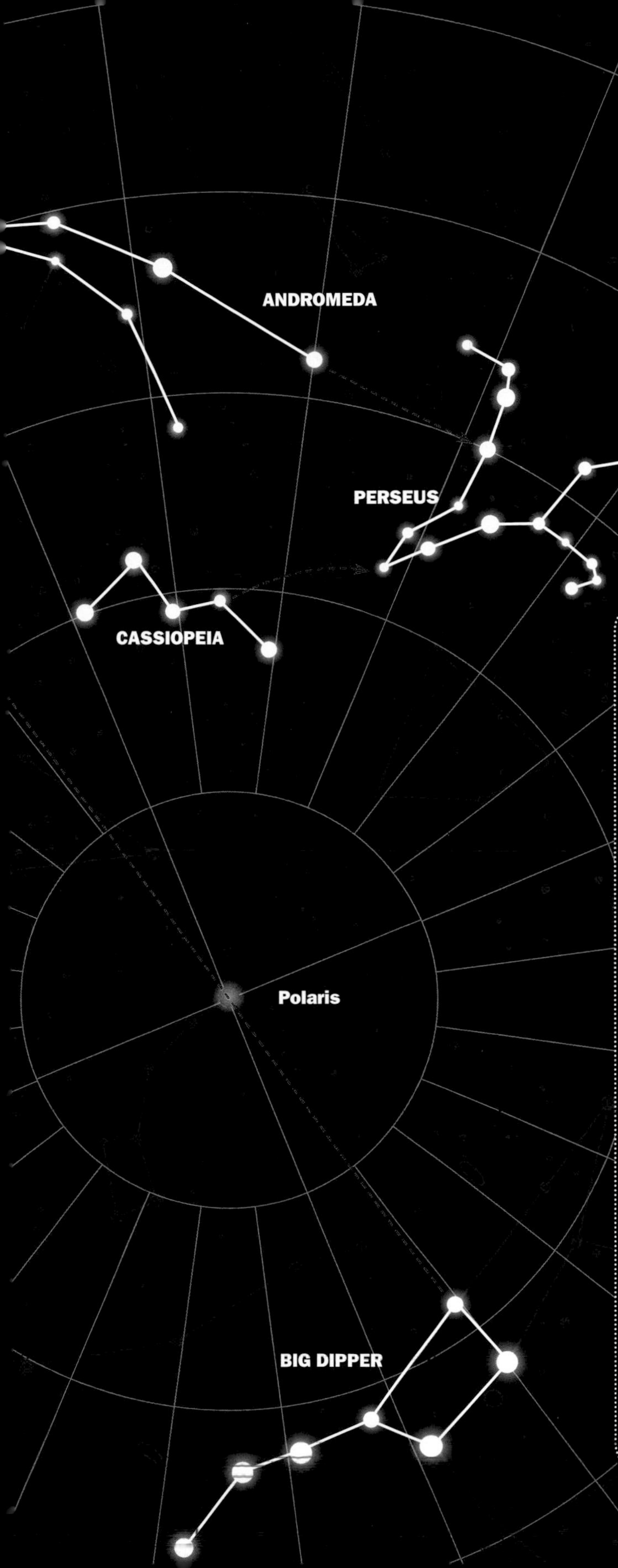

ANDROMEDA

PERSEUS

CASSIOPEIA

Polaris

BIG DIPPER

82 SCOUT PERSEUS THE HERO

No story of a princess about to be eaten by a monster is complete without a hero rescuing her. Perseus, having just beheaded Medusa, a snake-haired monster who could turn you to stone if you looked at her, rode in on the back of Pegasus just as the monster Cetus had reached Andromeda. Holding Medusa's severed head, Perseus forced Cetus to take a peek, turning the monster to stone and saving the princess. Perseus and Andromeda were married and lived happily ever after. The constellation depicts our hero carrying a sword in one hand and the head of Medusa in the other. If you can't see a man carrying a sword and a severed head, try imagining three fishhooks.

First, find Andromeda. Trace the back legs of Pegasus to locate the V shape of stars forming her legs. Using the two lines of stars as pointers, imagine extending both lines away from the Great Square of Pegasus. One line of stars will take you to the center of Perseus, while the other will reach the top of Perseus's head. To check your location using Cassiopeia, connect two of its center stars (left if they form a W, right if an M). Extend this line three times the distance between the stars, right to the head of Perseus.

83 ACQUAINT YOURSELF WITH FAINT ARIES

The ancient Babylonians, Egyptians, Persians, and Greeks all called this group of stars the Ram. In one version of the Greek legend, the king of Thessaly had two children, Phrixus and Helle, both abused by their stepmother. The god Hermes sent a ram with golden fleece to carry them to safety. Helle fell off the ram as it was flying across the strait that divides Europe from Asia. Once safe on the shores of the Black Sea, Phrixus sacrificed the ram, and its fleece was placed in the care of a sleepless dragon.

Aries is the zodiac's first constellation, since the Sun at one time was entering Aries on the day of the vernal equinox—the moment when it crosses from the southern to the northern half of the celestial sphere. Because of Earth's precession, the Sun is now in Pisces at the vernal equinox. Easy to find between Pisces and Taurus, Aries is made up of fairly dim stars, with Hamal (Alpha Arietis), Sheratan (Beta Arietis), Mesarthim (Gamma Arietis), and 41 Arietis being exceptions.

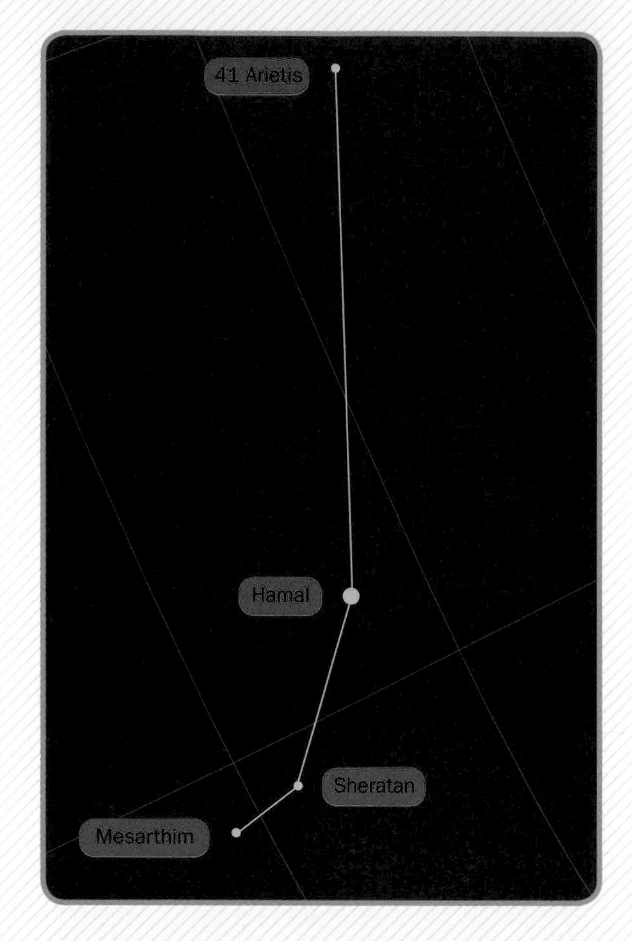

84 CHASE THE CHARIOTEER

Auriga the Charioteer is a lovely multisided figure that is easy to find in the sky, largely because of bright Capella, the she-goat star, and her entourage of three little kids. Ancient legends portray Auriga as a charioteer carrying a goat on his shoulder and two or three kids on his arm. The charioteer is also seen as Erechtheus, the son of Hephaestus (the Roman god Vulcan), who invented a chariot to move his disabled body about.

The sixth-brightest star in our sky, Capella (Alpha Aurigae) has been seen as the she-goat star since Roman times. Almost 50 light years away, Capella is similar to our Sun, only larger.

One extraordinary variable system within Auriga is Almaaz (Epsilon Aurigae). This supergiant star fades when its companion passes in front of it once every 27 years. During an eclipse, its brightness drops by two-thirds of 1 magnitude. The deepest phase of the eclipse lasts a full year, which may indicate that the companion star is surrounded by an enormous cloud of gas and dust.

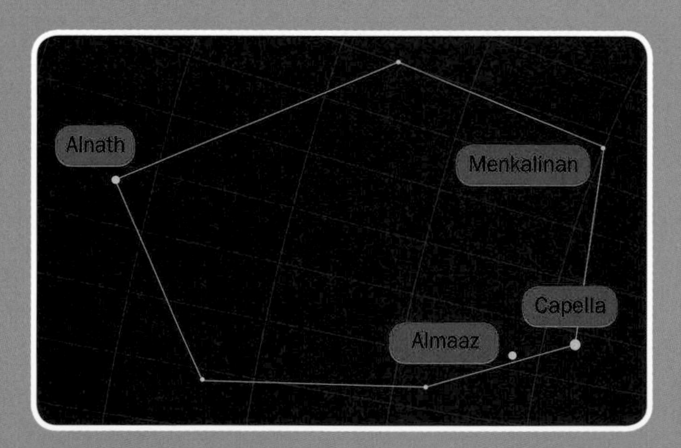

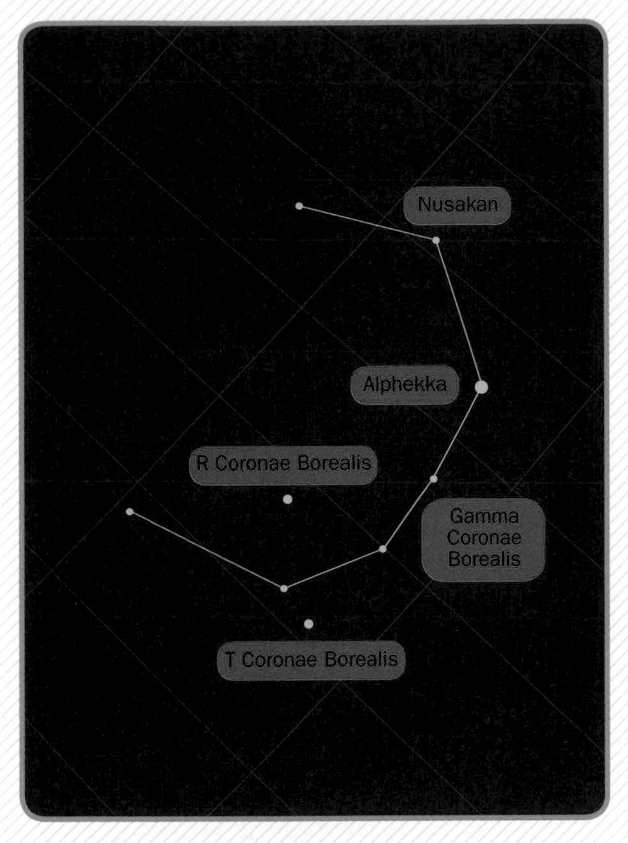

85 BEHOLD CORONA BOREALIS

Just 20 degrees northeast of Arcturus (Alpha Boötis) toward Hercules lies the Northern Crown, a small semicircle of faint but very distinct stars. The Greek myth claims the crown belongs to Ariadne, daughter of Minos, King of Crete. Since being deserted by the mortal Theseus, Ariadne was reluctant to accept a marriage proposal from Dionysus (in mortal form). To prove he was a god, Dionysus threw his crown into the heavens as a tribute to her. Satisfied, Ariadne married him, becoming immortal herself.

One of the more remarkable stars in the sky, R Coronae Borealis (or R Cor Bor), is a nova in reverse. Normally shining at magnitude 5.9, the star suddenly fades at completely irregular intervals—sometimes by as much as 8 magnitudes—as dark material erupts in the atmosphere. It then slowly recovers as the material dissipates. Another notable sight in Corona Borealis is T Coronae Borealis (also called the Blaze Star). Known as a recurrent nova, the star hosted large explosions in 1866 and 1946, and will probably do it again.

86 HAIL KING CEPHEUS

King of the ancient land of Ethiopia, Cepheus was the husband of Cassiopeia and father of Andromeda. Cepheus is an inconspicuous constellation whose five faint stars are easy to find only because they face the open side of the W shape of Cassiopeia, his queen. The constellation shape looks like a house with a pointed roof. Although the top of the roof does not really point to Polaris (the North Star, or Alpha Ursae Minoris), it offers the general direction of the pole at a time when the pointer stars of the Big Dipper aren't readily visible.

Delta Cephei is one of the most famous of the variable stars, and the prototype for *Cepheid variables* (variable stars with steady phases that help us predict their distance from Earth). Its variation was discovered by John Goodricke in 1784. It completes a cycle every 5.4 days.

Another notable star in Cepheus is the crimson supergiant Mu Cephei—a star so strikingly red that William Herschel called it the Garnet Star. It is one of the largest known stars in the Milky Way Galaxy.

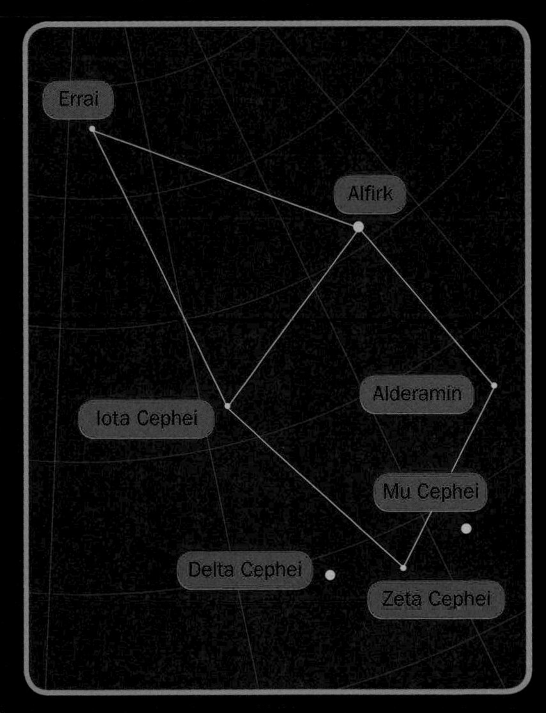

87 EXPLORE THE ZODIACS OF DIFFERENT CULTURES

What's your sign and what does it mean? While the zodiac constellations are neither the brightest nor the biggest clusters of stars, they do hold the distinction of lying on the *ecliptic*, the path that the Sun, Moon, and planets seem to take through the sky from our vantage point on Earth throughout the year. These storied constellations form the extremely distant background behind the plane of our solar system, traditionally splitting the sky into twelve 30-degree segments.

Even though the neat and tidy divisions of the zodiac don't quite match the messy reality of the constellations, your sign is *supposed* to indicate the location of the Sun along the ecliptic on the date that you were born. But the astrological system that we know today was invented many thousands of years ago, and since Earth's rotational axis has shifted over time, the original constellations no longer correspond exactly to where the Sun is when you were born. (It's about 20 days off.)

The Latin names used in the modern astronomical constellation definitions have the same origin as the astrological "signs" listed alongside your horoscope in the newspaper. Aquarius the Water Bearer, Capricornus the Sea Goat, Sagittarius the Archer, Scorpius the Scorpion, Libra the Scales, Virgo the Virgin, Leo the Lion, Cancer the Crab, Gemini the Twins, Taurus the Bull, Aries the Ram, and Pisces the Fish together make up the 12 traditional signs of the zodiac, all finding their origin in Mediterranean myths. But many other cultures have developed their own stories behind the zodiac. Here are a few of interest.

LATIN (ENGLISH TRANSLATION)	SUMERO-BABYLONIAN SYMBOL	MAYAN SYMBOL	CHINESE SYMBOL	SANSKRIT
CAPRICORNUS (SEA GOAT)	GOAT-FISH	BIRD	BLACK TORTOISE OF THE NORTH	MAKARA
AQUARIUS (WATER BEARER)	THE GREAT ONE, CARRYING A PITCHER	BAT		KUMBHA
PISCES (FISH)	GREAT SWALLOW	SKELETON	WHITE TIGER OF THE WEST	MINA
ARIES (RAM)	AGRARIAN WORKER	JAGUAR		MESA
TAURUS (BULL)	STEER OF HEAVEN	RATTLESNAKE		VISABHA
GEMINI (TWINS)	GREAT TWINS	OWL	RED BIRD OF THE SOUTH	MITHUNA
CANCER (CRAB)	CRAYFISH	FROG		KARKA
LEO (LION)	LION	PECCARY	BLUE DRAGON OF THE EAST	SIMBA
VIRGO (VIRGIN)	GODDESS OF SHALA'S EAR OF CORN	MOON GODDESS		KANYA
LIBRA (SCALES)	SCALES	BIRD		TULA
SCORPIUS (SCORPION)	SCORPION	SCORPION		VISCIKA
SAGITTARIUS (ARCHER)	SOLDIER	FISH SNAKE		DHANUS

HINDU/SANSKRIT ZODIAC In addition to the 12 "rasi" that correspond to the Greek zodiac, the Hindu tradition divides the ecliptic into 27 sections, or "nakshatras," rather than constellations. Nakshatras are identified by a prominent star or an asterism (group of stars), and are called lunar mansions. One example is Magha the Mighty One below (in Leo). The Moon appears in each nakshatra for one day.

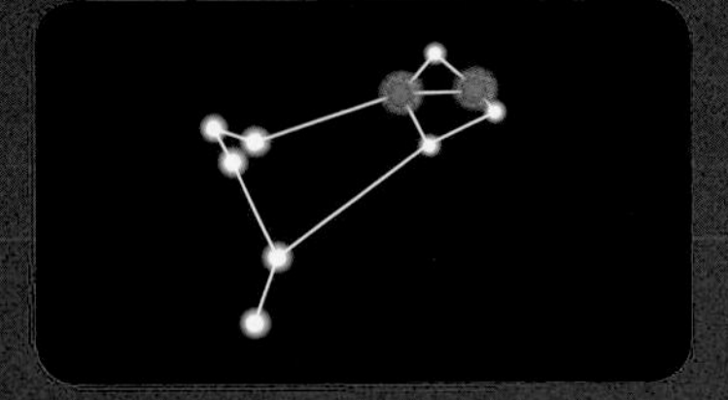

MAYAN ZODIAC The Mayans held that Earth exists on the back of a turtle, with each of the 13 animal constellations holding the Sun in its mouth for a period of the year. A sample sign: Instead of the stars in Orion's belt, the Mayans saw the sacred Three Hearthstones.

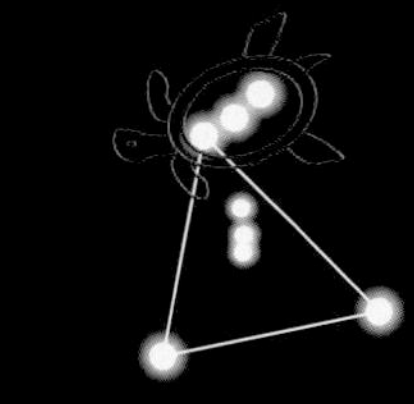

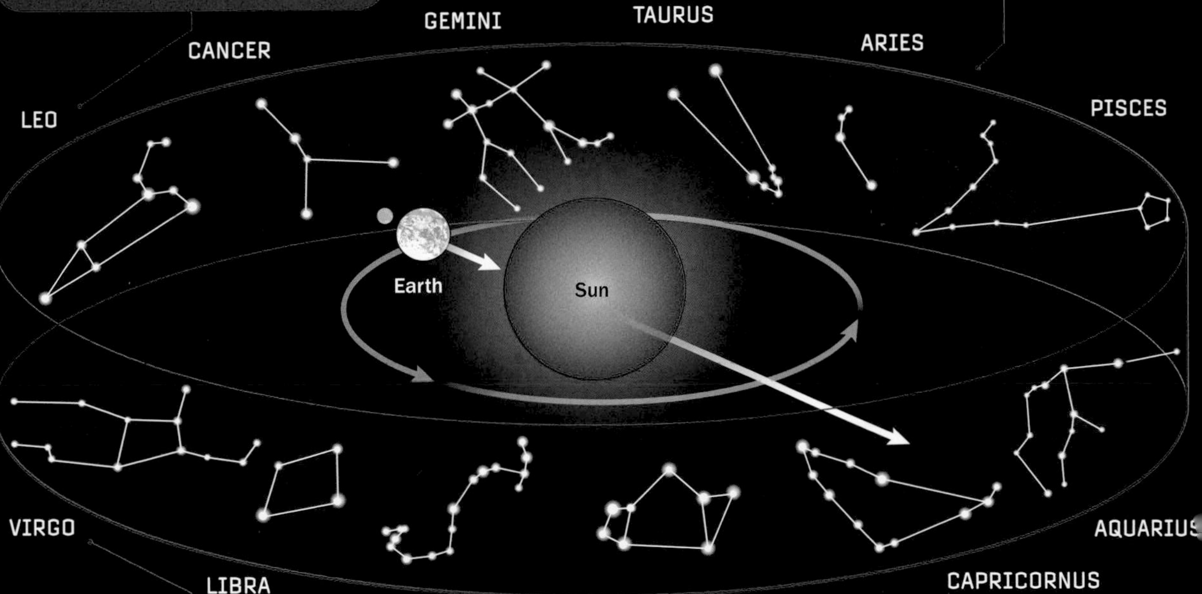

SUMERO-BABYLONIAN ZODIAC Babylonian cosmology dates back to 1000 BCE or earlier. While today's zodiac focuses on the Sun's path through the sky, Babylonians were more interested in the Moon's path. From an original 17 or 18 constellations along the lunar path, the list was reduced to 12 in the sixth century BCE and became the predecessors to today's 12 constellations (as well as our 12 months). Shala, the sign of the fertility goddess who began as an ear of corn in the sky, eventually became Virgo.

CHINESE ZODIAC You might be familiar with the animals used in the Chinese zodiac, each corresponding to a different year in a 12-year cycle: "Year of the Rat," "Year of the Ox," and so on. But they have nothing to do with the constellations along the ecliptic. Instead of focusing on the Sun, Chinese astronomers tracked the Moon through the constellations, dividing its path into quarters for each season: the Green Dragon of Spring, the Black Tortoise of Winter, the White Tiger of Autumn, and the Red Bird of Summer. The Blue Dragon corresponds to the western constellations Scorpius and Leo.

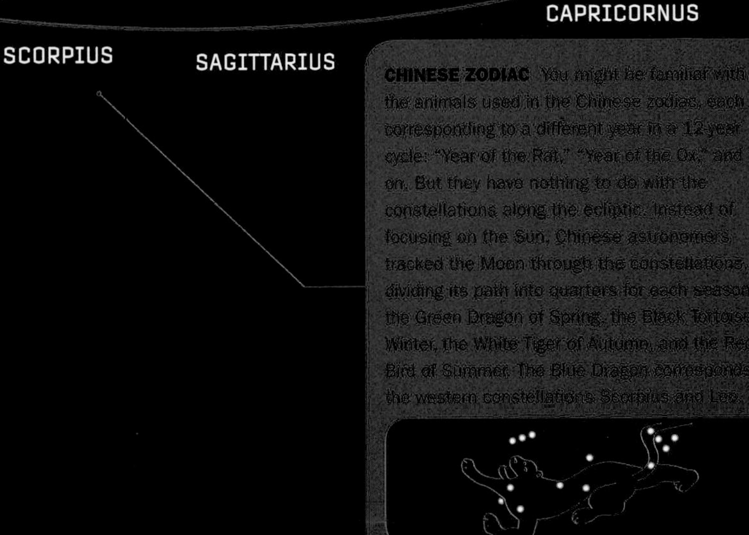

TELL A METEOR FROM A METEOROID

Meteor-what? You may have heard the terms *meteor, meteoroid,* and *meteorite* used interchangeably. While they do all describe small pieces of space rock, each refers to different points of a space rock's journey.

METEOROID As a piece of small rock or debris travels the millions of miles through space, it's called a meteoroid. Think aster-"oid," but smaller: Asteroids tend to be on the big side. (See #230–231 for asteroid info.)

METEOR Call them shooting stars or falling stars, meteors are the bright streaks that meteoroids make in the sky. These brilliant flashes are caused by atmospheric friction and pressure heating up the outer layer of rock so it's hot enough to glow. Much of the rock is vaporized as it falls to Earth. But the rocks aren't really "burning up"; the process is more akin to melting.

METEORITE When a space rock makes it through the atmosphere, "rite" to the ground, it's called a meteorite. Don't worry about being hit by meteorites, even during a meteor shower. Few people in recorded history have ever been hit—you're more likely to be bitten by a shark while winning the lottery!—especially since many space rocks never make it to the ground.

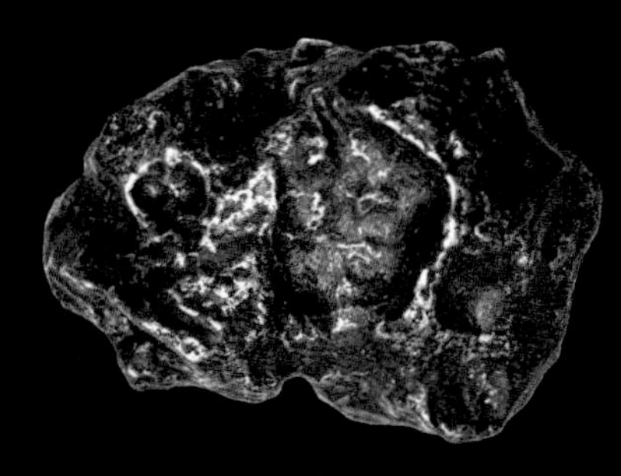

89 PICK THE BEST TIME TO METEOR-WATCH

We see annual meteor showers when Earth passes through the stream of debris left by a comet (see #232–234). Only a few comets cross through the path of Earth's orbit, but as they approach and are warmed by the Sun, they leave behind a stream of dust—bits of space rock and metal that make perfect meteoroids. They vaporize upon contact with our atmosphere and light up our night skies in the form of meteor showers.

The best time to see meteor showers is in the hours after midnight, when we're facing the direction in which Earth moves in its orbit. Think of it this way: When the field of debris left behind by a comet hits Earth's atmosphere, it's kind of like a swarm of lightning bugs hitting the windshield of a moving vehicle. After midnight, we are facing the same direction that the car is driving—looking straight at the swarm.

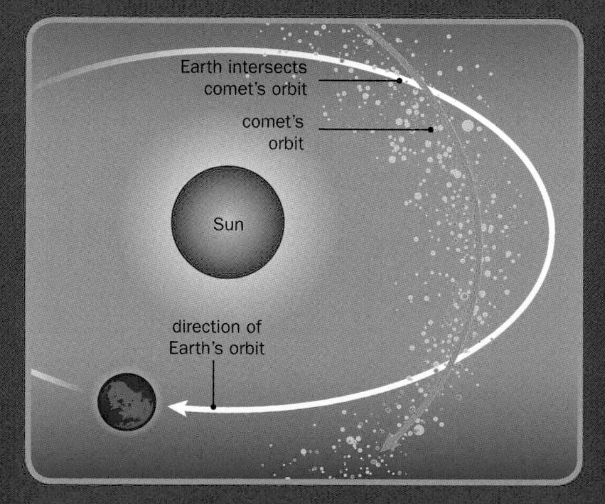

90

TOP FIVE
GO ON A METEORITE HUNT

There are actually a ton of meteorites on the ground, but they're not that easy to find—mainly because they look a lot like Earth rocks! Here are five tips for locating and IDing meteorites.

☐ **PICK THE PLACE** Search at a site with very few rocks so any rocks you do see really stand out. A sandy desert or an ice sheet are best, but a plowed field could work.

☐ **LOOK FOR DARK ROCKS** A meteorite's journey through the atmosphere likely gave it a dark fusion crust. It probably contains iron, so it will be heavier than most other rocks of its size, and chances are it is sturdy and nonporous.

☐ **TEST WITH A MAGNET** As a rule of thumb, meteorites' high iron content makes them stick to magnets. Unfortunately, magnets also attract other Earth rocks, so this is not a definitive test.

☐ **GIVE IT A GRIND** Scratch away a corner of the rock to see if it is metallic or contains metallic grains.

☐ **CALL THE EXPERTS** If a rock passes all these tests, take it to a local museum or university geology department for testing. (They may charge you, however.) Be aware that most small meteorites aren't worth much. The prize is knowing it came all the way from the asteroid belt!

91 NAME THAT METEOR SHOWER

Ever wonder where meteor showers get their names? If you watch for a while, you'll notice that during each particular shower, the meteors' trails radiate mostly from one general area in the sky. If you could trace them all back, they would point to a specific constellation. While this constellation actually has nothing to do with the shower's source, it is the direction the stream of particles is coming from—and so it informs the name of the shower. For example, the Perseids shower appears to come from the direction of Perseus.

92 SAVE THE DATE FOR METEOR SHOWERS

Want to catch the next meteor shower? Several recur roughly around the same time each year. Mark your calendar with these astronomical light shows.

- **QUADRANTIDS** *January 2–3*
- **LYRIDS** *April 22–23*
- **ETA AQUARIDS** *May 5–6*
- **DELTA AQUARIDS** *July 29–30*
- **PERSEIDS** *August 11–12*
- **ORIONIDS** *October 21–22*
- **TAURIDS** *November 4–5*
- **LEONIDS** *November 16–17*
- **GEMINIDS** *December 12–13*
- **URSIDS** *December 22–23*

93 MEET VENUS

Planning your next interplanetary vacation? Forget Venus. Second from the Sun and often called Earth's twin, this terrestrial planet has the hottest surface in our solar system, thanks to its dense carbon-dioxide atmosphere.

DIAMETER 7,520 miles (12,100 km), or about 95 percent of Earth's diameter

AVERAGE DISTANCE FROM THE SUN 67½ million miles (108 million km), or 0.7 times farther than Earth is from the Sun

LENGTH OF YEAR 225 Earth days

LENGTH OF DAY While Venus takes 225 Earth days to orbit once around the Sun, there are 243 Earth days between each Venusian sunset. So why is the Venusian day longer than its year? If you looked down on Venus over its north pole, you'd see that it rotates clockwise, unlike the other planets. In addition to making for a long day, the combination of a counterclockwise orbit with a clockwise spin means that the Sun rises in the west and sets in the east.

SURFACE TEMPERATURE 860°F (460°C), day or night. Venus's thick carbon dioxide atmosphere acts like a wool blanket, trapping solar heat energy. That's a temperature high enough to melt lead.

GRAVITY 0.9 times Earth's gravity. In other words, you would weigh about one-tenth less on Venus and could jump 10 percent higher.

MOONLESS Venus is one of two planets in the solar system without an orbital companion.

SUCCESSFUL MISSIONS
1970: *Venera 11* impacter (USSR)
1978: *Pioneer 1* orbiter (United States)
1989: *Magellan* probe (United States)
2005: *Venus Express* orbiter (European Space Agency)

ATMOSPHERIC PRESSURE While Earth's atmosphere pushes on every 1 square inch (6.5 cm²) of area with a force of 14 pounds (6 kg), on Venus the atmosphere is almost 100 times heavier. It exerts 1,200 pounds (540 kg) of force on the same area, making leisurely strolls impossible. On Earth, these pressures are only found in the deepest part of the ocean.

UNDER THE CLOUDS Most of Venus resembles a desert, with slabs of rock flattened by the high atmospheric pressure. The first spacecraft to land on the planet were the Soviets' *Venera 9* and *10*. They transmitted data for about an hour before losing radio contact. Venus features massive mountain ranges—including Maxwell, which stretches 540 miles (870 km) long and peaks at 7 miles (11 km) high—and coronae that resulted from volcanic activity. It also has two large continental plains: Ishtar Terra and Aphrodite Terra.

COMPOSITION Scientists have reason to believe that our sister planet is very similar to Earth in its makeup, with a core, a mantle, and a crust. The core is likely rocky, and the mantle might be partially molten, given the planet's rate of cooling. One point of difference: The crust of Venus is active, although it seems to have no tectonic plates.

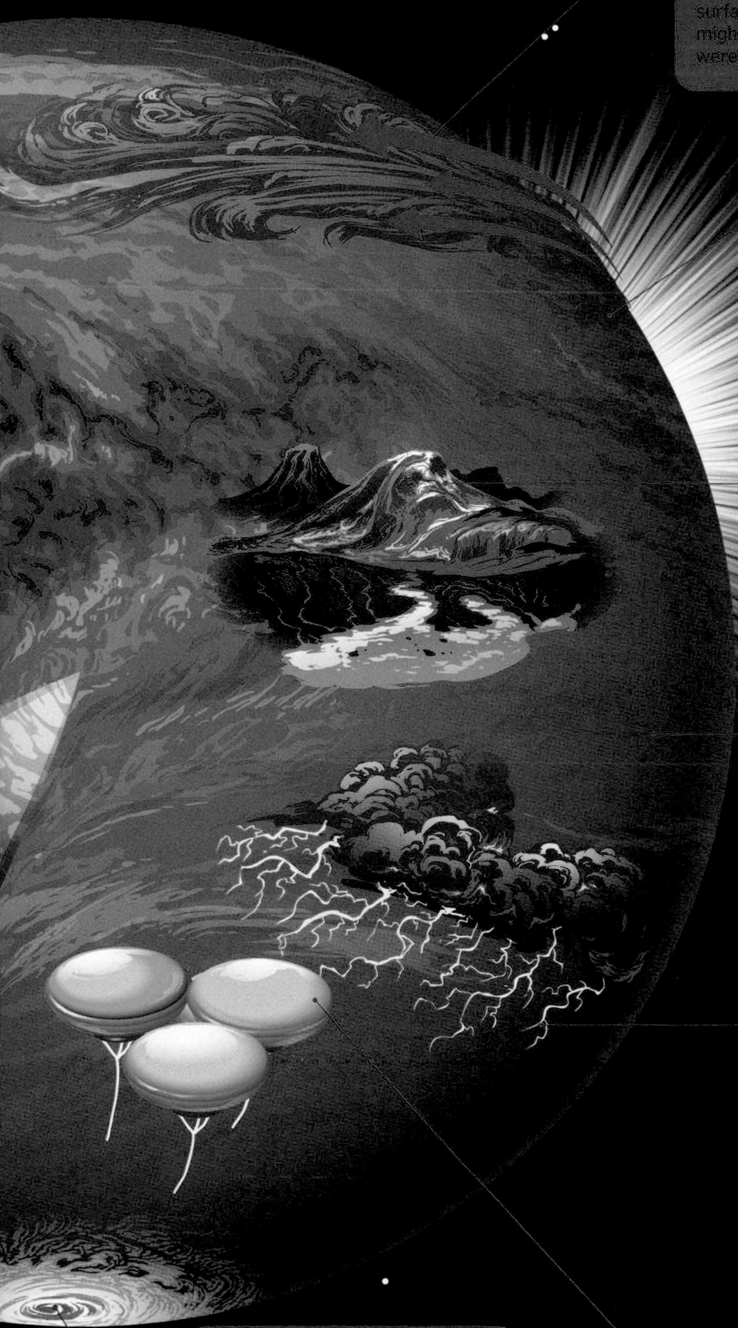

MIGHTY WINDS Winds in the upper part of the atmosphere move at hurricane speeds of over 215 miles per hour (345 km/hr), carrying cloud cover completely around Venus in about four days. Closer to the surface, the 3-mile-per-hour (5-km/hr) wind might feel more like a gentle breeze—if it weren't so incredibly hot, of course.

WHY THE Y? Scientists using ultraviolet technology have noted dark stripes in Venus's clouds that resemble a massive Y rotating around Venus with its stem at the equator. This shape may be caused by centrifugal forces and winds that swirl more quickly at higher latitudes, contorting the cloud.

VOLCANOES Venus is home to over 1,600 major volcanoes—more than any other planet in the solar system. There are 167 giant volcanoes, each at least 60 miles (97 km) across. Some flows are more than 3,000 miles (5,000 km) long. With a height of 5 miles (8 km), the tallest Venusian volcano is Maat Mons, named for the Egyptian goddess.

CLOUD COVER Venus is enshrouded in a blanket of thick white clouds. Sulfuric acid crystals in the cloud tops reflect most of the Sun's light back into space, making Venus the third-brightest object in our sky, after the Sun and Moon. It also provides for a runaway greenhouse effect.

LIGHTNING The clouds of sulfuric acid blanketing Venus do regularly shoot lightning down at the planet, putting it in the company of four other planets that experience lightning: Earth, Jupiter, Saturn, and Uranus. Venus may receive even more electrical discharges than Earth. It is the only planet to experience lightning created by sulfuric acid.

HAVOC The atmospheric pressure and temperatures measured at 31 miles (50 km) above the Venusian surface are very similar to those of Earth's surface. Some people at NASA have proposed building floating science stations in this part of the atmosphere (the High Altitude Venus Operational Concept, or HAVOC). To survive in these cloud cities, we would need to protect the spacecraft and its solar panels from the sulfuric acid found at these heights.

POLAR VORTEXES Venus has incredibly volatile vortexes—one at each pole. They change shape and size daily, and often there are even double vortexes!

DEMYSTIFY THE AURORAS' COLORS

Each shimmering, glowing, gaseous ring of the auroras borealis and australis (also called the northern and southern lights, respectively) is many thousands of miles in diameter and glimmers many miles above Earth's surface. These celestial light shows—which are sometimes bright enough to read by—occur when the stream of charged particles that constantly flow from the Sun (aka the solar wind) strikes Earth's magnetic field, generating powerful electric currents. These currents energize atoms in Earth's upper atmosphere, producing the beautiful colored lights we see.

Usually, the auroras are seen as luminescent curtains of green, but they can also appear in vivid red, pink, and blue. The colors we see depend on the kind of atom being energized, as well as the temperature and pressure at the height of those atoms. For example, while oxygen atoms make green light at 60 miles (97 km) above Earth's surface, the same atom at 200 miles (320 km) will give off red. Next time you catch a glimpse of an aurora, this list will help you make sense of its colors.

RED Emitted by oxygen atoms that are located higher than 120 miles (190 km) above Earth's surface.

BLUE Emitted by nitrogen atoms between 60 and 120 miles (97–190 km) above the surface of Earth.

GREEN Emitted by oxygen atoms between 60 and 120 miles (97–190 km) above the surface of Earth.

PINK Emitted by nitrogen atoms lower than 60 miles (100 km) above Earth's surface.

95 PICK A PRIME SPOT TO AURORA-GAZE

Here are some getaways that are front row center for nature's great light show.

SCANDINAVIA Norway and Sweden's extreme north positions make them shoo-ins for watching the aurora borealis. In particular, the town of Tromsø in Norway's Svalbard archipelago (the northernmost populated area in the world) boasts ample viewing opportunities. Meanwhile, Sweden's Abisko National Park in the Lapland area is also considered a magical viewing spot.

NORTH AMERICA The United States' and Canada's northernmost areas offer no end of packaged tours for those hoping to peep the northern lights. Try viewing them from Denali National Park, a two-hour drive from Fairbanks, Alaska, or travel by dog sled to the Yukon Territory's Whitehorse in search of neon skies.

SOUTH PACIFIC Stewart Island's Rakiura National Park translates from Maori as "land of the glowing skies," making this New Zealand island a dream destination for viewing the aurora australis. Australia's Tasmania is also a nice and accessible option.

DIGITAL VIEWS No travel plans on the horizon? Seek out an online live-stream view of the phenomena.

96 JOURNEY FOR THE AURORAS

If you want to see the auroras, your best bet is to take a trip to the extreme north or south—provided, of course, that you don't live there already. These areas (known as the auroral zones) are located, thin bands a few degrees in latitude and usually 10 to 20 degrees distant from the magnetic poles of Earth. Auroral zones exist because Earth's magnetic field traps charged particles from the solar wind and zaps them up and down toward both magnetic poles; these particles then strike the upper atmosphere long before they reach the surface.

While you can see the aurora from a plane at night if you happen to fly near one of the poles, one way to get the most out of your aurora-viewing expedition is to take a cruise to an auroral zone. This way, you can soak in many beautiful sights along the way. But the best method is to take a trip directly to these regions, spending a splendid vacation in a new land. Some towns specialize in aurora tourism, and there are also agencies that offer aurora-seeing trips.

Of course, there is a catch to visiting extreme northern and southern lands: It can be very hard to actually get there. (The aurora australis is particularly hard to reach, as it happens mainly over water or frigid Antarctica.) In fact, travel may not even be possible in the winter months, which are the best times to view the auroras.

While it may be easier to travel in the summer, you may have only several hours of night, at best, to view the aurora—that is, if the Sun sets at all. Regardless of how you choose to see an aurora, take this tip seriously and dress warmly!

97 TIME YOUR AURORA TRAVEL

The aurora borealis and australis are linked to an 11-year solar cycle. During years of peak activity, expect more drama in the sky and visibility much farther from the poles than usual. For shorter-term predictions—as little as three days out—consult the National Oceanic and Atmospheric Administration at www.swpc.noaa.gov. The best months to witness the aurora are August through April in the northern hemisphere and April through August in the southern hemisphere.

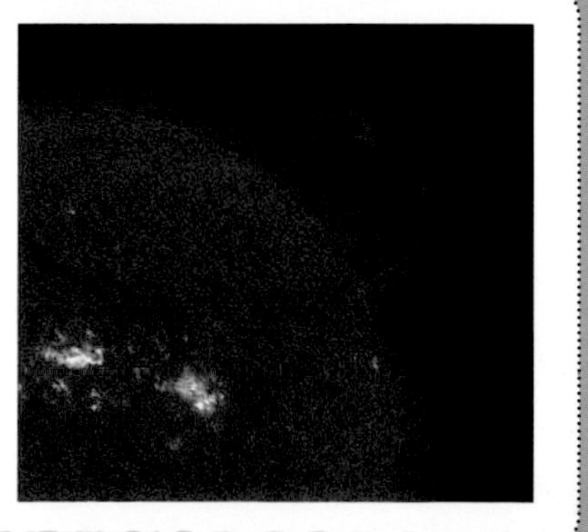

exosphere
(436–118,000 miles)

thermosphere
(51–435 miles)

mesosphere
(32–50 miles)

stratosphere
(7½–31 miles)

Karman line
(62 miles)

troposphere
(0–7½ miles)

ozone layer
(12½–18 miles)

98 PROBE EARTH'S ATMOSPHERE

While the above diagram is not to scale (in reality, Earth's atmosphere is about as thick as an apple's peel, if you shrunk our world down to that size), it shows us Earth's five layers. Each has its own unique chemical composition and temperature, and the layers become thinner the higher you go until eventually air dissipates into space at an imaginary boundary called the Karman line, located about 68 miles (110 km) from the surface of Earth.

TROPOSPHERE Stretching up to 7½ miles (12 km) above Earth's surface, the first layer is the most dense, containing half of Earth's atmospheric mass. It is mostly nitrogen (78 percent) and oxygen (21 percent), and it is home to most of the planet's weather.

STRATOSPHERE With a ceiling of about 31 miles (50 km) high, the stratosphere contains the ozone layer, which heats up the atmosphere and absorbs solar radiation. At this altitude, the air is extremely dry, and the air pressure is one-thousandth of what it is at sea level.

MESOSPHERE The mesosphere reaches about 50 miles (80 km) high in the sky and it is the layer where meteors burn up. The band at the top, called the *mesopause*, is the coldest part of Earth's atmosphere, with an average temperature of about –120°F (–80°C).

THERMOSPHERE Home to the International Space Station (ISS) and where space shuttles once flew, the thermosphere tops out at around 440 miles (700 km) above Earth's surface. Temperatures here reach up to a steamy 2,700°F (1,500°C). Auroras also occur in this layer when particles from space collide with and excite the thermosphere's molecules.

EXOSPHERE The layer where Earth's atmosphere merges into outer space, the exosphere contains extremely thin air composed of hydrogen and helium. It extends to about 118,000 miles (190,000 km) above sea level. At this altitude, atoms and molecules are so far apart that they can go hundreds of miles without running into each other, and they will routinely slip off into outer space.

99 CATCH A GLIMPSE OF GLORIES

To see a glory—which resembles the halos found in paintings of Christian saints—you must be positioned between the Sun and a cloud of fine water droplets. In a glory, light from the Sun behind you is *refracted* (or bent) by the droplets in front of you, creating a multicolor effect much like a rainbow—but visible only to you. The combination of the ghostly projection of your silhouette and the glory is called a *Brocken spectre*. Watch for glories on sunny yet foggy days, when you are above the fog line on a very tall building or mountain. On a plane? Look for a glory wrapped around the silhouette of your aircraft projected on the cloud deck below you.

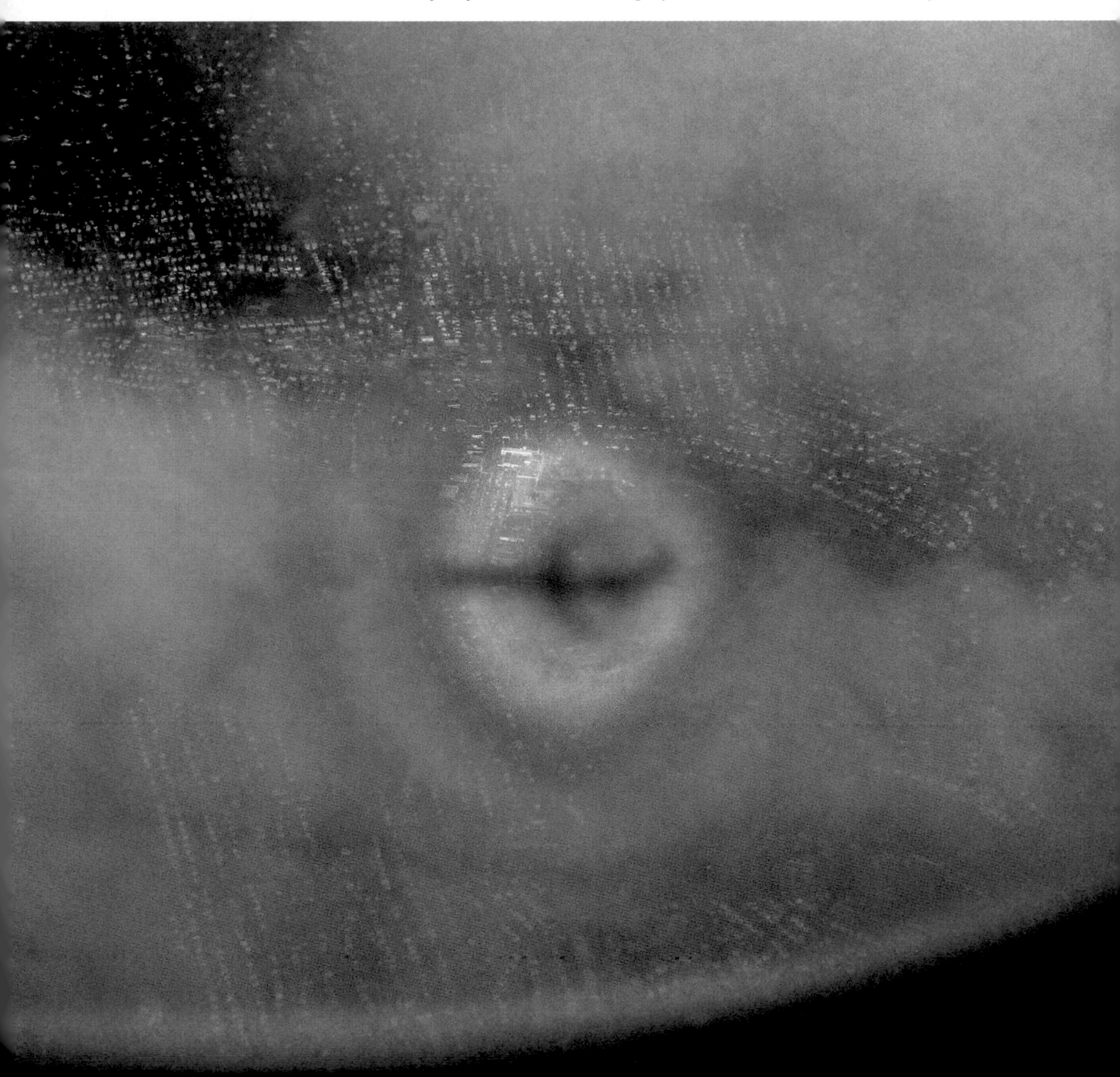

100 CHECK OUT SUN AND MOON DOGS

Our atmosphere is very dynamic, producing spectacular optical effects as the materials swirling above our heads filter light from celestial bodies. The specific culprits of these fantastic light displays? Fine dust particles, turbulence, water vapor, ice crystals, very fine and high clouds, light fog, and smoke. When these are present in high quantities, look up to witness extraordinary feats.

Moon and Sun dogs—just two of many fascinating effects caused by our atmosphere—are bright spots on either side of the Moon or Sun, usually part of a halo that seems to ring either object in an airy white circle. This phenomenon is caused by bright light from the Sun or a full (or nearly-full) Moon refracting off hexagonal ice crystals found in high, thin, and cold cirrus clouds, causing the appearance of extra-bright spots in the halo 22 degrees from the Moon's middle. A uniform distribution of crystals results in the pointed corners around the halo.

101 DEMYSTIFY TWINKLING STARS

Ever wonder what causes the stars to dance and shimmer in the nighttime sky? Before getting to your eyes, light from a star has to pass through our thick layer of atmosphere. Twinkling occurs when starlight passes through turbulent layers of air, becoming distorted before it finally hits your eyes. The twinkling effects diminishes near the zenith of the sky and is stronger at the horizon, simply because light travels through less air from higher in the sky.

To see some very violent twinkling, find a bright star near the horizon, like Sirius (Alpha Canis Majoris; see #59), which seems to flash many different colors. Note that while a lot of twinkling may be beautiful, it also means poorer visibility when using a telescope.

102 SPOT A SOLAR OR LUNAR HALO

Halos appear as bands of light around the Sun and Moon, forming when light strikes randomly arranged ice crystals in the atmosphere. While most halos are caused by ice high in cirrus clouds, you can also spot halos in the winter when ice crystals are suspended in the lower atmosphere. If conditions are just right, a halo around the Sun may look like a brilliant rainbow ring bursting with color. While halos around the Moon are always more subdued, you can still often see traces of red and blue. A related phenomenon is a Sun pillar (also caused by ice crystals), which resembles a bright streak of light shooting straight up from a setting Sun.

103 MEET MARS

Located fourth in line from the Sun, Mars is a small, rust-colored planet that you can easily see from Earth with nothing but your naked eyes. For this reason, it's long captured our imaginations, resulting in many missions, studies, and even hopes of finding intelligent life or establishing future habitation. Here are some fascinating facts about the Red Planet.

NAME Because Mars looks red in the night sky, many ancient civilizations associated it with blood and war. The Romans named it after their god of war, and today we use the same name.

DIAMETER 4,210 miles (6,775 km), or about half of Earth's diameter

MASS 639 sextillion kg, or about one-tenth of Earth's mass

AVERAGE DISTANCE FROM THE SUN Approximately 143 million miles (230 million km), or about 1.4 times farther than Earth is from the Sun

LENGTH OF YEAR 687 Earth days (almost two Earth years)

TEMPERATURE RANGE –225°F to 95°F (–145 to 35°C). Like Earth, Mars is hotter at its equator and colder at the poles. Because the Martian poles are tilted (25.2 degrees) and by almost the same amount as Earth's (23.5 degrees), Mars also experiences changing seasons. Because Mars takes twice as long to orbit around the Sun as Earth does, the Martian seasons are each twice as long. Using a telescope, you can watch the polar ice caps shrink and grow with the change in seasons.

MARS DAY Ancient Babylonians created the seven-day week, associating each day with the seven known moving objects in the sky: the Sun, the Moon, Mars, Mercury, Jupiter, Venus, and Saturn. Mars Day in Babylonia was the day we call Tuesday. Following the Babylonian tradition, Tuesday is named for the ancient Anglo-Saxon god of war, Tiw.

SUCCESSFUL MISSIONS

2001: *Mars Odyssey* orbiter (United States)
2003: *Mars Spirit and Opportunity* rover (United States)
Mars Express orbiter-lander (European Space Agency)
2005: *Mars Reconnaissance Orbiter* (United States)
2007: *Mars Phoenix Lander* (United States)
2011: *Mars Science Laboratory/Curiosity Rover* (United States)
2013: *MAVEN* orbiter (United States)

GRAVITY Since gravity on Mars is about 0.375 times Earth's gravity, you would weigh about two-thirds less on Mars and could jump three times higher.

3X

MARTIAN METEORITES Fewer than 200 pieces of Mars have been found on Earth. The result of collisions with comets and asteroids ejecting pieces of the planet into space, these rocky Martian invaders cost as much as US$1,000 per gram.

DUST STORMS Mars has the largest dust storms in the solar system, sometimes lasting for months and covering the entire planet.

COLOR Leave a piece of iron outside in the rain, and the water combined with oxygen in the air will cause the iron to rust. Mars is red because the iron in its rocks and dust rusted a long time ago—when Mars had more liquid water and its atmosphere contained more oxygen than it does today.

WATER On Mars, water ice exists in the polar ice caps and just beneath the surface. If it all melted, the entire surface would be covered in 115 feet (35 m) of liquid water. Scientists believe that Mars may have had oceans more than 3.5 billion years ago, when the atmospheric pressure and surface temperature were high.

MOONS The two known moons of Mars, Phobos and Deimos, are small and potato shaped. Phobos is 14 miles (23 km) across while Deimos is only 8 miles (13 km) across. Scientists believe each were probably wandering asteroids captured by Mars's gravitational pull.

OLYMPUS MONS The tallest mountain in the solar system, Olympus Mons is a dormant volcano 2.3 times taller than Mount Everest and almost the size of France. Scientists think it may still be active, since lava flowed 2 million years ago—"yesterday" for a planet more than 4 billion years old.

COMPOSITION Mars has no magnetic field, so it's uncertain if it has a molten iron core. Plate tectonics stopped long ago, as this small planet cooled and solidified faster than larger planets like Earth. Meanwhile, the Martian atmosphere is 96 percent carbon dioxide. Argon and nitrogen make up the remaining 3.8 percent, along with trace amounts of oxygen and water. The Martian atmosphere is about 100 times less dense than Earth's, making its surface atmospheric pressure very low.

PROXIMITY TO EARTH Mars and Earth are at their closest about every 26 months. Many Mars missions have taken advantage of this proximity to visit the red planet. That's why, depending on budgets, you'll often see that Mars missions launch about every two years.

104 CATCH A CONJUNCTION

A striking event that you can see regularly, a *conjunction* occurs when two or more orbiting objects pass each other in near alignment, making them appear as if they're close to each other (technically, within 1 degree), usually at right angles or parallel to the horizon.

The Moon will often appear to be in conjunction with the various planets as it travels through the ecliptic—for instance, here you see Venus and Jupiter in near alignment. On other occasions, different planets or bright stars will be in conjunction with one another. They seem to get close in the sky, then inevitably split apart as they move along their orbits. Of course, none of these objects is physically near any other—it is just an illusion caused by our perspective as we look across their orbits.

And, of course, there's always *alignment* itself, a phenomenon in which three or more celestial bodies appear to form a perfectly straight dotted line in the sky. A more specific form of alignment is *syzygy* (SIZ-eh-gee), when Earth and the Sun are aligned with another celestial body.

105 WITNESS PLANETS IN TRANSIT

Whenever you watch an object's orbit cross in front of another celestial body from our perspective on Earth (as Venus's orbit causes it to cross the Sun here), you're witnessing a *transit*. For a solar transit to be observable, the planet must orbit between Earth and the Sun, meaning the only dazzling glimpses we get of planetary transit are of Venus or Mercury crossing the solar disc. What's more, these transits may happen only once in a century—or even less frequently in the case of Venus. (Mercury is smaller than Venus, so observing it is a different story.) You can also see transits of other objects with a telescope—such as Jupiter's moons moving across Jupiter's disc. In the past, astronomers used the length of time it took for a planet to transit the Sun to help calculate the planet's size as well as its distance from both the Sun and Earth.

You can see Mercury transit the Sun on May 9, 2016, or November 11, 2019. However, you'll have to wait around until 2017 to see Venus transit again.

106 WATCH AN OCCULTATION

When a planet or smaller body (such as the Moon or an asteroid) crosses the path of another object in the sky, blocking it out, it is said to *occult* that object. The Moon can occult other stars and planets, planets can occult stars, and even a tiny asteroid can make itself known by briefly blocking the light from a star as it crosses in front. Here, Jupiter is occulting its moon Ganymede. Occultations can be very important scientifically, as the rate of the light's dimming as well as the length of time it takes it to fade and return can provide useful info about the size of the object and its atmosphere. However, it is also fun just to witness an occultation: It is a celestial "gotcha!" moment.

107 SCOPE PLANETS AT MAXIMUM ELONGATION

Have you ever caught a glimpse of Venus or Mercury higher in the sky than their usual positions? The planet was likely near its *maximum elongation*—the greatest angle from the Sun as seen from Earth. Maximum elongation is the best time to view these planets, as they are out of the Sun's glare, highest in the sky, and observable for the longest period of time for the year.

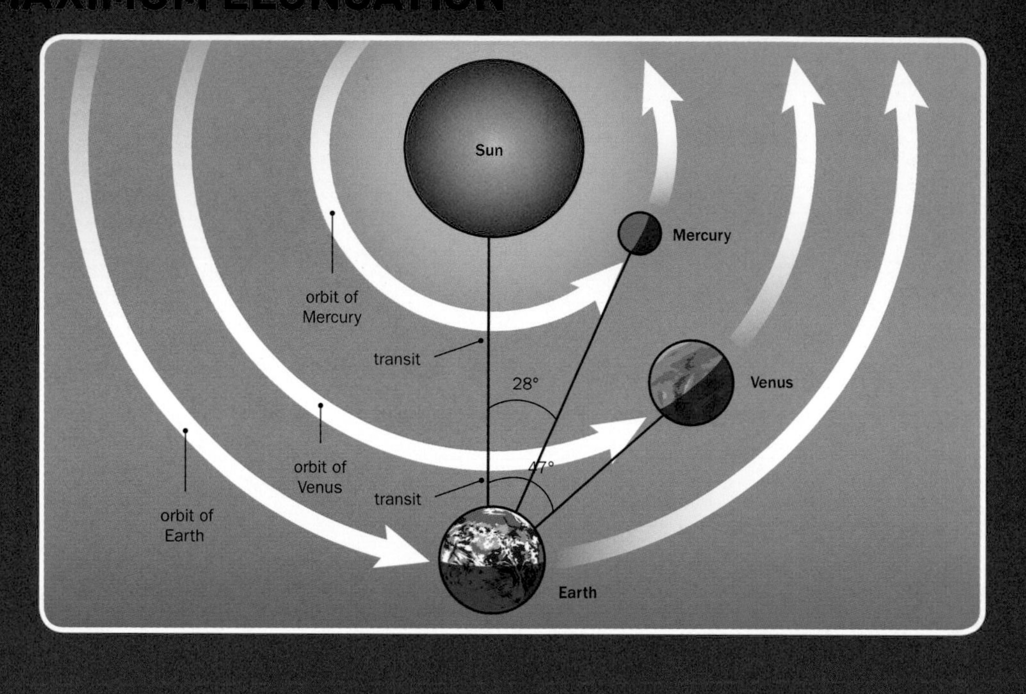

108 OBSERVE PLANETS IN RETROGRADE

From Earth, all the planets appear to wander from west to east in a straight line—and if you viewed them from above, you would see them all revolve in the same direction. But from time to time planets will drift "backward" (east to west in our sky) before reversing direction again. This backward motion is called *retrograde*. Astronomers using time-lapse photography have captured this looping motion, and the images are spectacular. But what's going on?

Planets don't actually change direction while orbiting the Sun. All planets, including Earth, orbit the Sun in the same direction all the time. The weird effect you're witnessing results from watching planets move from a vantage point that is also in motion—aka our Earth! You've likely experienced something very similar in a car: As you pass the slower car next to you, it appears to move backward, even though you are all moving in the same direction. The same thing happens when planets of different speeds pass each other: The slower one appears to move backward to observers on the speedier planet. As Earth continues to pull ahead, the other planet appears to correct course.

Want to witness a planet appear to back up in the sky? Mercury is your best bet: Every 116 days, it spends 21 days in retrograde.

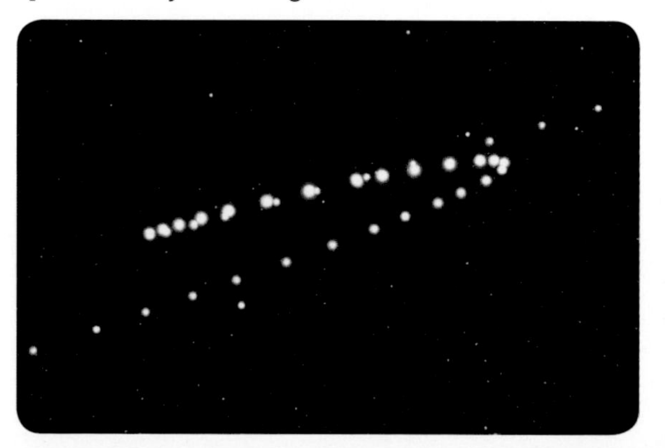

109 STAY UP ALL NIGHT FOR OPPOSITION

Watching a planet for the entire night in all its glory requires that it be in *opposition,* meaning it must be directly opposite the Sun as we see it from Earth. Thus, when the Sun sets, the planet rises, and when the Sun rises, the planet sets. It will also be fully lit by the Sun. Often an opposition can be favorable for viewing, as the planet can be much closer to Earth and thus appears larger and brighter through telescopes.

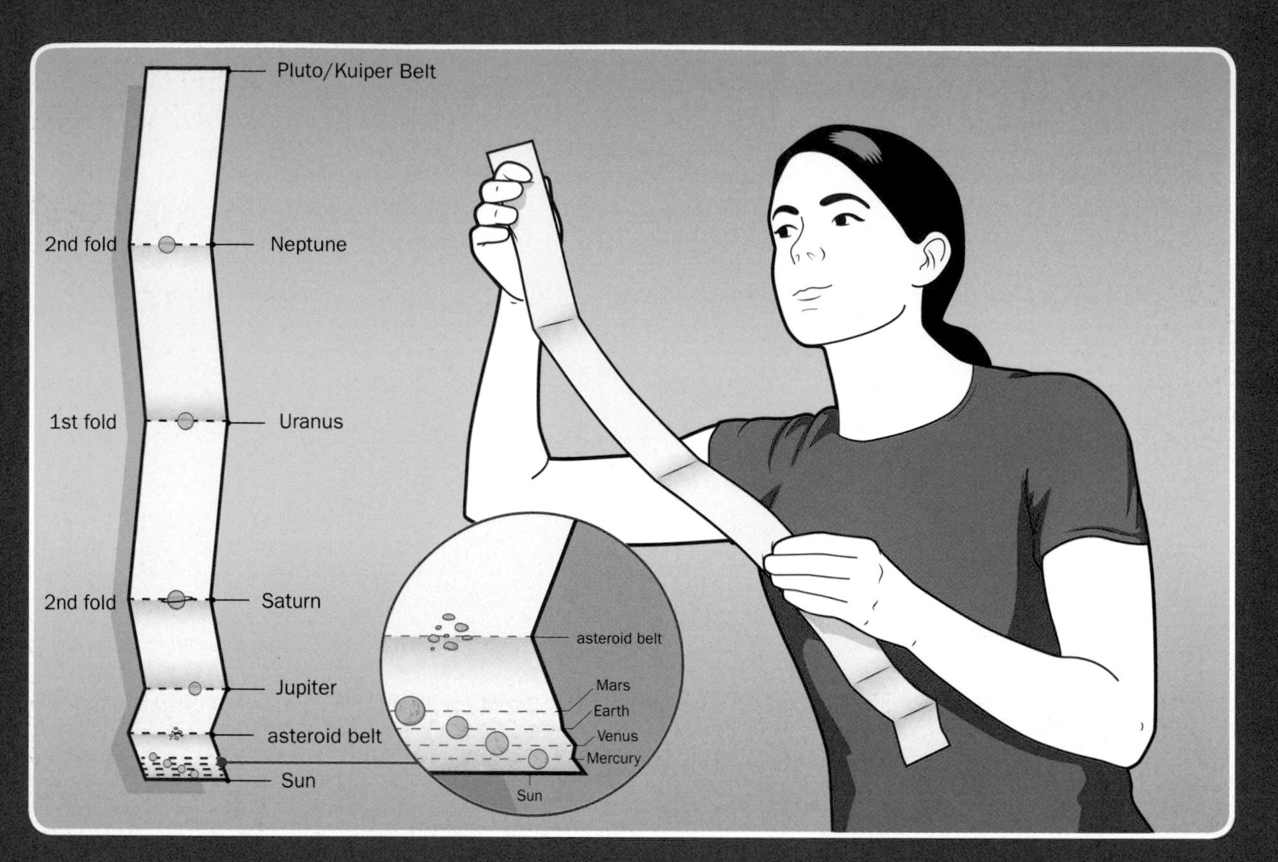

110 MODEL THE SOLAR SYSTEM WITH PAPER

Our solar system is so large and unimaginably empty that it is hard to model both its scale and the size of the planets at the same time without taking quite a hike. That said, creating this basic paper model can really put it in perspective.

STEP ONE Procure a strip of adding machine tape or similar paper, no shorter than 3¼ feet (1 m) and no longer than the length of your body.

STEP TWO On one end of the strip, write "Sun" in small letters; on the other end, write "Pluto/Kuiper Belt" (a group of bodies beyond Neptune's orbit; see #274).

STEP THREE Fold the paper in half and make a crease. Unfold the paper and write "Uranus" on the crease.

STEP FOUR Fold the paper back along the crease where you've written "Uranus." Now, fold the paper in half again, dividing the strip into quarters. Since Neptune is the only planet between Uranus and Pluto, write "Neptune" on the crease nearest Pluto. Label the remaining crease "Saturn."

STEP FIVE All the remaining planets must fit between the Sun and Saturn. Fold the Sun up to Saturn, make a crease, and open back up. That crease represents Jupiter's orbit.

STEP SIX Fold the Sun up to Jupiter and make another crease. Here's where the asteroid belt orbits (see #230).

STEP SEVEN Fold the Sun up to the asteroid belt and make a crease for Mars's orbit.

STEP EIGHT Here it gets a little tricky, so follow along closely. Fold the Sun up to Mars and leave it folded. Fold the crease side up to Mars again. This will leave you with three creases between the Sun and Mars: Mercury, Venus, and Earth.

111 UNDERSTAND PLANETS' RELATIVE SIZES

On the scale of a solar system the size of 3¼-foot (1-m) sheet of paper, the Sun would be smaller than a grain of sand and you wouldn't be able to see any of the planets. Still, that doesn't mean it's impossible to show an approximate scale of planets. Here's one way to visualize the size of major celestial bodies compared to one another.

STEP ONE Label sheets of paper with each of the eight planet names, plus Pluto for good measure.

STEP TWO Procure 3 pounds (1.4 kg) of dough or clay, dividing it into 10 equal parts. You may find it easiest to split the dough if you roll it all out into a cylinder first.

STEP THREE Once you've separated the dough into 10 parts, mush six of them together and place them on the Jupiter sheet. Then, mush three segments together and place it on the sheet for Saturn. You should have one piece left.

STEP FOUR Divide the remaining piece into 10 equal parts. Combine five with the clay on the Saturn paper; smush two together onto the Neptune paper; and put two more onto the sheet for Uranus. You should have one remaining piece.

STEP FIVE Divide your remaining piece into four parts, combining three of them with the clay on the Saturn sheet. Again, you should have one remaining piece.

STEP SIX Divide the remaining piece into five equal parts. Put one piece on the Earth sheet and one piece on Venus. Combine two parts with the clay on the Uranus sheet.

STEP SEVEN Divide the remaining piece into 10 equal parts. Put one piece onto the Mars sheet. Combine four pieces with the clay on Neptune's sheet, and another four with the mound on the Uranus paper.

STEP EIGHT Divide the remaining piece into 10 equal parts. Place seven of the pieces onto the Mercury sheet. Combine two pieces with the dough on the Uranus paper.

STEP NINE Divide the last piece into 10 equal parts. Take nine and combine them with Uranus. Place the last piece onto Pluto's sheet. Each of these piles of dough now represents a planet (and Pluto) by volume. Roll them into balls to best represent the planet shapes—and completely understand why Pluto is now considered a dwarf planet.

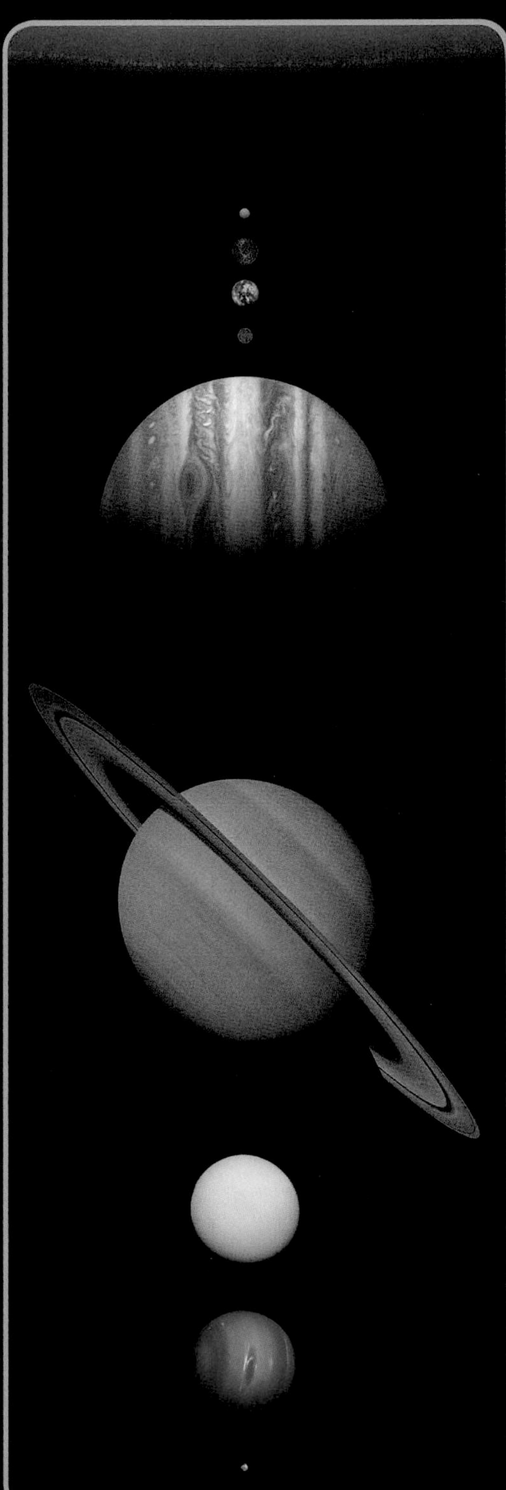

112 START AT THE SUMMER TRIANGLE

The Summer Triangle is not a constellation, per se, but an *asterism*: a pattern of stars forming a recognizable landmark. Three very bright stars (Deneb, Vega, and Altair—also called Alpha Cygni, Alpha Lyrae, and Alpha Aquilae) form the triangle's tips. For observers in the northern hemisphere, look for it from June through December. For southern hemisphere observers, August through October is best.

To find the Summer Triangle, wait a few hours after sunset and look almost straight up. The closer you are to the equator, the higher the Summer Triangle will be. The brightest of its stars is Vega. Altair marks the point with the sharpest of the three angles, leaving Deneb to close off the triangle. From there, other stargazing sites are a hop, skip, and a jump away.

Deneb
CYGNUS
SUMMER TRIANGLE
DELPHINUS
Altair
AQUILA

113 DETECT DELPHINUS

The group of stars forming adorable Delphinus looks so much like a leaping dolphin that you might notice it before you locate anything else in the sky. Despite this striking resemblance, some astronomers call the four stars of the dolphin's head Job's Coffin, because the shape looks a bit like a casket. In China, its stars are part of a larger animal (Genbu, the Black Tortoise of the North), while the Arabs named it al Ka'ud (the Riding Camel).

To locate Delphinus, find Altair (Alpha Aquilae) at the head of Aquila the Eagle and follow the bird's flight direction. The first dolphin-like constellation you hit will be Delphinus, which is visible for everyone except those in the very farthest latitudes south.

114 FLY WITH AQUILA THE EAGLE

To see the stunning Aquila the Eagle, notice that Altair (Alpha Aquilae) marks one of the four points of a diamond shape. This diamond forms the wings of the eagle. The long string of stars trailing from the side of the diamond opposite Altair make up the eagle's tail.

115 SOAR WITH CYGNUS THE SWAN

Deneb (Alpha Cygni) resides in the constellation Cygnus the Swan. Since many people see a cross in the sky with Deneb at one of the ends, this smaller pattern of stars inside Cygnus is sometimes called the Northern Cross. To see Cygnus, imagine a swan flying with its wings spread on either side of its body. The longest segment of the cross is the swan's extended neck, the shortest line of stars its tail (with Deneb at the very tip), and the stars forming a line perpendicular to the head and neck are outstretched wings. Cygnus is in the northern hemisphere; you won't be able to see it below 40 degrees south.

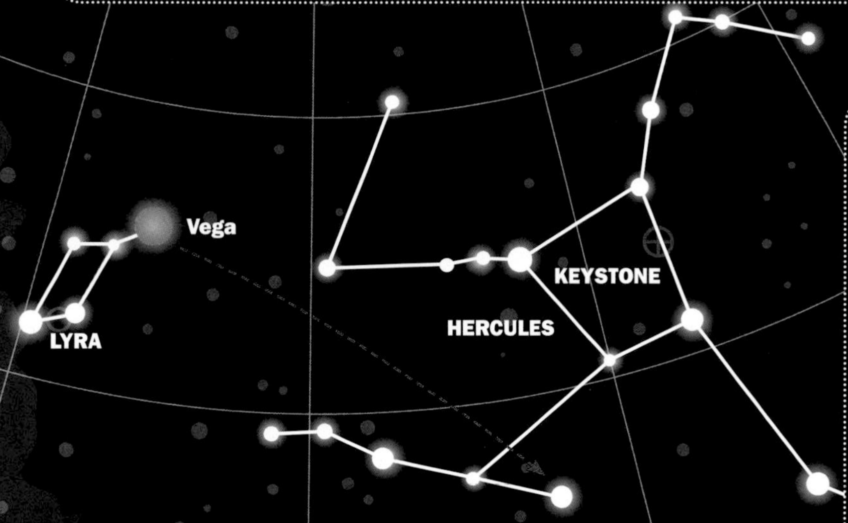

Vega

LYRA

KEYSTONE

HERCULES

117 LOCATE HERCULES

Hercules, usually envisioned as a kneeling warrior with a club over his head, is a large constellation. Like Orion, Hercules appears to be doing battle overhead. Unlike Orion, the constellation does not have a lot of very bright stars in it, making it sometimes difficult to see in cities or brighter night skies. One very recognizable feature inside Hercules is a four-star pattern making up a deformed box, which we call the Keystone (or Quadrangle). Hunting for this smaller group first may help you find Hercules.

To locate Hercules, connect Deneb (Alpha Cygni) and Vega (Alpha Lyrae) with an imaginary line, then extend that line past Vega. The first constellation you will pass through is Hercules. The Keystone asterism makes up the body of the hero. Hercules is visible throughout the northern hemisphere and north of 50 degrees south.

116 FIND LYRA

Vega (Alpha Lyrae) is the brightest star inside a small constellation that looks a bit like a squished cardboard box, or a shopping cart with a handle. The Greeks imagined it as a small harp or lyre—the name we still use today.

To find Lyra, simply look for Vega, which forms the end of the handle of this ancient stringed instrument.

118 SET YOUR SIGHTS ON HYDRA

Hydra—not to be confused with Hydrus the Water Snake—was the nine-headed serpent that Hercules had to kill as one of his twelve labors. Each time he lopped off one head, two others grew in its place. Hercules emerged from this nightmare by having his nephew burn the stump of each severed neck to prevent new heads from sprouting. In the midst of the struggle, Juno, the protector of the state, sent Cancer the Crab to attack Hercules and distract him. The crab nipped Hercules, who then stepped on it and killed it. For its bravery, Juno rewarded the crab with a place in the sky.

As in the case of other large constellations, some mapmakers have tried to break up Hydra's snaking form. In 1805, the French astronomer Joseph Lalande entertained himself by making up a constellation called Felis the Cat. Lalande formed his feline from stars of Hydra and Antlia, but it has not survived. Hydra remains, snaking a quarter of the way across the sky, bordered by Libra, Centaurus, and Cancer. Its brightest star is Alphard (Alpha Hydrae), whose magnitude is only 2.

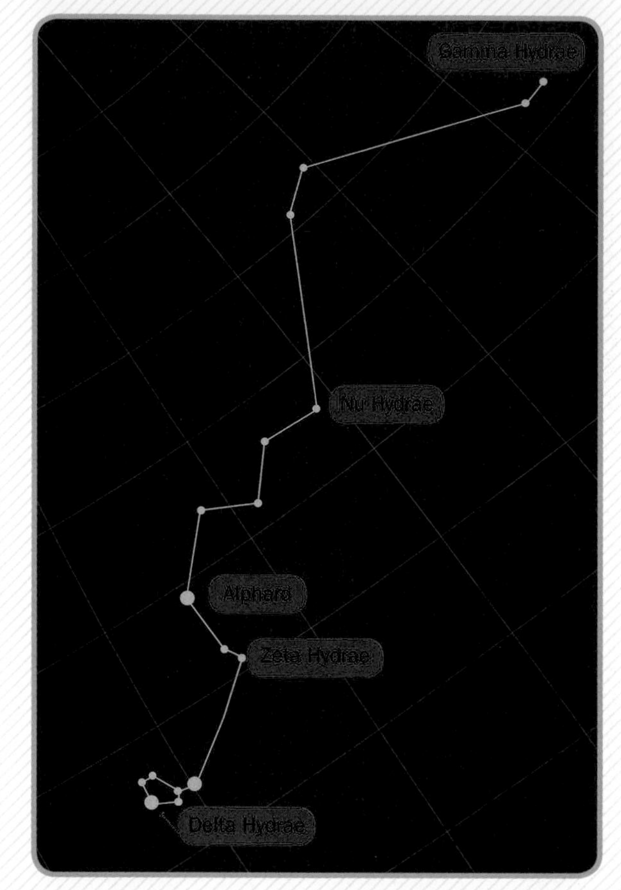

119 TRACK THE SERPENT BEARER

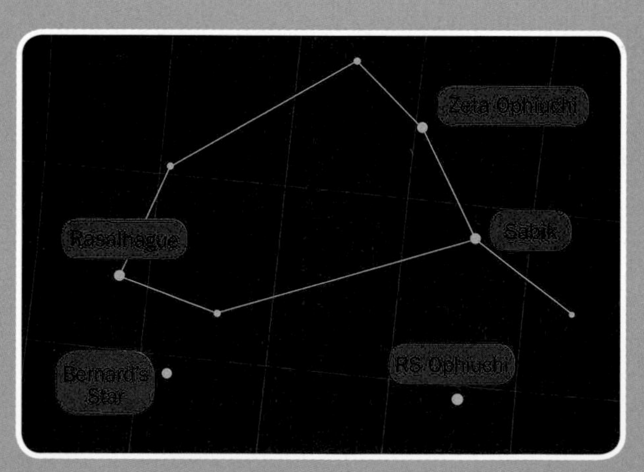

Ophiuchus, entwined with the constellation Serpens, covers a large expanse of sky and is filled with points of interest, including some of the Milky Way's richest star clouds. Greek for "serpent bearer," Ophiuchus is usually identified with Asclepius, the god of medicine. In one legend, Asclepius learned about the healing power of plants from a snake. His medical skills were so great that he could even raise the dead—a cause of concern for Hades, god of the underworld. Hades persuaded Zeus, his brother, to strike Asclepius dead. Zeus then placed Asclepius in the sky, in recognition of his healing skills, along with Serpens, his serpent.

Within Ophiuchus you can find the recurrent nova RS Ophiuchi; it had outbursts in 1898, 1933, 1958, 1967, 1985, and 2006. Its minimum magnitude is around 11.8, and during outbursts it rises as high as 4.3. Another notable sight is Barnard's Star (V2500 Ophiuchi). Discovered by E. E. Barnard in 1916, this 9.5-magnitude red dwarf star has the greatest *proper motion* (apparent motion across the sky) of any known star.

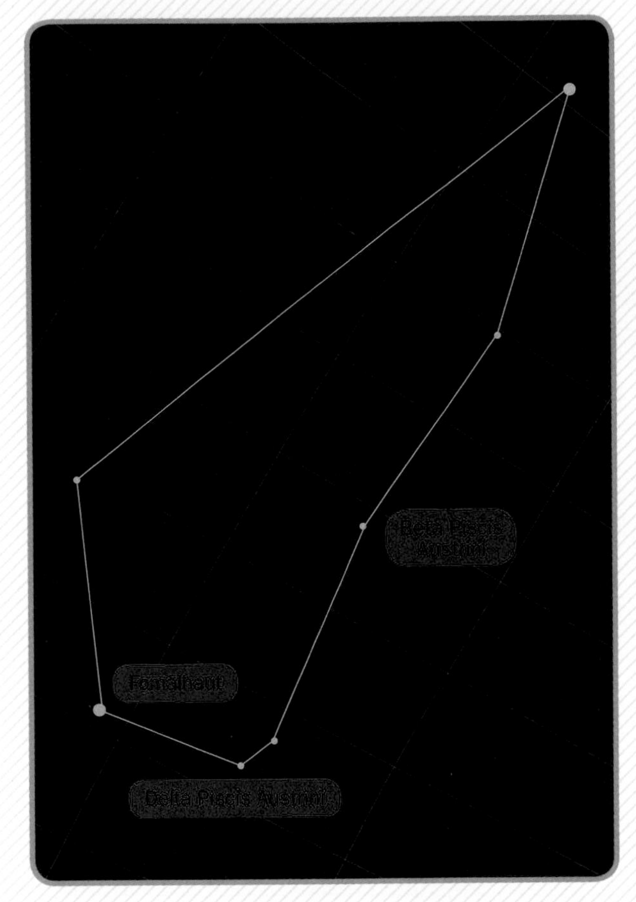

120 CATCH THE SOUTHERN FISH

Lying to the south of Aquarius and Capricornus, and to the north of Grus, the petite Piscis Austrinus, the Southern Fish—not to be confused with the larger constellation Pisces, the pair of fish—is relatively easy to spot because of its lone bright star, Fomalhaut (Alpha Piscis Austrini), often referred to as the Solitary One. For the Persians living 5,000 years ago, this was a royal star that had the privilege of being one of the guardians of heaven. Many early sky charts show the Southern Fish drinking water poured from Aquarius's jar.

At magnitude 1.2, Fomalhaut is 22 light years away—close by stellar standards. It is about twice as large as our Sun and has 14 times its luminosity. Some 2 degrees of arc southward is a magnitude 6.5 dwarf star that seems to be sharing Fomalhaut's motion through space. They are so far apart that it is hard to call them a binary system, but maybe these two stars are all that is left of a cluster that dissipated long ago.

121 SEE SAGITTARIUS

One of the 12 constellations of the zodiac, Sagittarius is located in the Milky Way in the direction of the center of the galaxy. Here the band of the Milky Way is at its broadest, although cut by dark bands of dust. It is a treasure trove of galactic and globular clusters, plus bright and dark nebulae.

The most distinctive aspect of Sagittarius is the group of stars within it that look like a teapot, complete with spout and handle. The handle also stands by itself as the Milk Dipper. Ancient Arabs thought of the western triangle as a group of ostriches on their way to drink from the Milky Way, while the eastern quadrilateral was ostriches returning from their refreshment. (See #60–64 for more Milky Way interpretations.)

Sagittarius is thought to be a centaur (half man and half horse)—often specifically Chiron, who is also identified with the constellation Centaurus. Sagittarius's drawn bow, however, is not in character with Chiron, known for his kindness. Some say Chiron created the constellation to guide Jason and the Argonauts.

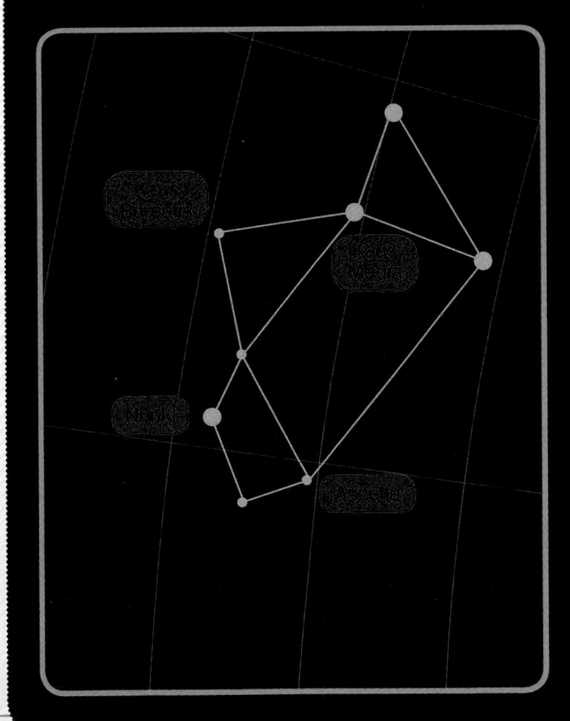

122 SAY HELLO TO THE FIRST GREAT SKYWATCHERS

Long before space telescopes streamed feeds online for anyone who cares to look, many people took educated shots in the dark about what makes the universe tick. Here are just a few of the figures who helped shape our understanding of our own tiny place in the cosmos, many years before we had much of a clue.

THALES
(624 BCE–526 BCE)

A Greek philosopher, traveler, and mathematician, Thales was one of the first to postulate that the movements of heavenly bodies and forces of weather were not, as commonly believed, controlled by gods but could be explained with scientific principles. According to Herodotus, Thales accurately predicted the occurrence of an eclipse around 585 BCE.

PYTHAGORAS
(570 BCE–485 BCE)

Possibly best known for the geometric theorem that bears his name, Pythagoras was also the first person to claim that Earth and the planets were spherical. While he got that right, he also got a lot wrong, like thinking that the enormous, planet-containing spheres rubbed against each other to create "music of the spheres."

ARISTOTLE
(384 BCE–322 BCE)

Doing one better than Pythagoras, Aristotle backed up his belief that Earth was round with observations, including the circular shadows that appear during eclipses, and the fact that stars shift position as a person moves north or south. Like Pythagoras, he got a lot wrong, too. He thought Earth was the stationary center around which everything else moved.

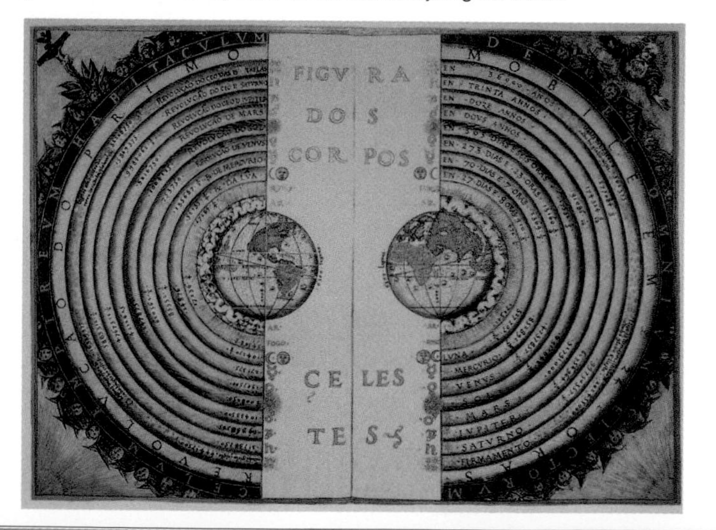

HIPPARCHUS
(190 BCE–120 BCE)

One of the greatest astronomers of ancient times, Hipparchus used relatively primitive astronomical tools—like the gnomon (see #38 to make your own)—to make his heavenly observations. Among his achievements are the creation of an enormous star catalog, as well as a scale comparing the brightness of stars.

ERATOSTHENES
(276 BCE–194 BCE)

Passionate for knowledge in general, Eratosthenes wore many hats. While this Greek thinker dabbled in everything from music theory to mathematics, poetry to astronomy, he essentially invented geometry and is best known for first calculating with great precision both the circumference of Earth and its tilt on its axis.

ARISTARCHUS
(310 BCE–230 BCE)

Ever in the shadow of Aristotle, Aristarchus was actually the first recorded person to propose a model of the universe with the Sun at its center and Earth and the other planets revolving around it. Far too radical an idea for the time, it would take almost 1,200 years for a heliocentric model to catch on in a serious way.

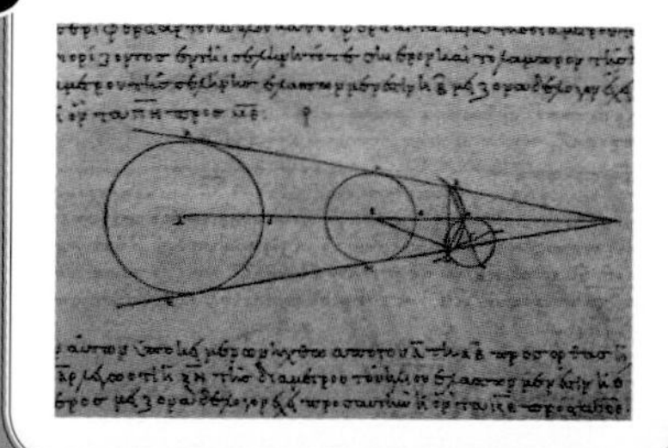

PTOLEMY
(90 CE–168 CE)

In keeping with our theme of getting some things very right and others very wrong, Ptolemy proposed the idea that all celestial bodies move at a constant speed and in a circular orbit. Besides orbits being not exactly circular, another major flaw in his model was the fact that Earth was at the center. His list of 48 constellations was the definitive register up until the Middle Ages.

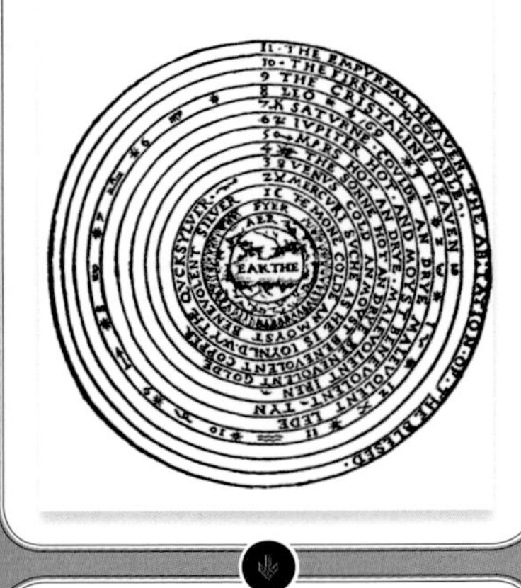

ARYABHATA
(476 CE–550 CE)

Indian mathematician and astronomer Aryabhata was the first to correctly postulate that our experience of day and night was caused by Earth completing a single rotation on its axis each day. He also claimed the movement of the stars to be only apparent motion caused by the rotation of Earth. Unfortunately, he, too, thought our planet was the center of the universe.

ABD AL-RAHMAN AL-SUFI
(903 CE–986 CE)

First to spot the Large Magellanic Cloud (which wouldn't be observed from Europe for another almost 800 years) and first to record seeing the Andromeda Galaxy, Persian astronomer al-Sufi was also the first person on Earth to spot galaxies other than the Milky Way. Further, he rightly noted that the ecliptic—the path our Sun takes along our equator—is inclined. In his entirely illustrated work *The Book of Fixed Stars*, al-Sufi also captured detailed information about each of the constellations, including position, brightness, and color.

NICOLAUS COPERNICUS
(1473 CE–1543 CE)

Nearly 1,200 years after Aristarchus guessed that Earth and other planets revolve around the Sun, Copernicus was the first to put forth a coherent model of a heliocentric solar system. While it wasn't perfect, it forever altered the way astronomers—and regular people the world over—would think about the universe. He also verified Aryabhata's theory that Earth's rotation is responsible for the appearance of celestial movement. He published his findings in *On the Revolution of the Celestial Spheres*.

NASIR AL-DIN AL-TUSI
(1201 CE–1274 CE)

Arabic astronomer al-Tusi built the most advanced observatory of his time—the Maragheh observatory in Iran—for the purpose of better predicting astronomical events. Most notably, al-Tusi was the first to say that the Milky Way was not a cloud, as was widely believed, but was made up of multiple clusters of individual stars. (Three hundred years later, Galileo Galilei was able to prove this premise with the aid of a telescope.) Al-Tusi also created the most accurate tables of planetary movements in his era.

TYCHO BRAHE
(1546 CE–1601 CE)

Based on seeing a new "star" appear (now known to have been a supernova and since named SN 1572), Danish astronomer Tycho Brahe challenged Aristotle's claim that everything outside the orbit of the Moon was forever unchanged. By observing the Great Comet of 1577, he also discovered that comets aren't phenomena that occur within our atmosphere but instead travel above it. After Brahe's death, Johannes Kepler analyzed his records of Mars's movement to develop the laws of planetary motion. And, in one of the more dramatic arguments in astronomical history, Brahe lost his nose in a duel with a relative over a mathematical formula. Thankfully, it is said they eventually made up.

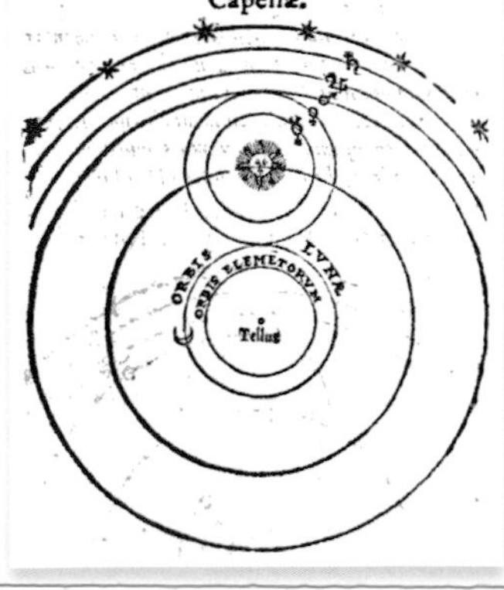

123 MEET JUPITER

Fifth from the Sun and the largest planet in our solar system, Jupiter is the fourth-brightest object in Earth's sky (behind the Sun, the Moon, and Venus). It's a gas giant, so if you were to attempt landing your rover there, you'd be crushed before finding solid footing. But its mysterious features and tremendous forces make it more than worth the visit.

DIAMETER 86,900 miles (140,000 km), or 11 Earths in a row

MASS 1.9 octillion kg, or the mass of 318 Earths (but less than one-thousandth of the mass of the Sun)

AVERAGE DISTANCE FROM THE SUN Approximately 484 million miles (779 million km)—about five times the distance from the Sun to Earth

LENGTH OF DAY A day on Jupiter is only 10 hours compared to tiny Earth's lazy 24 hours. It is the fastest spinning planet in the solar system, rotating at so great a clip that it bulges at the equator. Scientists call this shape an *oblate spheroid*. Clouds whip around the planet at more than 400 miles an hour (645 km/h)!

LENGTH OF YEAR It takes about 4,333 Earth days, or almost 12 Earth years, for Jupiter to orbit the Sun.

AVERAGE TEMPERATURE –160°F (–105°C) in the cloud tops

WHAT'S IN A NAME? In Greek mythology, Jupiter was known as Zeus, the king of all things on heaven and Earth, including the gods. Many of Jupiter's moons are named after related mythological characters.

SUCCESSFUL MISSIONS
1972: *Pioneer 10* flyby (United States)
1973: *Pioneer 11* flyby (United States)
1977: *Voyager 1* and *2* flybys (United States)
1989: *Galileo* orbiter (United States)
2011: *Juno* orbiter (United States)

AURORA Jupiter's tremendous magnetic field causes an aurora very similar to Earth's northern lights (see #041); if you could see it in the sky, it would be larger than five full Moons!

CORE The nature of Jupiter's core is still largely unknown. It is likely a solid core larger than Earth's mass, surrounded by a huge layer of metallic hydrogen. It's impossible to go very far into the Jovian atmosphere because the intense pressures would crush any spacecraft before it reached the core. When the *Galileo* spacecraft sent a probe into the atmosphere, it reported data for only 93 miles (150 km) before succumbing. Future orbiting space explorers will one day tell us more.

GREAT RED SPOT Jupiter's Great Red Spot—so big that two Earths could fit across the widest part—has been intriguing astronomers for more than 300 years. We know it is a huge storm, but scientists are uncertain as to why it has lasted for so long. It gets bigger and smaller and sometimes fades all together. Amateur telescopes of about 8 inches (20 cm) or more can see it.

ATMOSPHERE Jupiter's atmosphere is about 90 percent hydrogen and 10 percent helium (by volume), with trace amounts of methane, water ices, and ammonia. Its red and white stripes are clouds of chemicals. The light bands are called *zones*, each flowing in the direction opposite the adjacent dark regions, known as *belts*.

RINGS Though not visible with amateur telescopes, Jupiter has rings (far fainter than Saturn's) likely created by debris from its inner moons. All known rings exist inside the orbits of the Galilean moons.

MOONS Jupiter has 50 confirmed moons and counting, including Ganymede, the largest moon in our solar system, which is bigger even than the planet Mercury. The largest four Jovian satellites (Io, Europa, Ganymede, and Callisto) were discovered by Galileo and are called the Galilean moons. They were an important piece of evidence in proving that not every celestial body revolves around Earth. These are exceptional moons: Io is highly volcanic, while Europa and Ganymede have ice and oceans. Jupiter's other moons are dwarfed by comparison. Some have eccentric orbits and irregular shapes. Many are likely captured asteroids. (See #135 for info on tracking them!)

SUPERSUCKER Jupiter's gravity is about 2.5 times Earth's, which results in its acting like a cosmic vacuum cleaner, sweeping up solar system debris and likely saving Earth from many devastating impacts. Comets regularly dive into Jupiter kamikaze-style because its gravitational pull is so large. One of the most memorable was Comet Shoemaker-Levy 9 in 1994. Scars more prominent than the planet's Great Red Spot persisted for months.

TELESCOPES & OTHER TOOLS

124 PICK A PAIR OF BINOCULARS

If you're new to skywatching, it's a good idea to get to know the night sky with binoculars before purchasing a telescope. Binoculars have the advantage of being relatively inexpensive, easy to use, and portable. More important, they have a larger *field of view,* allowing you to see more of the sky than you would with a telescope. That means you're more likely to find what you're looking for, if it's visible. Now, how do you find just the right pair for night observing? Follow these tips.

KNOW YOUR NUMBERS Binoculars are often identified by two numbers—each of which can tell you a lot about what you'll see looking through a particular pair. Take 7x35, for example. The first number refers to *eyepiece magnification,* or *power*—the amount by which an object will appear larger over a certain distance—while the second number refers to the *aperture*—the diameter (in millimeters) of the *objective lens,* the lens farthest from the eyepiece that controls the amount of light. In our 7x35 binocular example, the Moon would appear seven times bigger than it does with the naked eye, while the binoculars' lenses measure 35mm across the middle. For most astronomical objects, a magnification between 5 and 8 is ideal—anything above that and you will need a tripod for stabilization. As far as aperture goes, the more light you collect, the brighter images will appear, so the larger the number the better. Often, the field of view is noted on the model's back.

CHOOSE LENSES Different coatings slightly affect the bending and transmission of light through the lens, which can carry more light to your pupils. Coatings are described by a series of codes, with FC meaning all the binoculars' lenses have a coating and FMC meaning all have multiple coatings. In general, multiple coatings are best but also priciest. For astronomical observations, avoid UV coating, which can dim the image.

STABILIZE YOUR IMAGE If your hands are shaky, consider binoculars with *image stabilization*—a feature that reduces the appearance of motion in viewing. While image stabilization provides amazing views, it can come with a hefty price tag.

ONE SIZE DOES NOT FIT ALL Smaller faces require smaller binoculars. For children or adults whose pupils are fewer than 2½ inches (6 cm) apart, you may want to choose compact binoculars.

125 GO WITH PORRO OR ROOF-PRISM BINOCULARS

Binoculars, like a pair of small refracting telescopes, have two main parts: the light-gathering, or objective lens, and the eyepiece with a magnifying lens. Binoculars also use two types of prisms, *Porro* or *roof*, to invert the image as it comes through the tube. A typical Porro prism yields sharp images with its zig-zag mirror placement, losing little light between lenses. Roof models (with their mirrors angled in a straight tube) are light and resilient but must be manufactured highly accurately to match the Porro models' sharpness and prevent distortion at the edges. Some expensive roof models can be quite sharp.

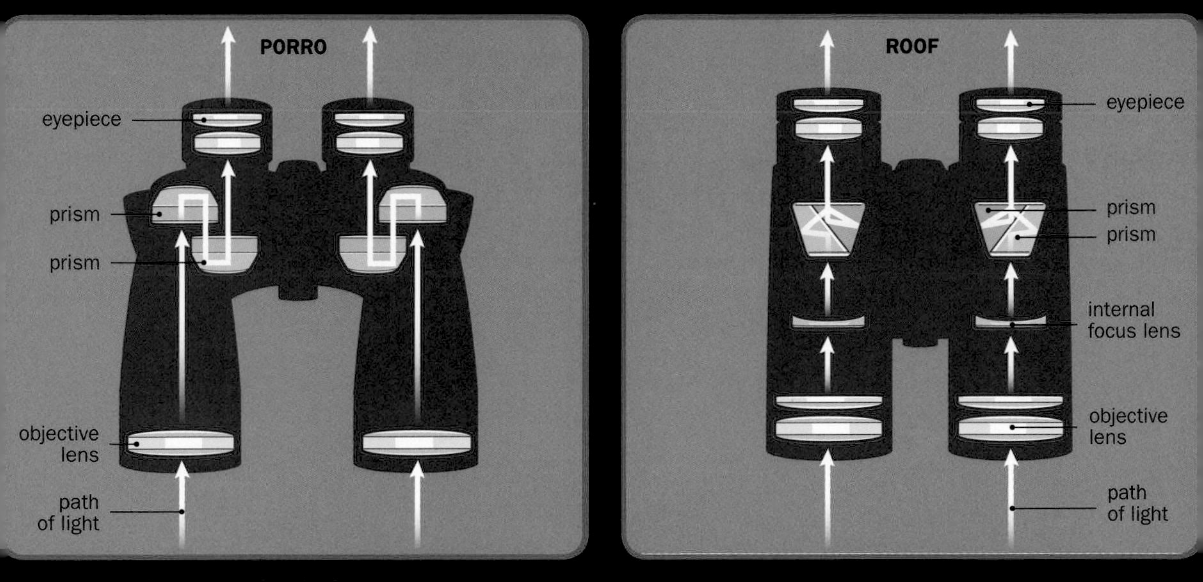

126 LEARN ALL ABOUT EYE RELIEF

Eye relief is the distance from the outermost edge of the eyepiece's lens to the *exit pupil,* the circle of image-forming light that binoculars and other optical tools deliver to your eye. For proper focus, your eye needs to be right at the exit pupil. You will often find, however, that higher magnifications result in shorter eye relief, which may force you to put your eye uncomfortably close to the eyepiece. Wide-field eyepieces often have shorter eye relief as well. If you wear eyeglasses, consider binoculars with adjustable eyecups or long eye relief.

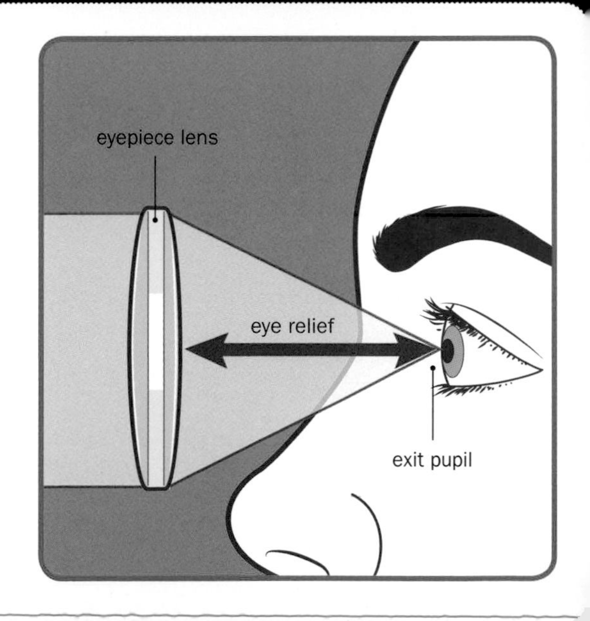

127 ADJUST BINOCULARS FOR VISION

Your eyes, like most people's, are probably not perfectly in sync: One eye's focus is probably ever so slightly different from the other. Despite this fact, people often overlook that binoculars are essentially just two small telescopes—one adjusted for each of your eyes. When you get a new pair of binoculars, it is essential to make sure you adjust them properly for your eyes, whether you wear glasses or not. Please note that these tips may vary according to different models of binoculars, so know your model—and its manual!

STEP ONE Identify the location of the *focus ring*; it will be in the center of the binoculars, between both oculars. Next, find the *diopter ring* near the right ocular. While the diopter allows you to focus individual oculars to compensate for the differences between your eyes, the focus ring adjusts the focus of both oculars at once.

STEP TWO Close your eye behind the diopter (usually your right eye). Now, with just your left eye open, adjust the focus ring until you see a perfect image of a distant object—say, a tree, the Moon, or a telephone pole.

STEP THREE Close your left eye. Open your right eye behind the ocular holding the diopter ring. The object you focused on before may be blurry; adjust the diopter ring until the object becomes perfectly clear to your eye.

STEP FOUR Open both your eyes and marvel at all you can see. You have a stereo view in crystal clarity.

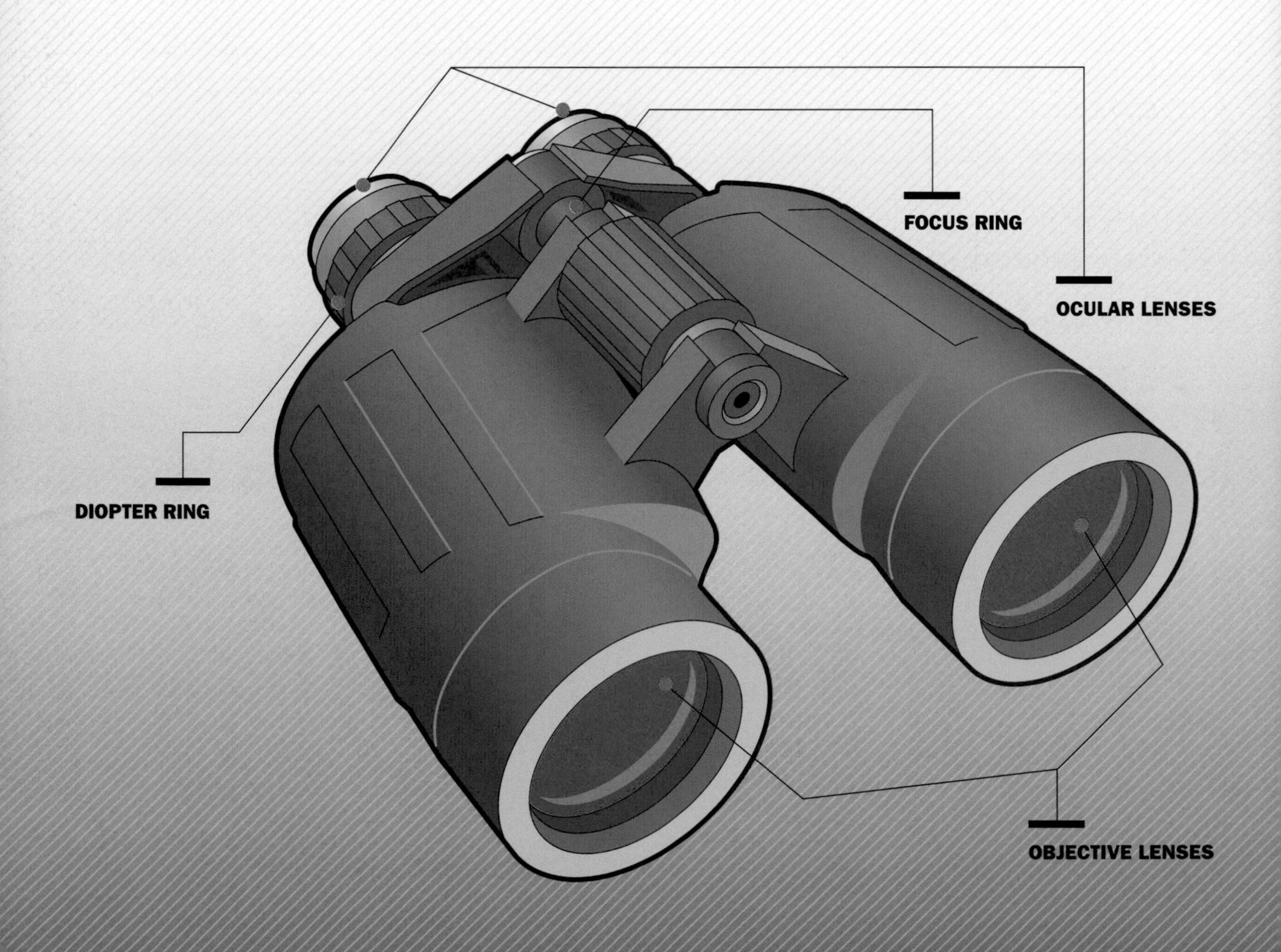

128 CLEAN YOUR BINOCULARS

Keeping binoculars clean means working sparingly and with great care. While you may be tempted to rub away any little speck of dust or smear on your lenses, resist this itch. Overzealous scrubbing can result in scratches or damaged optical coatings. Of course, you will eventually need to clean your lenses. To safely do so, follow these steps.

STEP ONE Check to see if there is dust or moisture inside your binoculars. If so, don't try to clean them yourself. Bring them to a specialist to clean and reseal them. If they were cheap binoculars, buying new ones that are waterproof (instead of water resistant) might cost less.

STEP TWO For minor cleans, start by removing the dust. Carefully blow away any dust from the lenses with compressed air or a special "air puffer." A soft brush can be used to help flick dust off as well.

STEP THREE Gently wipe away grease from accidental fingerprints or sticky pollen with a soft, lint-free cotton cloth. A genuine microfiber cloth is worth keeping in your cleaning kit. Specialty "Lens Pens" (shown below) are also great to use in this case, especially for eyepieces, which are smaller and harder to clean.

STEP FOUR Use cleaning fluid if your lenses are especially dirty—but choose carefully! Do not ever use alcohol or solutions such as Windex; they can strip away the coatings on your lenses and cloud them. A specialized photography cleaning fluid works well. Plain distilled water is much cheaper and also will do the trick.

129

TOP FIVE

TREAT BINOCULARS WITH RESPECT

When well kept, a good pair of binoculars should last a lifetime. Here are five tips to keep your binoculars in working order for the long haul.

☐ **STORE THEM PROPERLY** Make sure to always put your binoculars away after using them, ensuring the lens caps are back in place.

☐ **TRAVEL SAFELY** Store and transport your binoculars in a soft, padded case, like the precious cargo that they are.

☐ **PROTECT THEM FROM THE ELEMENTS** If you leave your binoculars outside, water will eventually seep in and spoil them. Even the Sun can hurt the binoculars. Constant heating and cooling can cause the seals to expand and contract, shortening the lifespan of the optical tubes and rubber eyecups.

☐ **DON'T NEGLECT THE EYECUPS** To keep rubber eyecups (the pieces that rest against your face) in good shape, store them popped up. Wipe them with rubber or vinyl treatment once in a while to help prevent them from cracking, too.

☐ **SPORT A STRAP** Always wear your binoculars' strap if using them freehand. No matter how confident you are that you won't drop them, at some point, you will. (Just look at your cracked phone screen!)

130 PEEP CASSIOPEIA'S PACMAN NEBULA

In the fall and winter, the sky tells a tale of vanity and valor. Recall the legend of Cassiopeia—a queen so enchanted with her image that she infuriated the gods, who sent the sea monster Cetus to devour the people of Ethiopia. The king sacrificed his beautiful daughter, Andromeda, tying her to a rock to await the sea monster. Luckily, the hero Perseus, with wings on his feet, saw Andromeda's compromised state. Fresh from beheading Medusa, Perseus used Medusa's eyes to turn the serpent Cetus to stone. The two young lovers then rode into the sunset (the cosmos may be more appropriate) on a winged horse, Pegasus.

When skyhopping with binoculars, try beginning with the W of Cassiopeia, high in the northern sky for those in the northern latitudes. While you're there, if you're under dark skies, aim your binoculars at Schedar (Alpha Cassiopeiae). Nearby, you'll see an *emission nebula* (a cloud of ionized gas that emits light in different colors): NGC 281, better known as the Pacman Nebula. Named for the star character in the eponymous arcade game, the Pacman Nebula was discovered in 1881—long before any pixelated yellow jaws munched on virtual ghosts or cherries. The Pacman Nebula is visible in many amateur binoculars and telescopes.

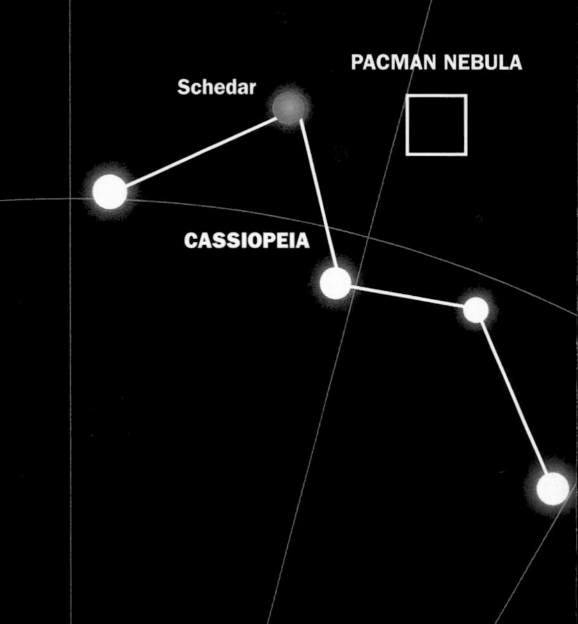

131 NAB THE DOUBLE CLUSTER WITH BINOCULARS

Following Cassiopeia, as you may recall, is the fainter, sweeping Perseus (see #82). From Perseus, you can find a few other special objects. On your way to the hero, be sure to check out the famous jewels of the Double Cluster: NGC 869 and NGC 884. These open clusters are only a few hundred light years apart. Spot them about halfway between the center star of Cassiopeia's W and the brightest star in Perseus, Mirphak (Alpha Persei).

At times of the year when the Double Cluster appears closest to the horizon (particularly spring and summer), it can be hard to spot. Your best bet is fall or winter. A favorite of many for its beauty and accessibility, the Double Cluster can be seen with your naked eyes under dark skies, or very easily with binoculars. While you're in that neck of the sky, check out the loose group of stars around Mirphak—great to view in binoculars.

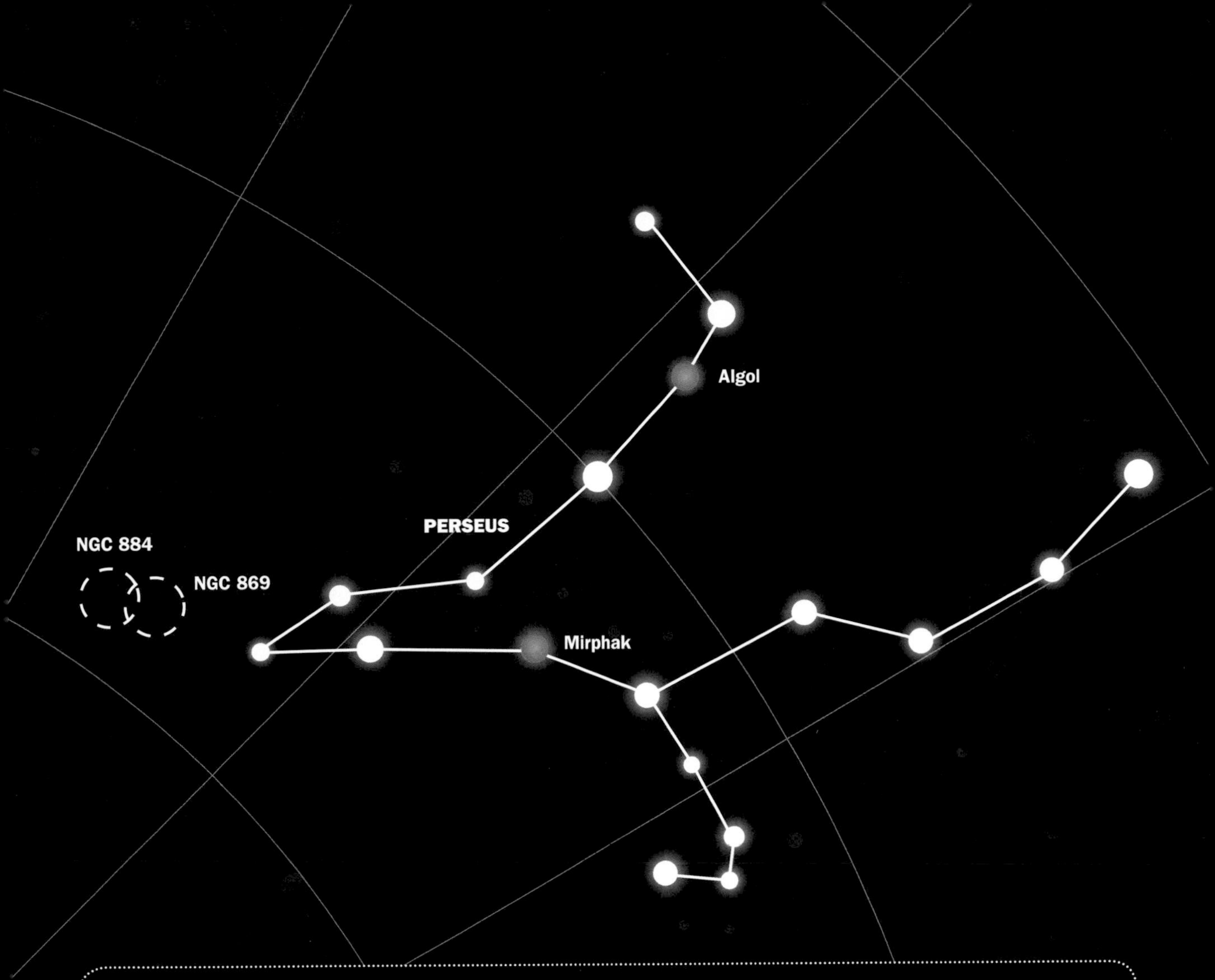

132 ALIGN YOUR SIGHTS WITH ALGOL

One star in Perseus has been known since ancient times. Algol (Beta Persei), from the Arabic "ghoul," is often referred to as the demon star. In the legend of Andromeda, this star is the eye of Medusa—the snake-haired monster that turns Cetus to stone. Formally discovered by Italian astronomer Geminiano Montanari in 1667, Algol has been associated with violence across a number of cultures.

A binary star system, Algol eclipses from our perspective about every three days (it's a triple star system, but the third star is much smaller and doesn't play a part in the dimming). These two main stars orbit each other closer than the orbit of Mercury around our Sun. Its primary star, Algol A, has much more mass than its neighbor Algol B, even though Algol B is much older. Known as the Algol Paradox, this discrepancy is likely due to the stars' proximity to each other. The larger-massed Algol A—and its resulting gravitational pull—tears a steady stream of matter from its older companion.

The cycle from bright to dim and back to bright takes nearly 10 hours, but you can see a difference quickly. In just 5 hours, this bright, 2.1-magnitude star will dim to a magnitude of 3.4. For reference, use binoculars or a telescope to better compare its various stages to the brightness of nearby stars.

133 SEE STAR CLUSTERS WITH BINOCULARS

You can catch the intricate, amazing details of star clusters through binoculars.

Ⓐ OPEN CLUSTERS These scattered groups make for the best binocular viewing, as binoculars' wide field of view offers great glimpses. Clouds of stars born at the same time and from the same matter, these clusters tend to be young, bright, and blue. The teenage sisters of stars, they've blown off the rest of their parental dust and hang together but will eventually drift apart. Examples of open clusters are Pleiades (M45) in Taurus (see #202), the Beehive (M44/ NGC 2632) in Cancer (#215), and the Wild Duck (M11/NGC 6705) in Scutum (#238).

Ⓑ GLOBULAR CLUSTERS Meanwhile, globular clusters are much more distant and will appear tighter and more compact. Through binoculars, you often see only a small fuzzy point. These tight spheres are bound together by gravity, and their stars tend to be among the oldest in our galaxy—and there are more than 150 in our Milky Way! Their distribution once gave astronomers clues that we are not, in fact, at the center of our galaxy. Examples include the Scorpius Cluster (M4/NGC 6397; see #204) and the Great Cluster in Hercules (M13/NGC 6205; see #192).

134 FEAST YOUR EYES ON DISTANT GALAXIES WITH BINOCULARS

The most distant objects we can see through binoculars are galaxies. You can even see a few with low-powered models, if you stretch your observing muscles. While each galaxy consists of millions to billions of stars, don't expect to see grand, swirling pictures like those from the Hubble Space Telescope (see #262). Instead, galaxies viewed through binoculars will look like faint smudges. In the northern hemisphere, try the Andromeda (M31/ NGC 224; see #35) and Ursa Major's interacting Cigar galaxies (M81/NGC 3031 and M82/NGC 3034), and don't miss the Magellanic Clouds in the south. These nearby dwarf galaxies are so large they won't fit inside a telescope field of view—they're ideal for binocular viewing.

135 USE BINOCULARS TO TRACK JUPITER'S MOONS

A great observing project is to study Jupiter's moons—Io, Europa, Ganymede, and Callisto—over a weeklong period. If you observe at the same time each night with your binoculars, you'll likely be able to tell which moon is which. The inner three—Io, Europa, and Ganymede—have a 4:2:1 resonance. In other words, for every one time Ganymede goes around Jupiter, Europa goes around twice and Io four times. When Galileo made this observation more than 400 years ago (see his original chart at right), it proved that not every object orbited Earth—one of the final nails in the coffin of an Earth-centered universe. Prove it to yourself!

Create a simple chart like the one below, drawing the moons that you see and where they're positioned in relationship to Jupiter. (Each tick is roughly the width of Jupiter.) You may not see all four each night. Since you are viewing from the side, they will appear to zip back and forth in a straight line (rather than appear to go in circles).

STEP ONE Check a skytracking app to be sure that Jupiter will be visible.

STEP TWO Using binoculars, observe Jupiter at a regular time during the course of at least seven nights.

STEP THREE See if you can tell which moon is which during the one-week period. Do you notice any patterns? Can you discern a moon's *period*, the time it takes for it to go all the way around Jupiter and back to the same spot? Compare your findings to an online resource and see if you guessed correctly.

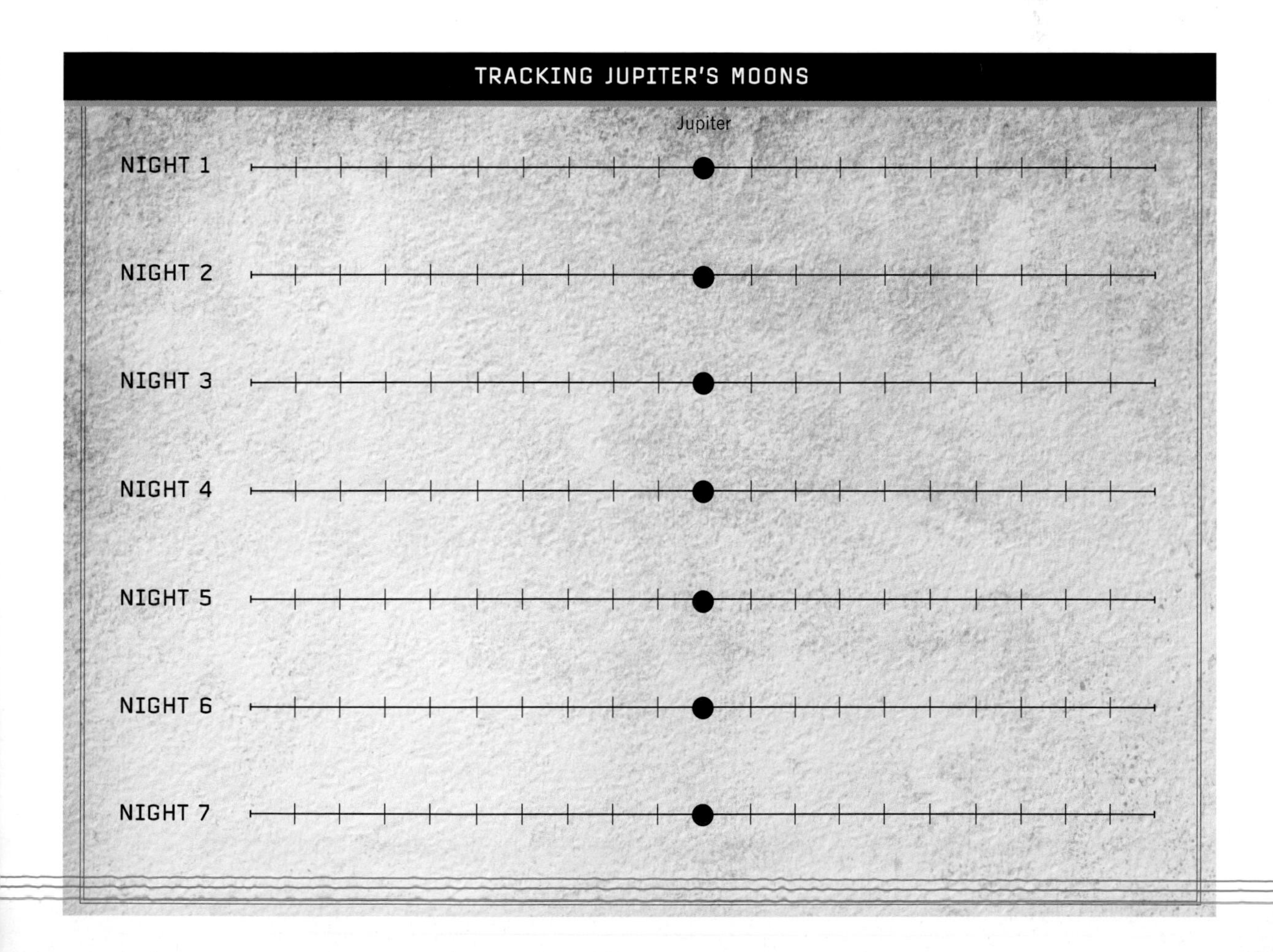

136

TOUR THE MILKY WAY WITH BINOCULARS

There are many binocular gems in the Milky Way Galaxy: clusters, nebulae, and dark patches of gas and dust, all poised for the finding. While we highly recommend exploring it on your own at your leisure, here's a starter tour that's great for low-powered binoculars (say, 7x50 or 10x50).

STEP ONE During summer in the northern hemisphere and winter in the south, begin by looking toward Sagittarius, in the direction of our galaxy's center. Find its famous Teapot, an asterism made up of eight stars that gives off "steam"—that so-called steam is an edge-on view into our Milky Way.

STEP TWO Train your binoculars above the steam to glimpse two beautifully bright nebulae: the Lagoon (M8/NGC 6523), which is between 4,000 and 6,000 light years away from Earth and looks like a sunken oval pond with a dark channel cutting across it, and Trifid (M20/NGC 6514), which is estimated to be about 5,000 light years away from us and is famous for its three-lobed appearance. These two objects are close to each other in our sky. See if you can fit them both into your binoculars' field of view!

STEP THREE Look even farther up to make the acquaintance of the Eagle (M16/NGC 6611), 7,000 light years distant and the home of the famed Pillars of Creation (see #269), which you would need a high-powered telescope to see. You can also spy the lovely Omega Nebula (M17/NGC 6618—also called the Swan or Horseshoe).

137 CATCH CANES VENATICI

Just south of the Big Dipper's handle, Canes Venatici are the hunting dogs Asterion and Chara, held on a leash by Boötes as he hunts for the two bears Ursa Major and Ursa Minor. They were described as such by Polish astronmer Johannes Hevelius in 1687. Besides the stars Cor Caroli and La Superba (Alpha and Y Canum Venatoricum), Canes Venatici hosts deep-sky objects viewable with a telescope.

A rare gem of the northern sky, M3 (NGC 5272) is a globular cluster midway between Cor Caroli and Arcturus (Alpha Boötis). Located on the fringe of Canes Venatici close to Ursa Major, the Whirlpool Galaxy (M51/NGC 5194) is a famous Messier object, appearing as a round, magnitude-8 glow with a bright nucleus. A 12-inch (30-cm) scope will show its spiral structure. The Sunflower Galaxy (M63/NGC 5055) is another galaxy worth spotting; with an 8-inch (20-cm) scope, you can begin to see its spiral structure. These spiral arms were first glimpsed by Lord Ross in the 1800s.

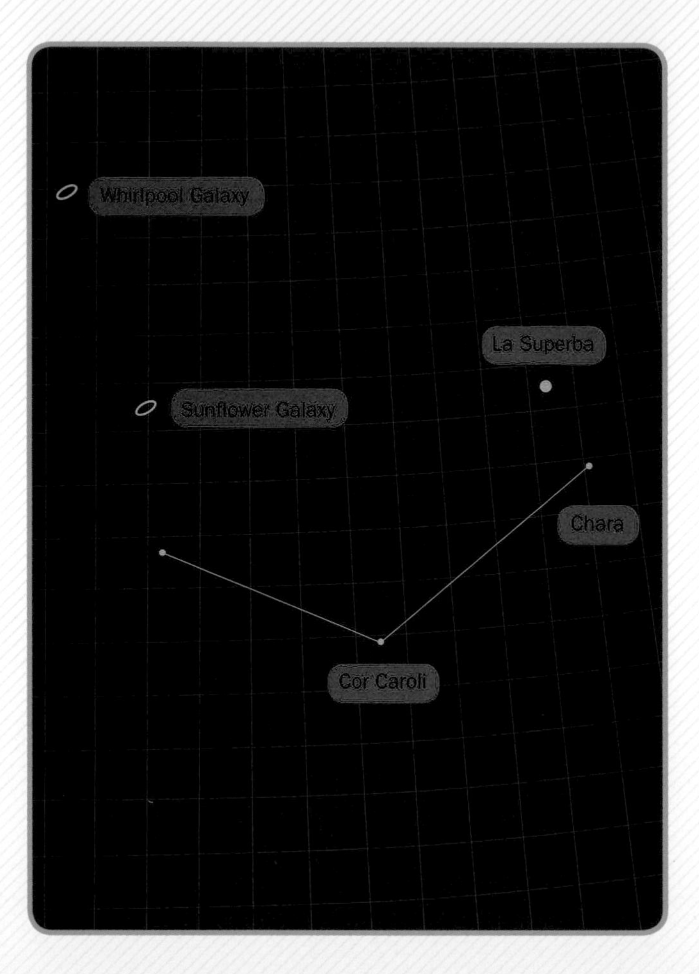

138 VIEW VULPECULA

This constellation, invented by Hevelius in 1690, is without an exciting story or moral tale. Hevelius's name for it was Vulpecula cum Anser, the Fox with the Goose, but now it simply goes by the Fox. A fairly faint constellation, you can find it in the northern sky near the middle of the Summer Triangle (see #112).

One of the finest planetary nebula in the sky, Vulpecula's Dumbbell Nebula (M27/NGC 6853) was also the first ever discovered. Bright and large, it is easy to find just north of Gamma Sagittae. Having a magnitude of 7, it may be visible through binoculars, but only as a faint nebulous spot. With a small telescope, you can make out its odd shape. A larger model that's at least 10 inches (25 cm) in diameter will reveal the almost magnitude-14 central star—a white dwarf, the remnant of the star whose gases now compose this beautiful nebula. Although the nebula's gases are expanding at 17 miles per second (27 km/s), there will be no noticeable change in the nebula's appearance within a human's lifetime.

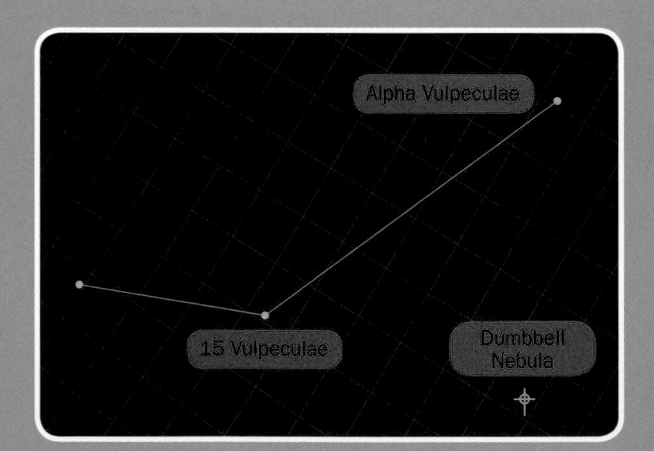

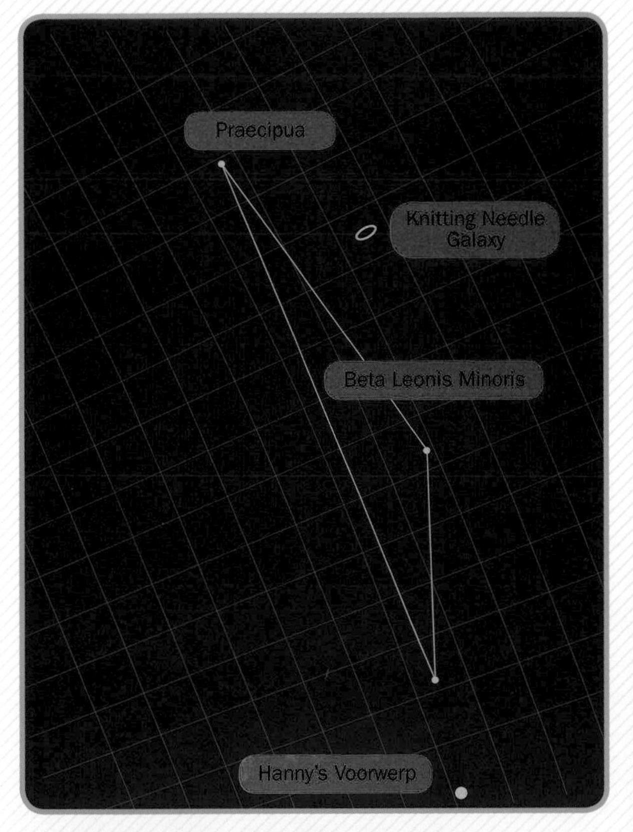

139 LOOK FOR LEO MINOR

The Little Lion was not always viewed as its own distinct constellation. Named in 1687 by the astronomer Johannes Hevelius, Leo Minor is situated in the northern sky, between Ursa Major to the north and Leo to the south. Cancer and Lynx form the constellation's western border.

Leo Minor has a handful of noteworthy stars, many best seen through binoculars or a telescope. The star Praecipua (46 Leonis Minoris), at a magnitude of 3.8, is the brightest in Leo Minor. An orange giant, its color becomes especially apparent when seen through a pair of binoculars.

Leo also houses a number of deep-sky objects easily spotted with a decent-size telescope. The Knitting Needle Galaxy (NGC 3432) can be seen in a lot of amateur telescopes. Seen nearly edge-on, the Knitting Needle Galaxy is slowly moving away from our galaxy.

Discovered by a Dutch schoolteacher in Leo Minor in 2007, and thought to be the light echo of a now inactive quasar, Hanny's Voorwerp is a rare deep-sky object worth seeking out. It is about the size of the Milky Way.

140 PINPOINT PISCES

For thousands of years, Pisces has been seen either as one fish or two. In Greco-Roman mythology, the monster Typhon chased Aphrodite and her son Heros. To escape, they turned themselves into fish and swam away, tails tied together so they would not be parted.

The ring of stars in the western fish, beneath Pegasus, is called the Circlet. The eastern fish is beneath Andromeda. A rare cycle of triple conjunctions, in which Jupiter and Saturn appear close in the sky three times in a single year every 800 years, began in Pisces in 7 BCE.

Other sights include Zeta Piscium, a beautiful double star of magnitudes 5.6 and 6.5, separated by 24 arc seconds. M74 (NGC 628) is a large spiral galaxy, seen face-on, close to Eta Piscium. Rather faint, it requires a dark sky and an 8-inch (20-cm) telescope or larger to be seen. Van Maanen's Star is a white dwarf star you can see in an 8-inch (20-cm) telescope.

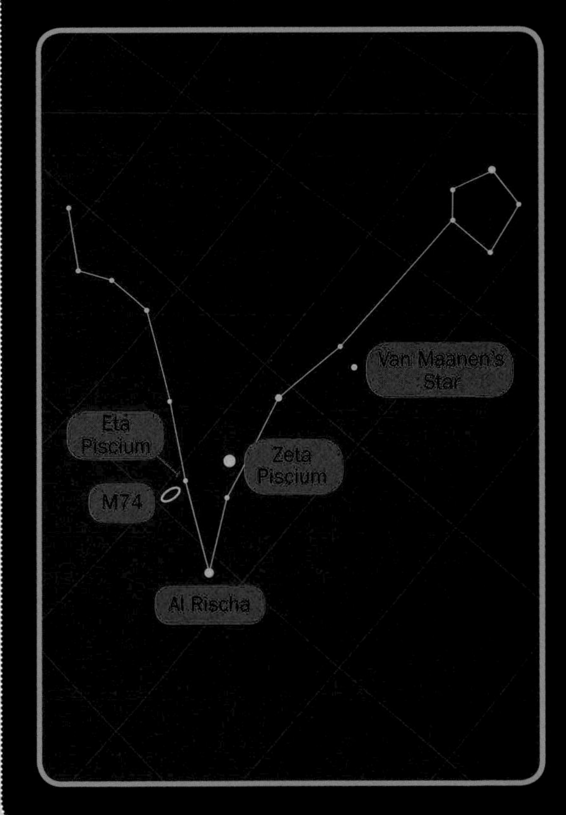

141

DISSECT A BASIC TELESCOPE

There are many types of telescopes, but they all perform the same basic function: They gather an immense amount of light—far more than our eyes could ever hope to collect—and focus it to create clear images of objects in the universe's darker corners. The two oldest and most basic types are refractors and reflectors, but most models share some basic parts.

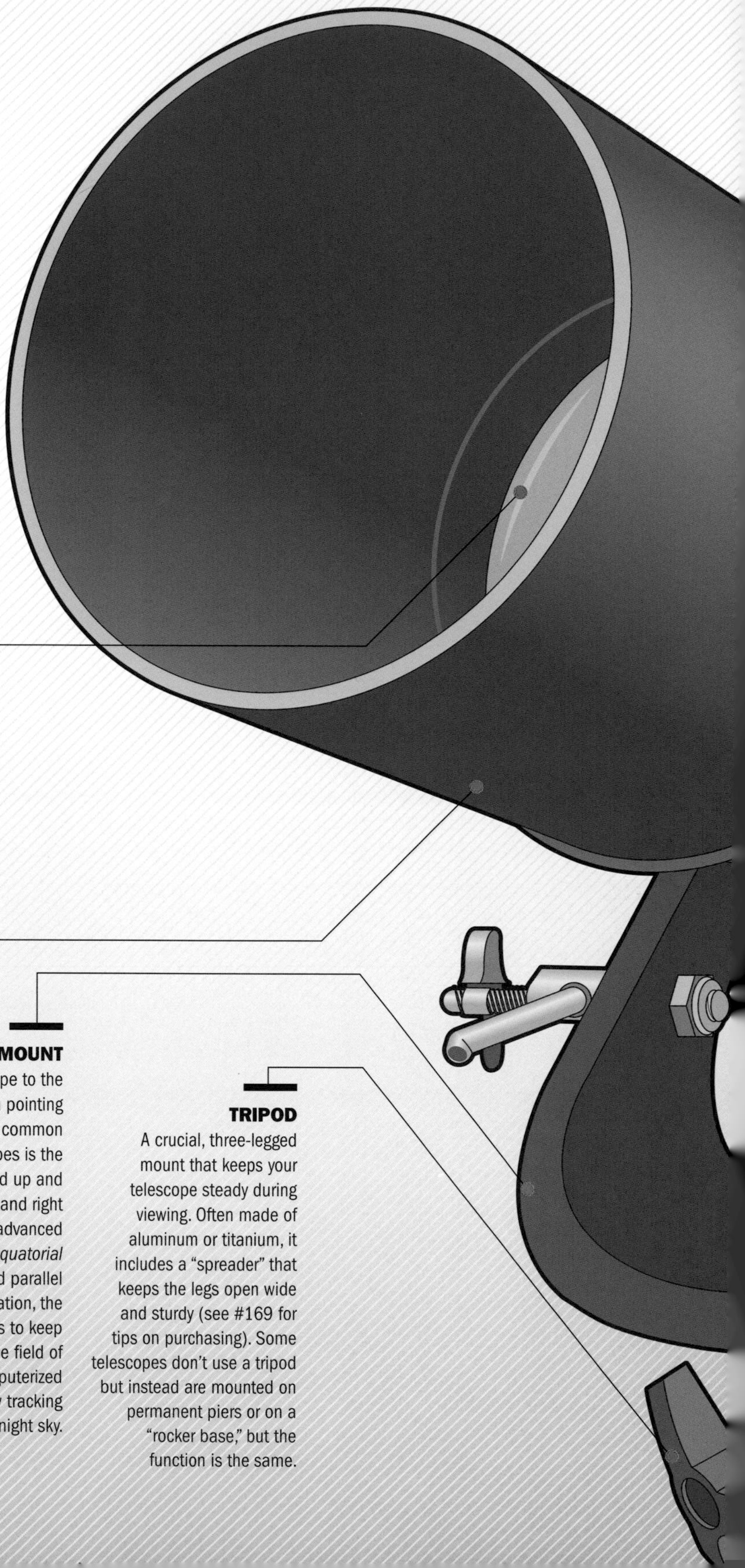

OBJECTIVE LENS
The lens that gathers light from the observed object and focuses the light rays to produce a real image.

DEW SHIELD
An accessory extension of the main tube that minimizes the amount of cold air that reaches your objective lens, preventing condensation from forming. As a bonus, the dew shield will also guard your lenses from stray light. (See #180 for specific dew shield solutions.)

MOUNT
The piece that fixes the telescope to the tripod and allows for smooth pointing and tracking. One of the most common mounts for amateur telescopes is the *altazimuth*, which can be aimed up and down (altitude) and left and right (azimuth)—see #158. More advanced telescope setups tend to use *equatorial mounts*: With one axis aligned parallel to the axis of Earth's rotation, the motorized mount rotates to keep objects fixed and in focus in the field of view. (Check out #160.) Computerized versions of both mounts allow tracking of objects in the night sky.

TRIPOD
A crucial, three-legged mount that keeps your telescope steady during viewing. Often made of aluminum or titanium, it includes a "spreader" that keeps the legs open wide and sturdy (see #169 for tips on purchasing). Some telescopes don't use a tripod but instead are mounted on permanent piers or on a "rocker base," but the function is the same.

FINDER
The telescope's aiming device. It is often a small telescope with a smaller magnification and a much larger field of view than the telescope's eyepiece, which helps you locate objects before getting a closer look with the main viewer (see #151). Some finders are not telescopes at all but project a red dot or crosshairs (see #165–166).

EYEPIECE
The lens closest to your eye when you're looking at the night sky. Positioned at the focal point of the objective lens or primary mirror, the eyepiece receives and focuses the light to create a magnified scene. There are many on the market with varying fields of view, and these are interchangeable so you can achieve the best focus for the specific object you're hunting (see #161).

ADJUSTMENT KNOBS Found on the finder to allow the aim to be fine-tuned.

ALSO

CAPS Cover the telescope's openings when not in use. Prevent the buildup of dust.

COLLIMATION KNOBS Found on reflectors, they allow the mirrors to be fine-tuned, as they may sometimes fall out of alignment.

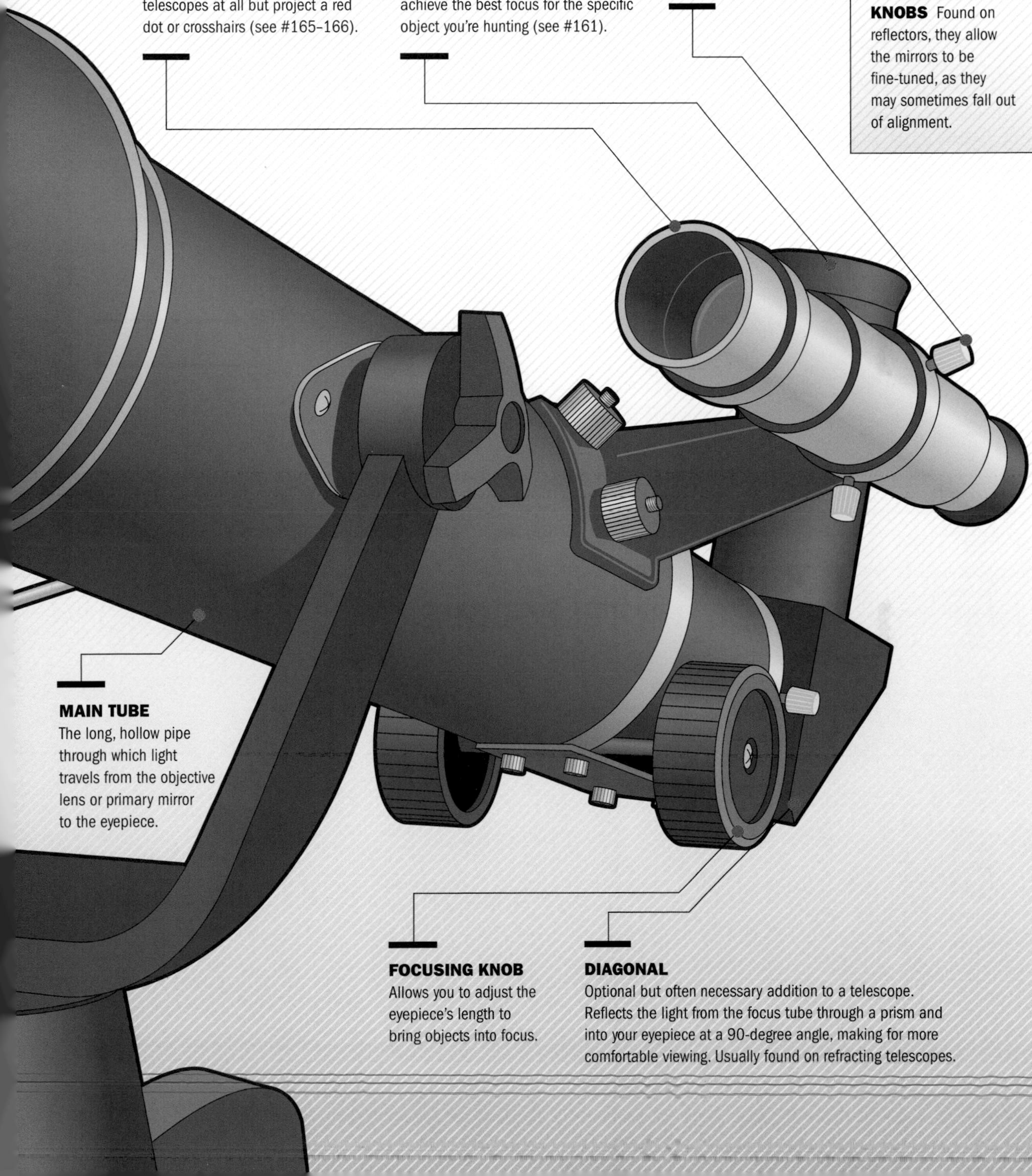

MAIN TUBE
The long, hollow pipe through which light travels from the objective lens or primary mirror to the eyepiece.

FOCUSING KNOB
Allows you to adjust the eyepiece's length to bring objects into focus.

DIAGONAL
Optional but often necessary addition to a telescope. Reflects the light from the focus tube through a prism and into your eyepiece at a 90-degree angle, making for more comfortable viewing. Usually found on refracting telescopes.

142 DISCOVER THE CLASSIC REFRACTOR SCOPE

The most recognizable telescope (and the oldest, dating back to the 1600s when Lippershey invented it and Galileo improved upon the original design), the refractor is very simple, long-lived, and requires barely any maintenance. It's just a lens sturdily mounted inside a tube, and it works by using the lens to *refract* (bend) light from distant objects into focus, then deliver that light to the eye via the eyepiece. The larger the refractor and the more refined its optics, the more expensive it will be.

One thing to watch out for with refractor scopes: Objects can be slightly haloed by other colors. This is called *chromatic aberration*, which means that all wavelengths of light (red, orange, yellow, green, blue, and violet) can't be focused at the same point, which can cause an inaccurate fringed or rainbow effect. More pricey models are *apochromatic*: They have extra lenses and special glass to correct for aberration, resulting in a pristine image (see #291).

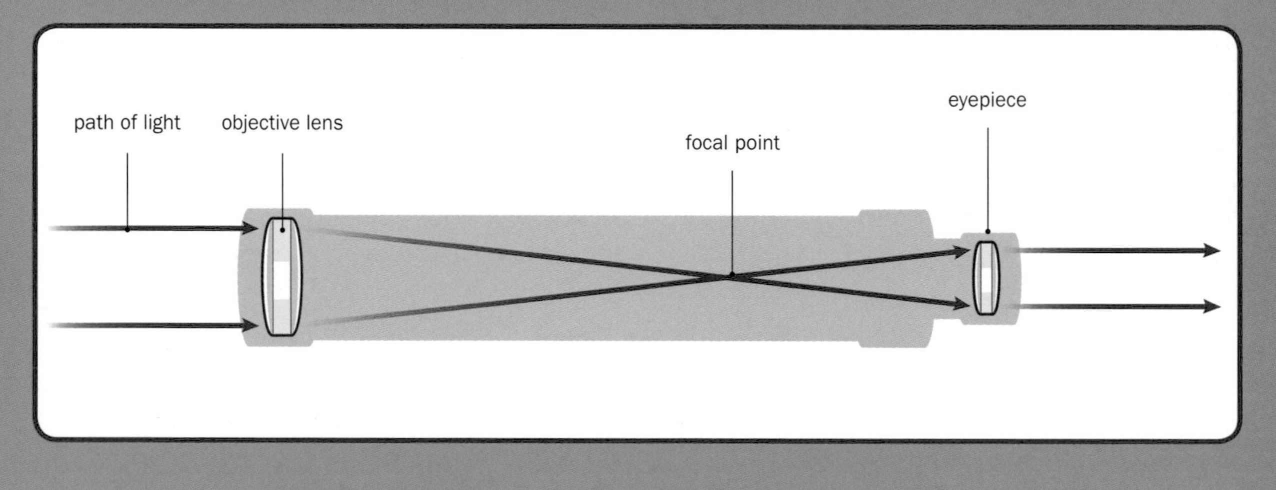

143 CONSIDER A REFLECTOR TELESCOPE

A reflector telescope bounces light from an object via a specially curved mirror and up into your eye. Reflectors tend to be shorter, cheaper, and easier to use than refractors—you can, with some determination and skill, even make one yourself. They tend to be larger than a refractor as well.

Reflectors do not suffer from chromatic aberration, but they have their own limits: A reflector can have problems resolving objects caught near the edges of its mirror, which causes visual defects based on *spherical aberration*. Their mirrors can also require more maintenance. Still, like refractors, they provide a much finer view of objects in our night skies than our own unaided eyes can provide.

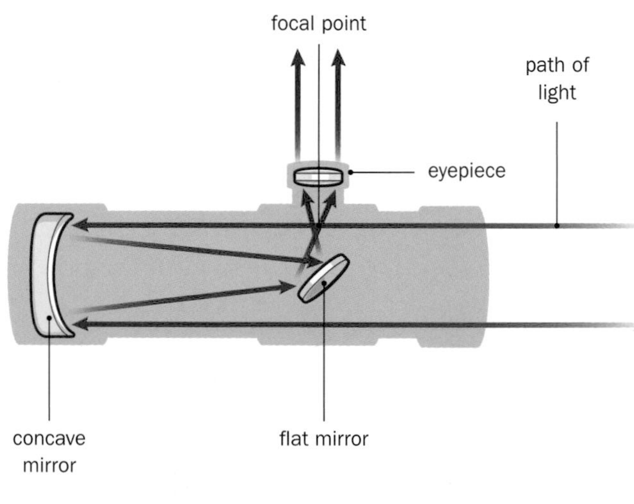

144 LEARN ABOUT OTHER TELESCOPE TYPES

While refractor and reflector models are the most common on the scene, there are a few other telescope models to be aware of and consider.

SCHMIDT-CASSEGRAIN TELESCOPE
Venture out to a star party some night, and there's a fair chance this design will be in attendance. Usually operated with computer controls or a camera system, the Schmidt-Cassegrain combines the refraction power of the Schmidt corrector plate with the Cassegrain's mighty reflector to reduce both chromatic and spherical aberration, as well as reduce the overall size of the telescope tube.

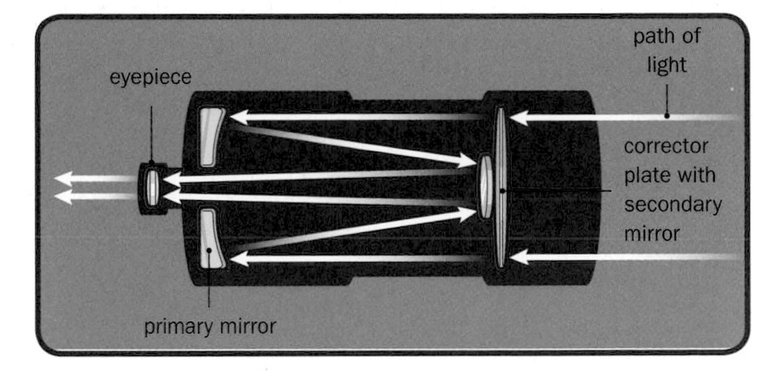

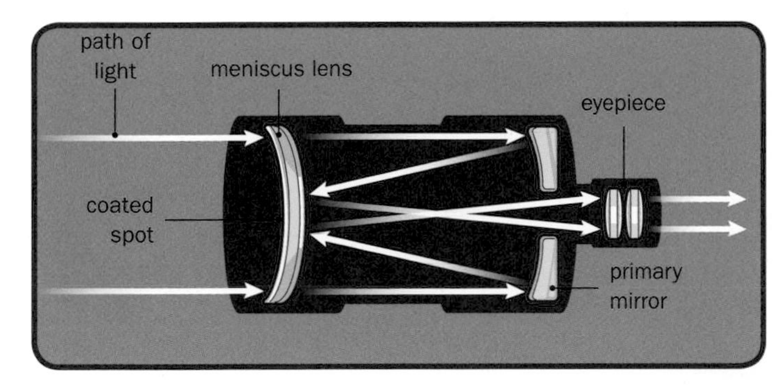

MAKSUTOV TELESCOPE Often used in spotting scopes but also in smaller portable astronomy-focused designs. The Maksutov (shown below) is great for terrestrial observation and bird-watching, and it can double as a serviceable astronomical instrument at night. Many campers are fond of this design, as they tend to be rugged and portable. Its thin, convex lens differentiates it from the Schmidt-Cassegrain's corrector plate.

145

TOP FIVE SELECT A TELESCOPE

Buying a telescope for the first time can be intimidating. There are many companies selling a ton of models, all claiming to be the best! Relax. You aren't picking the best telescope in the world; you're picking the one best suited to you right now.

☐ **LOOK FOR DIAMETER, NOT POWER** Ignore telescopes claiming to have 800x power. Instead, look at the size of a telescope's optics. Most good first scopes have mirrors between 4½ to 6 inches (11–15 cm) in size. Scopes with mirrors 10 inches (25 cm) or larger are generally not well suited to beginners.

☐ **GET THE RIGHT SIZE** Don't get a huge telescope. Look for something lightweight that won't strain your back and allows for easy travel. "Will it fit in my trunk?" is a great prepurchase question. Above all, make sure the telescope is comfortable to use.

☐ **VALUE SIMPLICITY** Get a simple, sturdy telescope with a good mount, a couple of eyepieces (see #161), and a good finder (see #165). You can always add more items later as you gain experience.

☐ **TRY BEFORE YOU BUY** Some local astronomy clubs have telescope-lending programs. If you can, borrow a telescope to try. You can also check out many telescopes at a local star party.

☐ **PAY A FAIR PRICE** Don't spend too much, but don't be too stingy either. While you can get a good starter telescope for a couple hundred dollars, telescopes costing less than US$100 will likely be poor quality.

146

GO COMPUTERIZED OR STICK WITH MANUAL

Tempted by a computerized GoTo telescope that promises to show you thousands of objects with the simple push of a button? Hold on! Computerized telescopes are a lot more complicated to set up than they may seem (check out #149), often frustrating beginners out of the hobby. Learning the night sky can be challenging enough without having to learn a new computer.

Begin your telescopic observing with an electronics-free manual telescope. You will learn the sky much better, starting with the Moon, planets, and other bright objects. Since a manually operated telescope needs no power, you don't need to worry about batteries or extension cords. It is also ready to use in a couple of minutes.

Of course, computerized telescopes can be useful for more advanced skywatchers—once you're familiar with the night sky and telescope operation. Still, keep in mind, astronomers who show off their computerized scopes often still have a manual telescope in reserve!

147 BUY A STARTER SCOPE FOR A KID

Kids love space. If you're searching for a scope for young ones, look for something fairly cheap, small, and sturdy. While there are many good starter scopes for kids (some under US$100), avoid toy scopes; they aren't great for much other than seeing the Moon and can lead to quick boredom. Small reflectors are cheap and offer reasonable views of the night sky. It's also important to find something that can be carried around by young hands without being easily damaged by dropping or misuse. (Bonus: These scopes are great for camping or travel.)

A 3-inch (7.6-cm) reflector is a fine first choice, offering a nice wide view and being strong enough that the craters of the Moon, moons of Jupiter, and rings of Saturn can be seen—all popular choices for kids and adults. The real secret for buying a scope for kids is to make sure an adult could use it and not be bored or frustrated. A so-called kid's scope can make a fine starter scope for adults, too. Because, let's face it, once the kids go to sleep, parents may want to take a peek themselves!

Keep a small notebook and a couple of pens and pencils with the scope so your child can start his or her own observation log, complete with simple sketches of what he or she sees. (See #254–256 for tips.) Wet wipes are also a good idea—not to clean your scope, but to clean sticky little hands before they touch it!

148 SET UP YOUR TELESCOPE

Nobody opens up a telescope for the first time and knows exactly what goes where. Follow these simple steps to safely and easily set it up.

STEP ONE If it's cloudy, rainy, or overly humid, forget it. Your telescope's mirrors and lenses will get coated with dew, ruining any attempts to observe the sky.

STEP TWO Set the tripod down on a flat, sturdy, and dry area. Grass or dirt is better than pavement, which retains heat that can hurt your seeing. Also avoid decks, as boards will lightly vibrate. Make sure the view is free of obstructions, then extend and lock your tripod's legs.

STEP THREE If the mount is not already built in, attach and secure it to the top of the tripod.

STEP FOUR Lock the mount's two directions of motion (up and down, and left and right) with the mount's locking knobs. If you use an equatorial mount, attach and secure your counterweights now.

STEP FIVE Do you need to align your scope with the North or South Celestial Poles? Some telescopes require you to align the mount first, with the help of a polar alignment scope built into the mount. If so, do that now.

STEP SIX Gently raise your telescope and set it on the mount. Make sure every knob and latch has been secured and that the scope is firmly locked into place.

STEP SEVEN Attach accessories, such as your finder scope and eyepiece. Double-check your polar alignment.

STEP EIGHT Make sure your finder is aligned with your telescope's own field of view. You can do this in the daylight or by focusing on a lightsource after dark. The Moon works very well, as does a dim flashlight.

STEP NINE Let your scope cool until it's the same temperature as the air around it. This could take a few minutes to more than an hour.

STEP TEN Pick a target (such as the Moon, a bright star, or a planet) and train your scope on it. Focus your eyepiece carefully until you see a crisp image in view. You are now observing with your telescope!

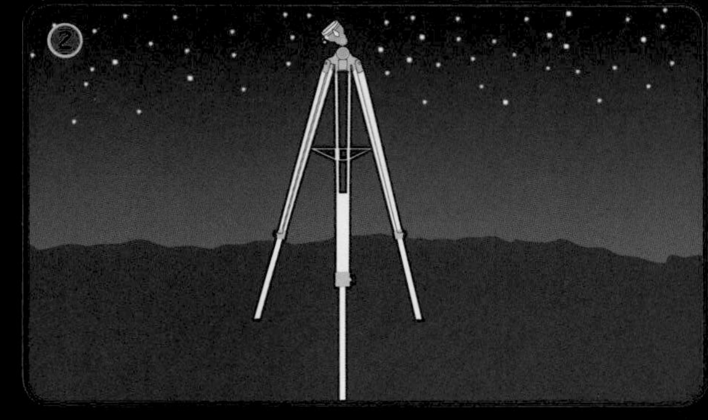

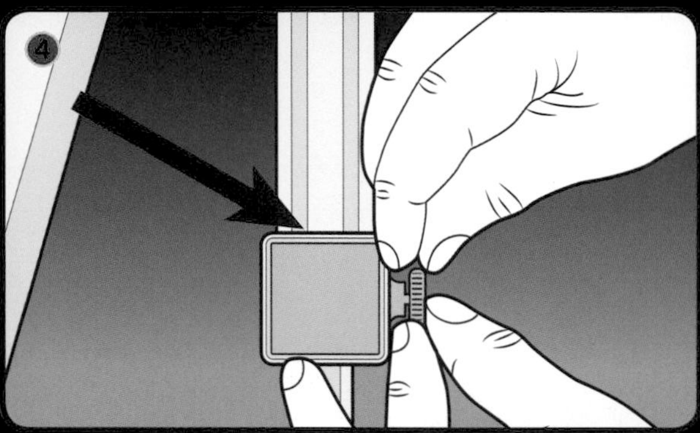

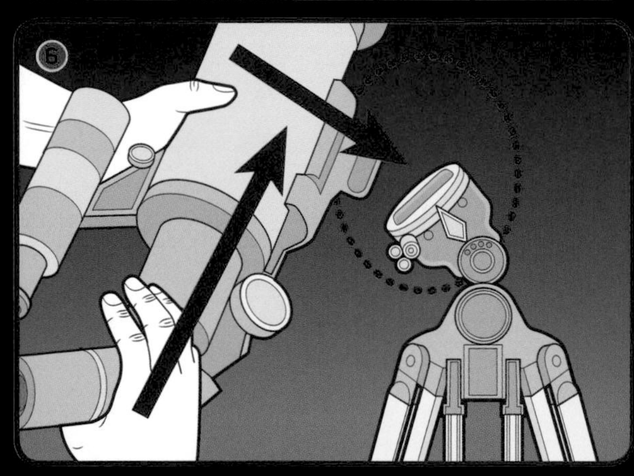

149 VIEW THE SKIES WITH A COMPUTERIZED SCOPE

While a computerized or GoTo telescope promises easy observing, you still need to prepare before taking advantage of features like searching and tracking. Don't be frustrated if at first your telescope doesn't seem to work as advertised. Even a computerized scope takes practice, and each model has its quirks. Follow these tips to get the most out of a computerized scope.

STEP ONE Read through your scope's manual for basic setup instructions beforehand. Keep the manual handy.

STEP TWO Make sure you have power for your scope. Batteries should be fully charged, or your scope should be plugged in.

STEP THREE Put your scope in HOME position. (This usually means making sure your telescope is level with the ground and pointed north). You may need to twist the telescope around its mount clockwise or counterclockwise a certain number of times to ensure freedom of movement. Consult your telescope manual carefully on this step—it's the most important one!

STEP FOUR Check the speed of your scope's motor. Most GoTo scopes offer several settings, since the planets, Moon, and Sun travel at different speeds from both each other and the background stars. The different modes are: sidereal (stars), lunar, solar, and planetary. You'll likely almost always use sidereal, but when observing a solar system object for longer than a half hour or doing astrophotography, your scope may start to drift. Plus, most scopes know when you're tracking, say, the Moon and will automatically adjust motor speed.

STEP FIVE Make sure the time, date, and location are correct, as well as whether or not you are observing daylight saving time. While some telescopes do this automatically via GPS, others require manual entry.

STEP SIX Start the telescope's alignment process. The telescope will usually pick a couple of bright prominent stars to set up its alignment. You will need to confirm each star's identity as the telescope targets and slews toward them. If you are unsure about the identity of a star, consult a star chart. If a chosen guide star is behind an obstruction, skip it; the telescope will pick another star.

STEP SEVEN Adjust the telescope's aim slightly at each guide star. Use the controller to align each guide star to the center of your eyepiece before continuing to the next star. The telescope will beep and confirm it is aligned. This can take a few minutes.

STEP EIGHT Pick another object from the telescope's controller. To make sure your scope is truly aligned, try an object that you will instantly recognize. Another very bright star is a good choice.

STEP NINE See your target in your eyepiece? Congratulations: You are set for a night of discovery! You may need to try a few times before you get the hang of aligning your computerized scope.

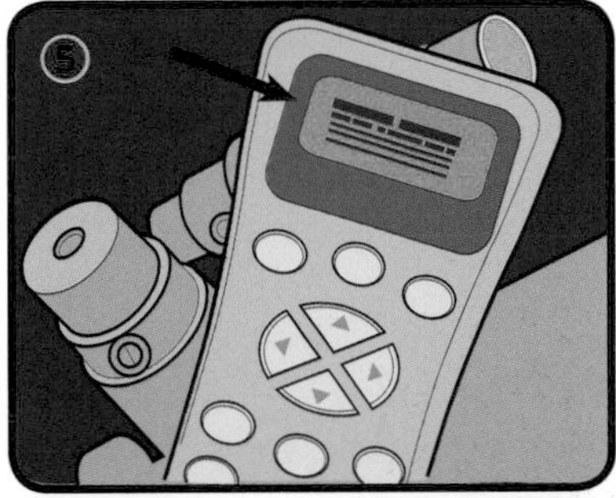

150 COLLIMATE YOUR TELESCOPE TO FINE-TUNE YOUR VIEW

Get the most out of your telescope's mirrors by making sure that they're aligned, or *collimated*. Any telescope that uses mirrors—such as Newtonians (a common type of reflector telescope) and Schmidt-Cassegrains—will require collimation, but reflectors will require the most adjustment—especially after a long, bumpy ride to the viewing spot! Collimation makes sure that the primary mirror is centered on your eyepiece and that the secondary mirror is aligned with your eyepiece and primary mirror.

STEP ONE Open up your telescope and point it at a bright light, such as a streetlight or lightbulb.

STEP TWO Remove your eyepiece (if there is one in place) and replace it with a collimation cap. This is a small cap that fits snugly to your eyepiece holder with a tiny hole in the center. (You can easily make one by poking a small hole in the end of a film canister, or taping foil over your eyepiece holder and poking a hole in it.)

STEP THREE Adjust the secondary mirror if you see warped black edges on either side of your field of view (which are a reflection of the inside of your eyepiece tube). Locate the three tiny screws at the top of the primary mirror and, using a small screwdriver or hex key, slowly adjust them until the secondary mirror is centered and you don't see any black edges.

STEP FOUR Move onto the primary mirror. Peer through the hole and locate the tiny ring at the mirror's center. If your scope is aligned properly, the ring will be centered in your field of view, with the dot of your collimation cap centered in the ring.

STEP FIVE Locate the three large collimation knobs at the bottom of your telescope's mirror cell. (Some setups have locking knobs that you'll need to loosen first.) While looking through the collimation cap, slowly adjust all three knobs until the ring is centered in your view and the collimation cap's dot fits like a bull's-eye in the ring's center.

STEP SIX Double-check your primary alignment again and do any needed fine-tuning. Once everything is centered and you can see the collimation cap's dot inside the ring, you're done. Lock down the knobs, if applicable.

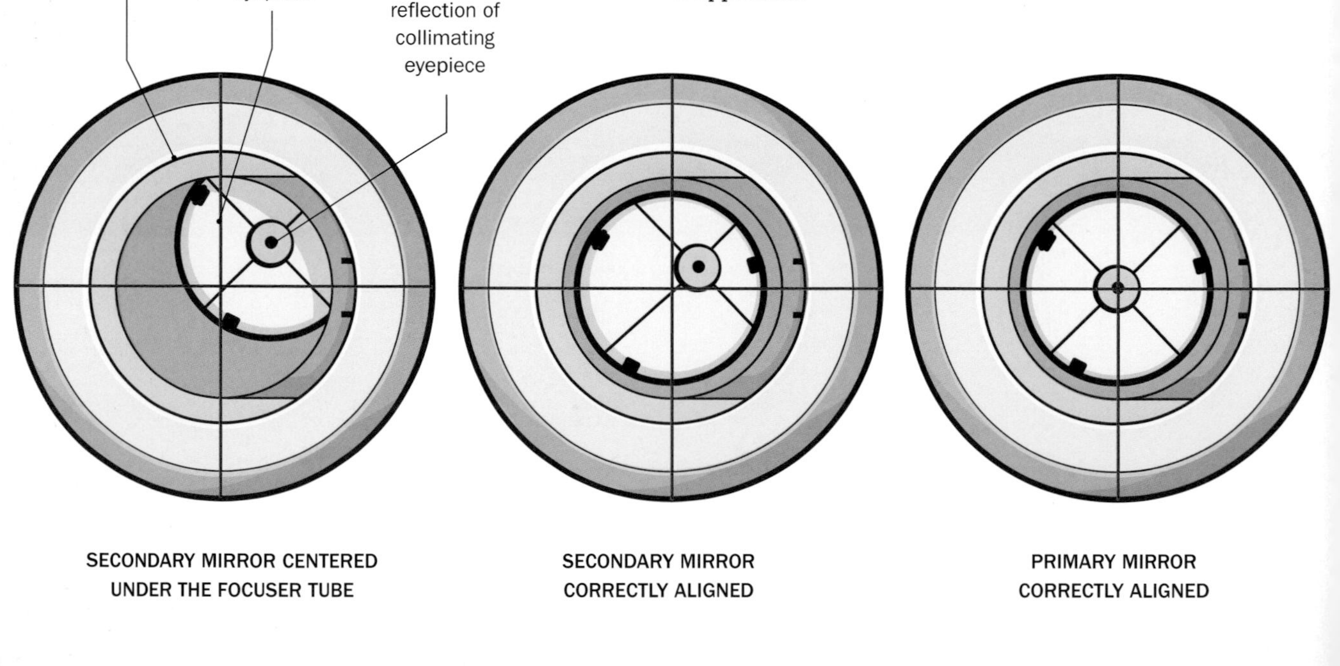

151 AIM YOUR SCOPE WITH A PIGGYBACK FINDER

When it comes to aiming your telescope, your finder is your friend. As you would a good friend, treat your finder with the care and respect it deserves.

A finder essentially piggybacks on top of your telescope, allowing you to target objects with a wide field of view before zooming in closer with your scope. A finder can be a smaller telescope on top of your main telescope, a device that projects a red dot over a field of view, or a glass "bull's-eye" style reflex sight (see #165–166). If your finder uses batteries, like a red dot finder, shut it off after every observing session—and keep a spare battery handy just in case.

To make the most of your observing session, you'll need to align your finder with your telescope's field of view, or *calibrate* it. Properly aligning your finder before viewing can save you a lot of time finding objects.

STEP ONE Aligning your finder in the daytime will save you a lot of frustration the first few times you do it. First, aim your finder at a distant object, such as a treetop or telephone pole (but not the Sun). If you can't align it in the daytime, pick a bright nighttime object, such as a distant streetlamp.

STEP TWO Check in your telescope's eyepiece, being careful not to touch your scope. (The vibrations will blur the image). Your object is probably off center from what your finder shows—or the object may not even be visible at all. No worries. Aim the telescope at the object (your finder should have gotten you close).

STEP THREE With your telescope aimed at the object, go back to your finder. Use the adjusting screws located on the finder's mount to carefully fine-tune its aim until the object is centered.

STEP FOUR Double-check to make sure the same object is centered in both your finder and your scope. You should be set.

152 FOCUS AND POINT YOUR TELESCOPE

So your telescope is assembled, cooled down, and calibrated with your finder scope. Next up? Focusing.

STEP ONE Pick the right target to start observing. Practice in the daytime to get a feel for your scope and how to operate it smoothly. Remember, in the dark you won't be able to see the scope very well. For your first few times at night, the Moon is a great target, as is a bright star like Sirius (Alpha Canis Majoris—see #59 for how to find it).

STEP TWO Once you have the target in your (hopefully calibrated) finder, look in your telescope's eyepiece. If you have aligned the finder properly, you will see your target; however, it may seem blurry.

STEP THREE Focus by slowly adjusting the knobs until you have the best image. Keep in mind that fog and turbulence will affect how finely you can focus.

STEP FOUR If you see the star as a bright sharp point of light—with no halos, double images, or blurring—you have perfect focus.

153

VIEW THE MOON WITH A SCOPE

So you've gotten your first telescope, and now you're ready to observe your first astronomical object. We suggest starting with Earth's nearest neighbor, the Moon. You may think that other astronomical objects are far more interesting than our Moon, but the lunar surface is absolutely stunning and surprisingly interesting in a telescope. The Moon is also immediately recognizable and easier to find using a telescope than the more elusive planets, nebulae, or galaxies.

During a full Moon, the intensity of the sunlight reflected by the Moon washes out a lot of surface detail. While the full Moon is certainly bright and beautiful, you'll see lunar craters and mountains in far more detail if you use your telescope to examine the lunar surface at other times of the month.

This diagram shows you the phases that are visible in the evening hours, just after a new Moon. For each phase, we also provide locations of some of the Moon's most interesting craters, basins, and mountains.

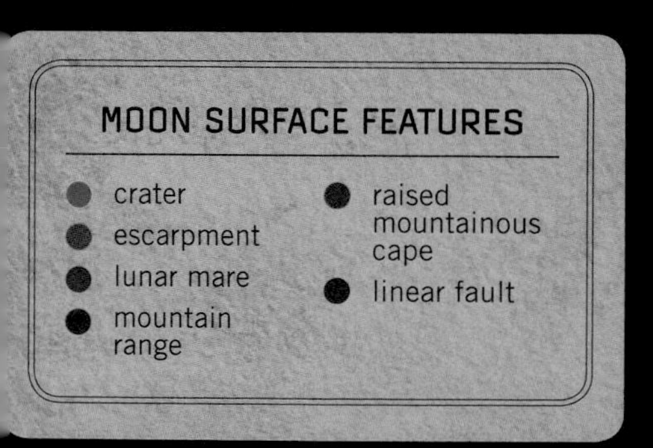

IDEAL PHASES FOR VIEWING THE MOON WITH A TELESCOPE

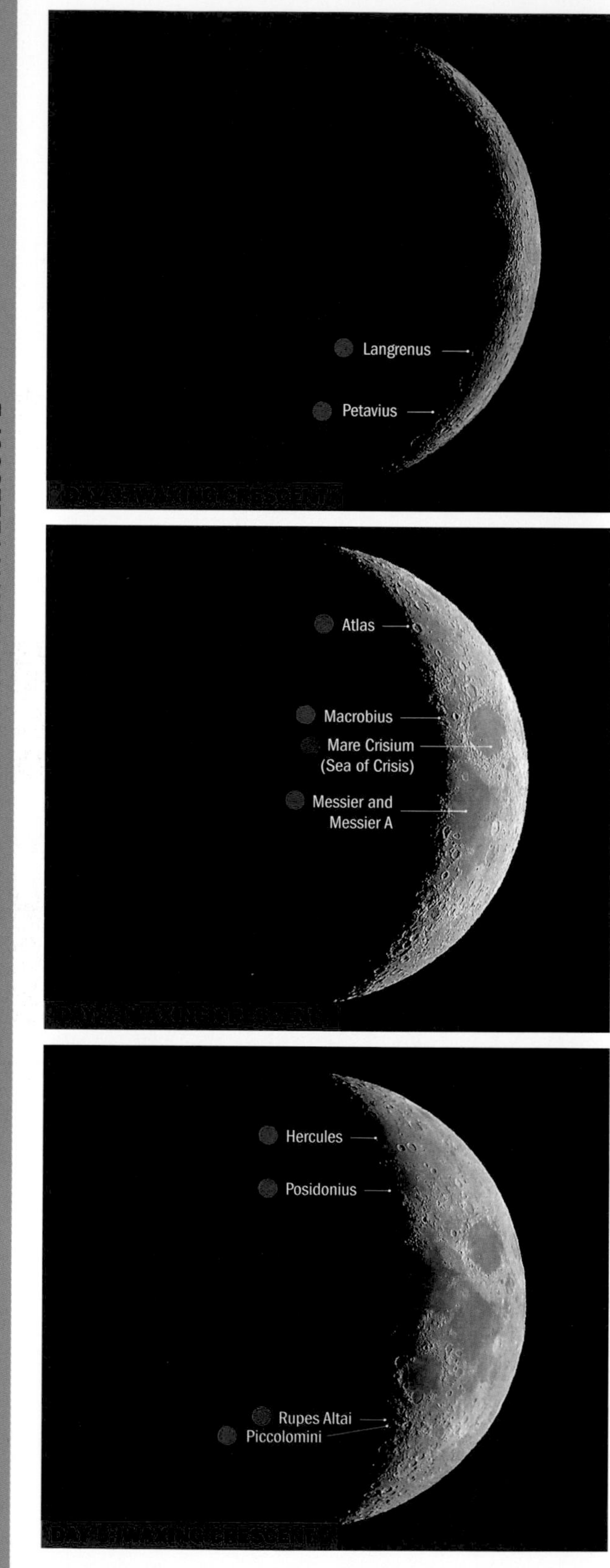

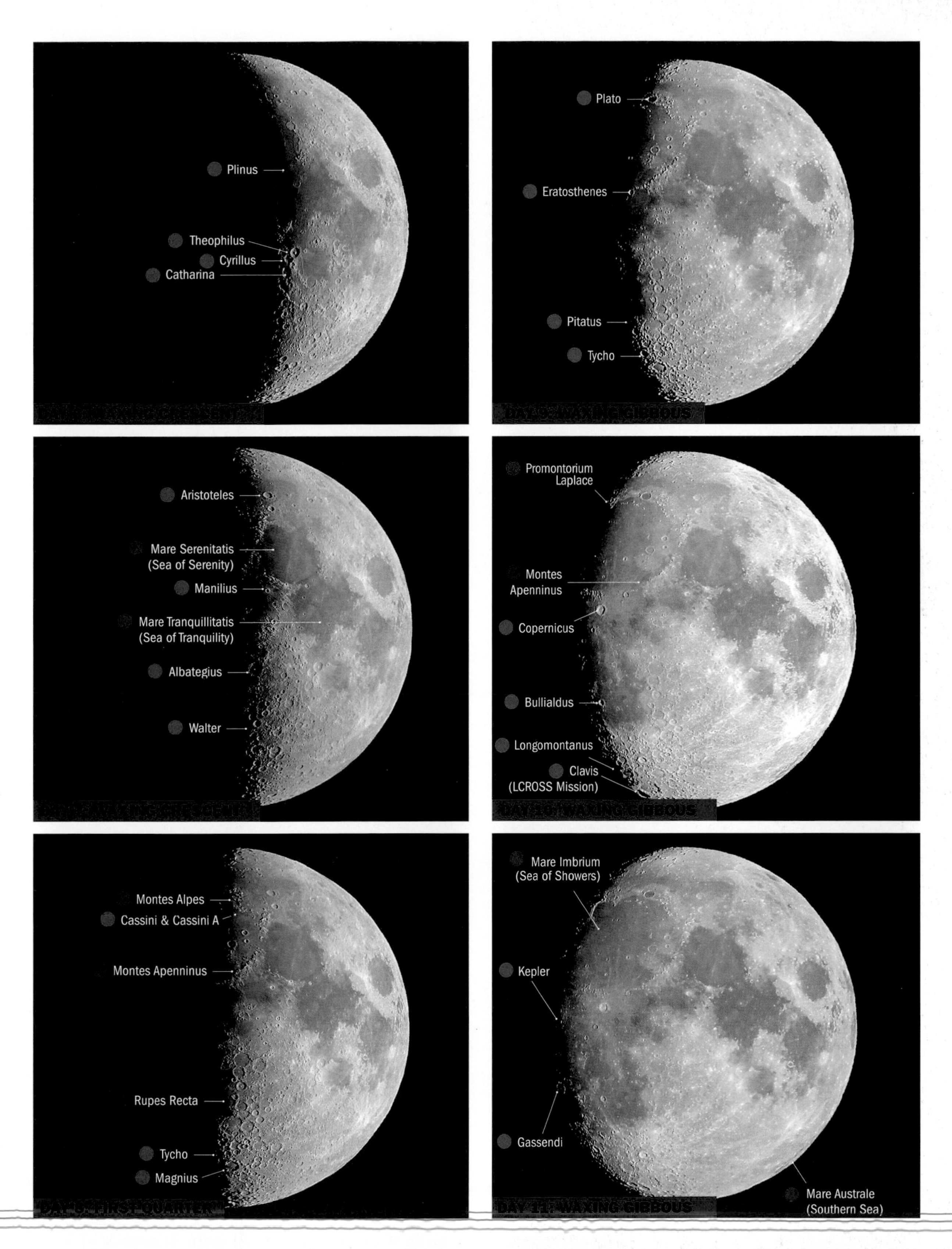

Plinus
Theophilus
Cyrillus
Catharina
Aristoteles
Mare Serenitatis
(Sea of Serenity)
Manilius
Mare Tranquillitatis
(Sea of Tranquility)
Albategius
Walter
Montes Alpes
Cassini & Cassini A
Montes Apenninus
Rupes Recta
Tycho
Magnius
DAY 8: FIRST QUARTER
Plato
Eratosthenes
Pitatus
Tycho
DAY 9: WAXING GIBBOUS
Promontorium
Laplace
Montes
Apenninus
Copernicus
Bullialdus
Longomontanus
Clavis
(LCROSS Mission)
DAY 10: WAXING GIBBOUS
Mare Imbrium
(Sea of Showers)
Kepler
Gassendi
Mare Australe
(Southern Sea)
DAY 11: WAXING GIBBOUS

154 AIM FOR ANTLIA

Antlia Pneumatica, the Air Pump, named after 17th-century physicist Robert Boyle's invention, is a small, faint constellation just off the bright southern Milky Way. While ancient Greek astronomers could likely make out the stars of Antlia, they were far too dim to have taken a position in any prominent constellation.

The Air Pump was given its somewhat unpoetic name by French astronomer Nicolas-Louis de Lacaille during the time he spent working at an observatory at the Cape of Good Hope, from 1750 to 1754. As a result of his observations of some 10,000 southern stars, de Lacaille divided the far southern sky into 14 new constellations, of which Antlia is one.

Not far from Vela and Puppis, Antlia is also bordered by Hydra the Snake, Centaurus the Centaur, and the compass Pyxis. The alpha star of Antlia is just barely the constellation's brightest star and has been given no proper name. Quite red in color, it possibly varies slightly in magnitude.

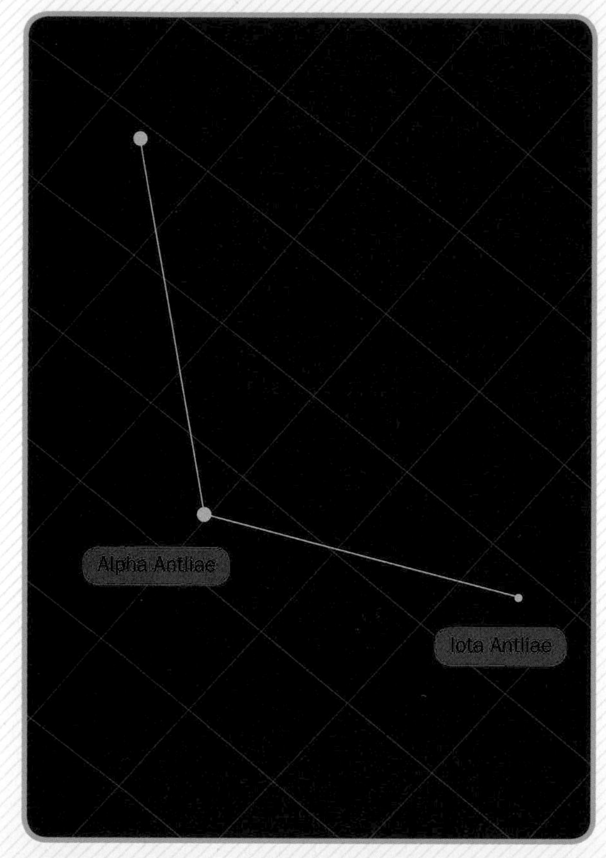

155 DRAW THE SOUTHERN TRIANGLE

A simple three-sided figure deep in the southern sky, Triangulum Australe first appeared in 1603 in Johann Bayer's great *Uranometria*—an atlas illustrating dozens of constellations as well as overviews of the northern and southern hemisphere skies. It lies just south of Norma the Level and east of Circinus the Drawing Compass—tools used by woodworkers and navigators on early expeditions to the southern hemisphere.

One of several of the constellation's *Cepheid variable stars* (stars that regularly pulsate and vary in temperature and diameter around relatively stable periods), R Trianguli Australis has a brightness that alternates from 6.0 to 6.8. Because it is a Cepheid variable, we know its period precisely: 3.39 days. For Cepheids with a variation this rapid, it's fascinating and worthwhile to compare magnitude at least once a night. Another bright Cepheid variable is S Trianguli Australis, which shifts from magnitude 6.1 to 6.7 and back during a period of 6.3 days.

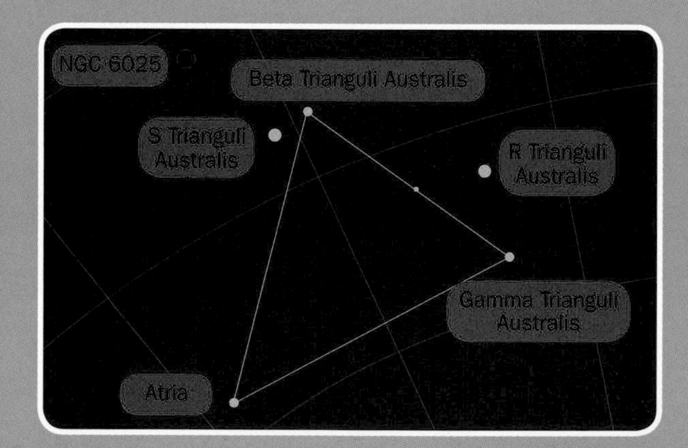

The Southern Triangle also contains NGC 6025, a small open cluster of about 30 stars of magnitude 9 and some fainter background stars.

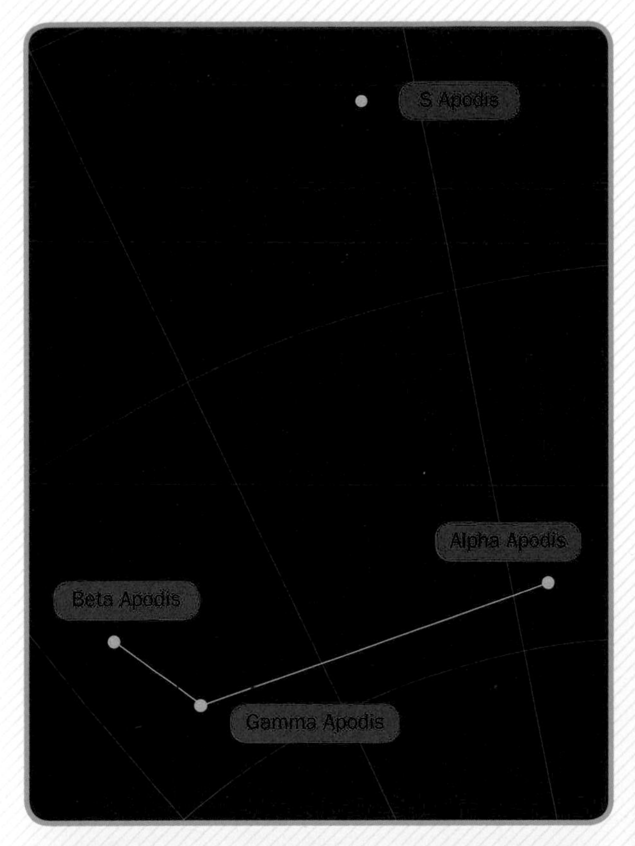

156 CATCH THE BIRD OF PARADISE

Apus is an ancient Greek word that means "footless." It is derived from *apus indica*, the Latin name given to India's Bird of Paradise by Europeans who erroneously believed the bird lacked feet. (The first dead specimens had their feet removed when shipped to Europe in 1522.) In the 16th century, Dutch astronomer and cartographer Petrus Plancius named this series of stars, basing his documentation on the detailed observations of Frederick de Houtman and Pieter Dirkszoon Keyser, who had traveled to the southern hemisphere to chart its unfamiliar skies. Today, we know Apus as a faint constellation directly below Triangulum Australe, the Southern Triangle. Being close to the southern pole, it cannot be seen from most northern latitudes.

The most variable of the stars in Apus is S Apodis—a "backward" nova. Usually, it shines around a magnitude of 10—bright enough to be seen through a small telescope—but at irregular intervals, it erupts, possibly sending dark, soot-like material into the atmosphere. It then fades dramatically, only returning to its original brightness after several weeks.

157 NOTICE NORMA

East of Centaurus and Lupus is a small constellation called Norma the Square. When he named this group of stars, our pal Nicolas-Louis de Lacaille decided to call it Norma et Regula, the Level and Square, after a carpenter's tools. Since those days, however, the Regula has been forgotten; the Square stands alone.

The constellation lies alongside Circinus the Drawing Compass, which de Lacaille named at the same time. Set in the southern Milky Way, Norma is interesting when viewed through binoculars. One notable open cluster found farther out past Kappa Normae, NGC 6067 reveals hundreds of stars within a stunning field when viewed through large binoculars or a telescope. Near NGC 6067, telescope-assisted observers can glimpse NGC 6087, another of Norma's striking open clusters. It is made up of some 40 stars centering around the variable star S Normae.

Norma has had some luck with novae: huge nuclear explosions that cause sudden star brightening. Its IM Normae is one of only 10 known recurrent novae in the Milky Way. It erupted in 1920 and 2002, and possibly in 1961.

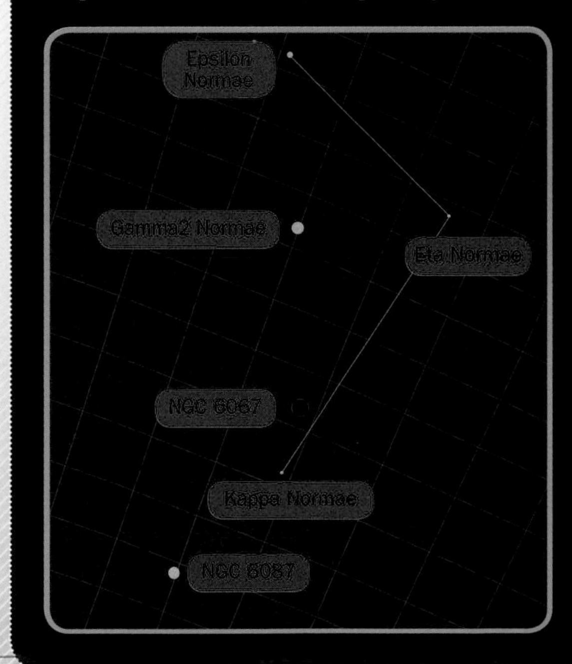

158 SCAN SKIES WITH AN ALTAZIMUTH MOUNT

So your telescope is equipped with an *altazimuth mount*—a simple, two-axis mount that allows you to track objects by adjusting the vertical and horizontal controls. Luckily, setup is very simple. Once you've picked a sturdy, level place and assembled your tripod, these steps should have you stargazing on two axes in no time.

STEP ONE Place your telescope into the mount. Make sure you pick the right height for the most comfortable viewing.

STEP TWO Keep your scope balanced by first mounting it parallel to the ground. Add your eyepieces and accessories, and see which way the scope tends to tip. Then adjust it accordingly by sliding it back and forth along its clamp rings until the scope remains level.

STEP THREE As the sky rotates above you, follow your target by slowly turning the knobs that control the horizontal and vertical motion of your telescope. If you want to hop to a nearby object, turn the knobs faster. If you are switching over to another section of sky altogether, unlock your telescope and manually swing it over.

159 SET UP A DOBSONIAN

Beloved by amateur astronomers, Dobsonians—named after inventor John Dobson—are light, simple Newtonian reflector scopes with large apertures that collect tons of light. They revolutionized amateur astronomy because you can build an inexpensive one out of ship glass and plywood. Follow these pointers when setting up yours.

① **BREAK IT DOWN** You may be tempted to move your telescope as one unit—tube and mount combined. But breaking it down for transport is best.

② **TRAVEL BARE** If you are traveling with your Dob, take off accessories. Nobody likes to pull up to a campsite only to notice a finder scope snapped off its mount.

③ **CARRY THAT WEIGHT** If you are planning on viewing objects near the horizon, bring a few small weights with you. By adding a small weight or two to the bottom of your scope—even just with duct tape—you can help keep the scope balanced when viewing at extreme angles.

160 TRACK STARS WITH AN EQUATORIAL MOUNT

Compared to Dobs or altazimuth mounts, telescopes with *equatorial mounts*—which follow the rotation of the sky, putting tracking on autopilot—are much fussier. For these mounts, leveling is critical, and most good mounts include a built-in bubble level to help you do this with ease. Here are a few more tips to keep you running smoothly.

A SAVE YOUR TOES Meet the most important part of your mount: the toe saver. A small knob that screws into the end of your counterweight bar, it prevents any counterweights from falling off and crushing your toes.

B GET ORIENTED Make aligning your telescope to the north or south easier by marking directions on the ground. Just don't create any tripping hazards.

C MATCH LATITUDES Remember to set your latitude scale to match the latitude of where you will be observing. If you don't, your tracking will be wildly off.

D GET ALIGNED Achieve as accurate a polar alignment as possible by using a mount with a polar alignment scope. Built into your mount, these scopes help you zoom in on the pole star when setting up—allowing for even more accurate alignment.

E RUN THE NUMBERS Learn the number system on your declination (dec) and right ascension (RA) axis. These numbers correspond to the coordinate system in our night sky (see #49–51 for more info). Use these numbers to manually locate hard-to-find objects from very detailed star charts or planetarium programs.

F TRACK MANUALLY The telescope traces the celestial sphere along the lines of north-south and east-west. In other words, one knob moves the telescope from north to the south (dec), and the other from east to west (RA). If everything is properly aligned, you need only to gently turn the RA knob to track and keep an object in your sights.

161 PICK A PROPER EYEPIECE

Telescopes are a lot like cameras in that you can adjust their capabilities with a variety of attachments. When outfitting your telescope for ideal skywatching, eyepieces are arguably one of the more essential accessories—mostly because a lot of telescopes only come with one eyepiece, often of average quality. Since celestial bodies are visible in a number of different ranges of magnification, one eyepiece usually won't cut it. Here are a few solid options to get you pointed in the right direction.

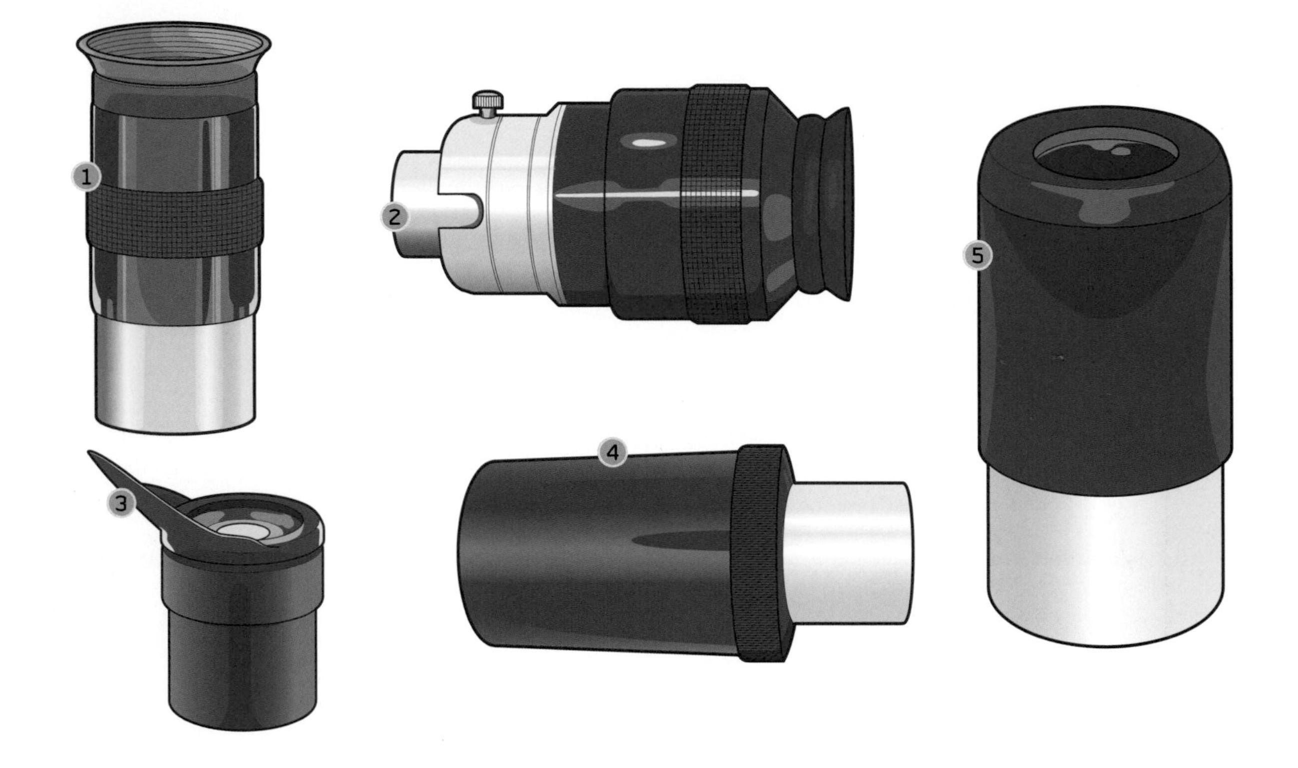

1 PLÖSSL Also called a symmetrical lens, the Plössl was designed by Georg Simon Plössl in 1860. Made up of two sets of two lenses, it allows for a wide field of view and is useful for many skywatching purposes, from peeping planets to exploring the deep sky. They are currently among the most popular eyepieces (pretty much every telescope comes equipped with a Plössl), with quality varying depending on the manufacturer.

2 NAGLER Designed by Albert Nagler, these eyepieces include a comparably large number of optics parsed out into a number of different groups. While Nagler eyepieces excel in high magnification and provide an ultrawide field of view, their weight—sometimes more than 1 pound (0.5 kg)—can actually tip smaller scopes. These tend to be very expensive but provide fantastic image quality—especially for deep-sky objects.

3 ORTHOSCOPIC Designed in the 1800s by Ernst Abbe, the original aim of orthoscopic eyepieces was to allow distance on microscope slides to be measured accurately. Their narrow field of view makes them best suited to applications where they'll stay in one place, such as on a telescope set up to track one object across the sky for detailed observing.

4 KÖNIG The original König eyepiece was designed by Albert König as something of a simpler version of the orthoscopic eyepiece. Whereas the original produced high magnification with a field of view comparable to a Plössl, more modern versions adapt the original design by adding sets of lenses, improving both magnification and field of view. These are good for viewing deep-sky objects with a high level of detail. Combined with filters, they are excellent for gazing at emission nebulae and galaxies.

5 KELLNER/ACHROMAT Among the least expensive eyepieces for practical use, Kellner uses three optical elements in two groups. While not as strong as a Plössl or with as wide a field as a Nagler, Kellner does a pretty good job in terms of viewing planets, the Moon, and deep-sky objects.

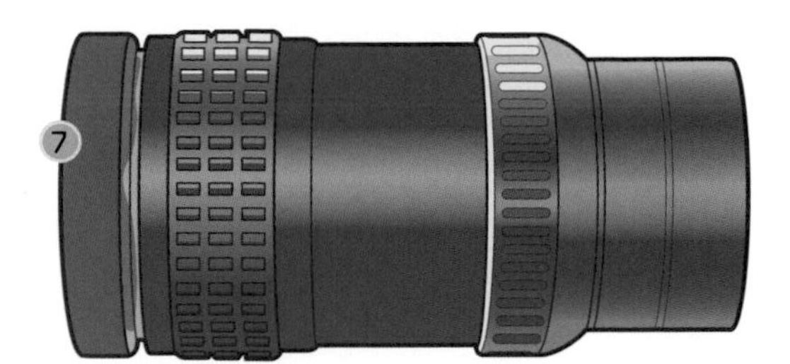

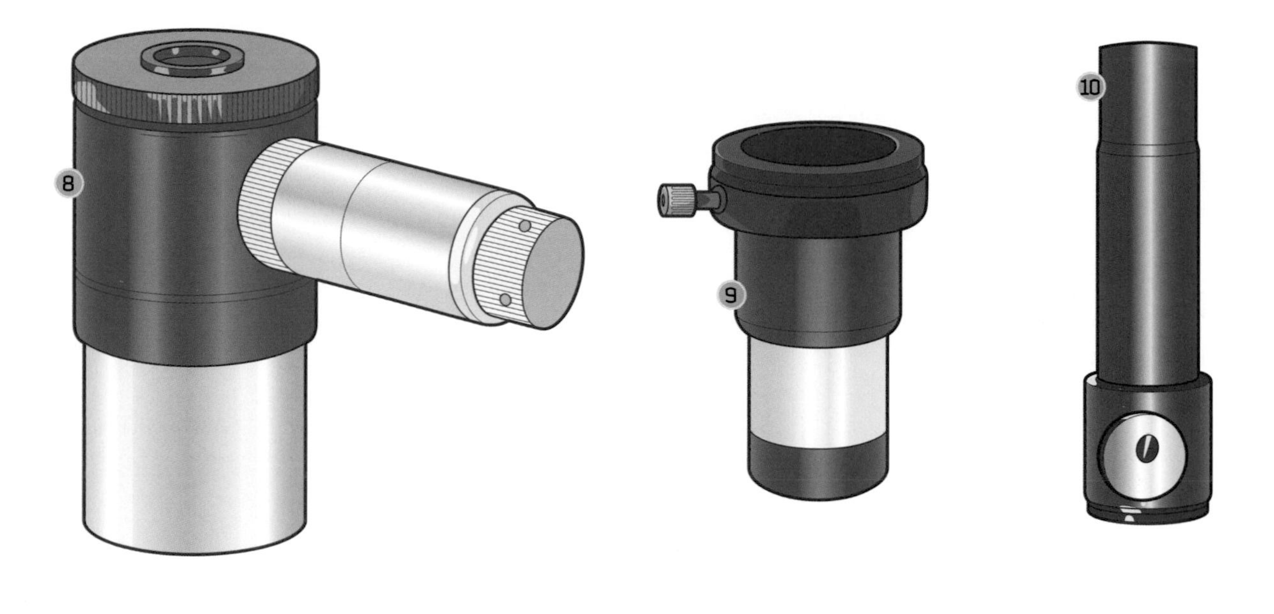

6 **MONOCENTRIC** Closer to the expert level, monocentric eyepieces are somewhat unusual in construction—having three glass elements fused together—and narrow both in field of view and use. Designed by Adolf Steinhall in 1883, the monocentric gives a bright, high-contrast image and is most useful for timing transits of moons around Jupiter, star occultations by our Moon, or changes in the Martian surface. Note: It has bad eye relief (see #126).

7 **ZOOM** As you might infer from the name, zoom eyepieces allow you to adjust the power of your eyepiece, often within a range of 8 to 24x, enabling you to tweak your piece to fit your viewing conditions. Of course, what a zoom eyepiece contributes in terms of convenience does translate to a loss of image quality.

8 **ILLUMINATED RETICLE** These eyepieces are powered with an illuminated crosshair or grid (*reticle*), as might appear in a hunting scope. They are especially helpful when trying to photograph the night sky because they allow you to precisely center your eyepiece on the target. Illuminated reticle eyepieces can also be useful for collimating your telescope or measuring an object's apparent size.

9 **BARLOW LENS** While technically not an eyepiece, a Barlow lens, when used with a telescope and an eyepiece, increases the magnification of an image, commonly by two or three times. With this one piece (and a small loss of light), you can essentially double the apparent number of eyepieces you have, allowing each to function at different powers.

10 **COLLIMATION** To see the sharpest image possible through your telescope, it's important that your optics are aligned as precisely as possible. Collimation refers to that alignment process, and a collimation eyepiece is a tool that makes aligning the optics of your scope much easier than it is without a similar tool. (See #150 for a collimation tutorial.)

162 UNDERSTAND EYEPIECE MAGNIFICATION

When it comes to telescope eyepieces, more magnification is not always better. The more you magnify an object, the dimmer your view will be. Zoom in too much during a night of bad seeing—or beyond the limit of your scope—and the view will be crummy. Highest magnifications are best for bright objects such as the planets, the Moon, and double stars, while lower magnifications are often best for views of distant, dimmer objects such as galaxies and nebulae.

For starters, it helps to know the highest practical magnification for your telescope. This is usually about 50x for every 1 inch (2.5 cm) of glass. For example, a 6-inch (15-cm) scope would have a highest practical magnification of 300x. Beyond that, you can calculate the magnification of your scope and eyepiece with this easy formula:

focal length / eyepiece diameter = magnification

For example, if you have a 10mm eyepiece and a telescope with a 1,000mm focal length, your magnification is 100x. You can get good views of objects starting at about 30x. Generally, the useful limit of magnification in an amateur telescope stands at 400x (due to the limits of seeing through our atmosphere). If you are worried about finding these numbers for your equipment, rest easy: Almost all manufacturers put them on the sides of their eyepieces and scopes.

low magnification

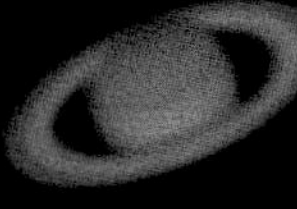

high magnification

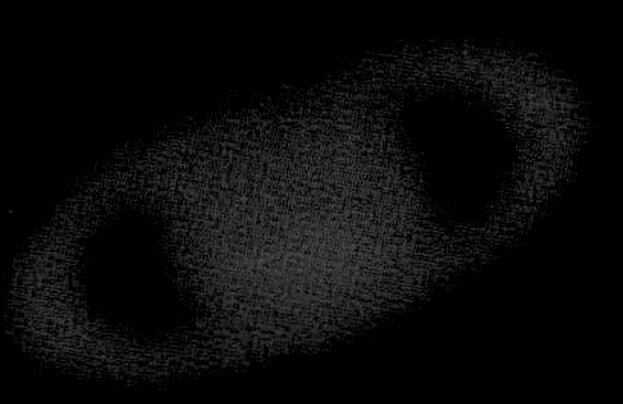

excessive magnification

163 FIND THE RIGHT FIELD OF VIEW

Getting the most out of your telescope also means understanding how different types of eyepieces show different fields of view at similar magnifications. While some eyepieces offer a narrow field of view, akin to looking through a tunnel or pipe, others offer a wide view of the heavens, more like looking through a porthole.

It is helpful to know the difference between the *actual field of view* of an eyepiece versus its *apparent field of view*. The actual field of view is how much of the sky is shown through your eyepiece, while the apparent field of view is how large the section of the sky appears to your eyes. For example, you may have an eyepiece that has a 35-degree apparent field of view. The actual field of view may be 0.5 degrees.

You can figure out your actual field of view by dividing your eyepiece's apparent field of view by its magnification with your scope (see #162 above to figure out your magnification).

apparent field of view / magnification = actual field of view

Say your telescope and 10mm eyepiece have a 100x magnification. If your eyepiece has a 50-degree apparent field of view, divide 50 by the magnification (100x) to arrive at 0.5 degrees of the sky as your actual field of view. That is the size of the full Moon in the sky, so the Moon would fill your entire eyepiece (A).

Another 10mm eyepiece may have an apparent field of view of 80 degrees, giving you an actual field of view of 0.8 degrees: a larger section of the sky at the same magnification. You would then be able to have the Moon centered with a good amount of space around it (B)—which is great if you are looking for it to occult a star.

164 BOOST RESOLUTION WITH THE RIGHT EYEPIECE

Look at an older TV and then a new high-definition TV. Notice how much clearer the HDTV image is? That's because the HDTV has greater *resolution*: how sharp or finely detailed the image is. Picking the right eyepiece helps bring out your telescope's best possible resolution. While a large telescope brings in more light and offers more detail, mating that telescope with a poor-quality eyepiece will destroy much of those gains. Like having your HDTV on the wrong settings: The signal may come through, but the picture won't be as good as it could be.

Next time you're at a star party (see #177), try looking at the same object with a few different eyepieces. Notice the maximum amount of fine detail that each of these eyepieces brings out, especially if you are choosing eyepieces with higher magnification. Also pay attention to how different sizes of telescopes and different eyepieces show more or less resolution on these objects. Ask the astronomers about their setup and choice of eyepieces; they will almost certainly like to share their thoughts on resolution!

165 PICK THE RIGHT FINDER SCOPE

These small, auxiliary telescopes connect to your main telescope, giving you a wider field of view to guide you in searching the heavens. From traditional piggybacking scopes to finders with lasers, there are several types.

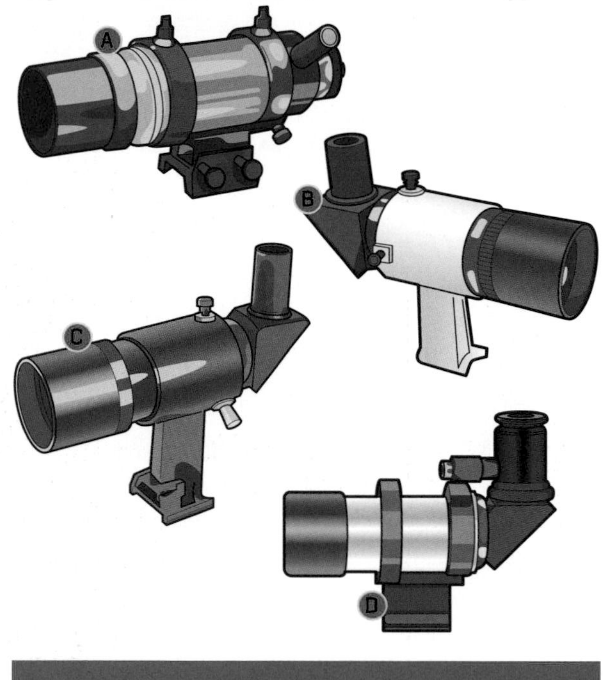

A STRAIGHT-THROUGH FINDER SCOPE The most common form of finder, this tiny scope should have good field of view in order to be useful. While many starter scopes feature this type of finder, they are often nearly useless due to a small field of view. Look for a finder that is at least 6x30. In this model, the image is displayed upside down and is reversed from what you see in the sky.

B RIGHT-ANGLE FINDER SCOPE The more advanced version of the straight-through finder is the right-angle finder, handy because the end of its scope is tilted 90 degrees toward you, much like a diagonal eyepiece in a refracting telescope. Since you don't need to tilt your head at an uncomfortable angle under the scope, you can spare your neck some strain. The image displayed is a mirror image of what you would see.

C RIGHT-ANGLE CORRECT IMAGE FINDER Unlike finder scopes that display an image in reverse for you, this type of finder scope shows you the sky in the same orientation as it would appear to you normally.

D ILLUMINATED FINDER SCOPE This finder scope has an extra feature: Glowing crosshairs are displayed inside its field of view, allowing you to more accurately target objects in the night sky. You can get them dead center.

166 FIND A RED LINE SIGHT

This family of devices projects a red light in the direction in which your telescope is pointed.

A RETICLE SIGHT They're great for smaller scopes, and their large field of view helps you aim roughly and quickly with the projected crosshairs.

B RED DOT FINDER Also called a reflex sight, these project a small red dot against a piece of clear glass or plastic, helping you find your target. They often come attached to small beginner telescopes and are not notably accurate.

C TELRAD A larger version of the reticle sight, Telrads are better suited to larger telescopes and are often used in conjunction with a finder scope. Many star charts and star-hopping guides, as well as astronomy software, use Telrad circles to help point out where to look in the night sky.

D LASER HOLDER This tool is made up of two clips that resemble the rings that hold a finder scope in place, only they're smaller and hold a green laser pointer. To use, turn on your laser pointer and aim the laser at a target after aligning it with your scope.

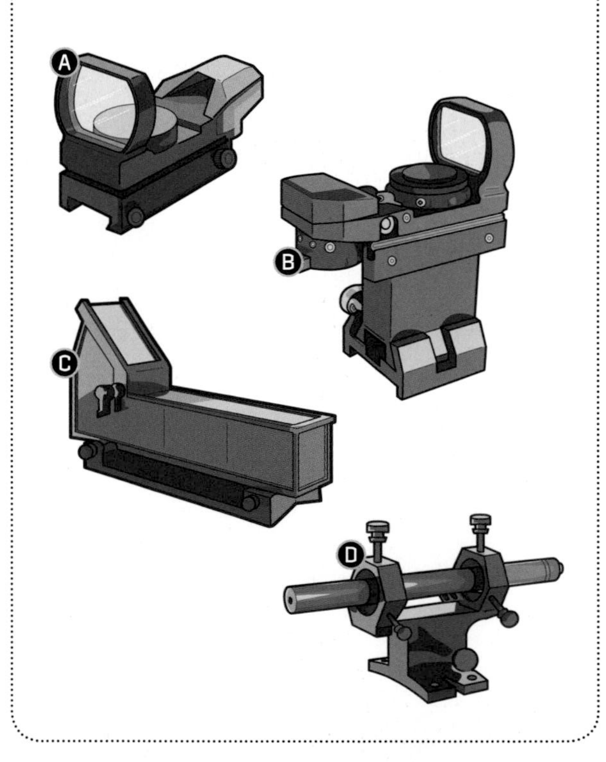

167 LIGHT THE WAY WITH LASERS

Green lasers let you point out the heavens to others—their green beams will seem to shine straight up to touch the stars. Of course, this is just a trick of perspective; the laser's coherence is only for a few hundred or thousand feet. Still, it is good enough for a dazzling—and educational—trick!

There are very important safety laws to follow with green lasers. First, never point them at an aircraft. Besides being a good way to get yourself arrested, it can cause confusion in the cockpit and ruin a pilot's night vision. Similarly, never shine them in anyone's eyes: Lasers can cause permanent damage, including blindness. The legality of these valuable tools is ensured by responsible use.

168 MOUNT A WEBCAM TO YOUR TELESCOPE

Webcams are fantastic tools for sharing the view from your telescope. They work best with brighter objects, making Mars and Jupiter popular targets, as well as the Moon or Sun (when using a solar filter, of course).

To start, use an adapter to attach a regular desktop webcam onto your telescope's eyepiece holder or a Barlow lens. (Just be aware that you won't be able to view through your telescope's eyepiece while the webcam is attached.) Then hook up the webcam to your red-light-filtered laptop (see #248). You can live-stream what you see or record and share it later.

Many amateurs remove filters from the webcam to enhance the image, or they use special planetary cameras. Generally pricier than regular webcams, they're worth it if you want the built-in support and software. Many stunning webcam videos are made by recording and then stacking each individual frame using software (see #246).

169 TOP FIVE PICK A SOLID TRIPOD

The rewards for choosing a good tripod are comfortable viewing and a hassle-free setup for your equipment. Here are the top five things to consider when making your selection. Take the time—it's worth it.

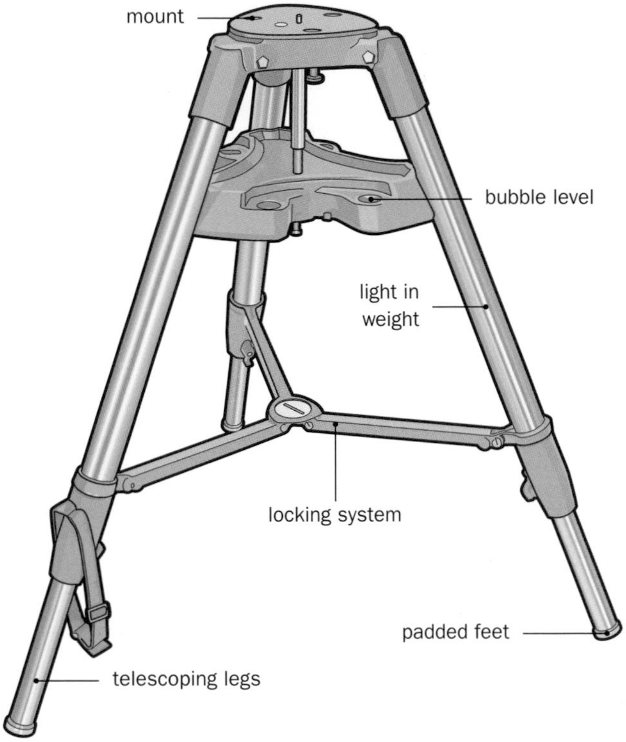

☐ **LOCKING LEGS** Make sure your tripod has adjustable telescoping legs for a wide variety of heights. Also, look for a tripod with a locking leg–spacer system. Not only do they keep the legs from spreading too far, but spacers often have compartments for holding eyepieces, too.

☐ **STURDINESS** Pick a lightweight but stable tripod—one that is rated to hold the weight of your equipment. Add padded feet or studs to the bottom of your tripod's legs to get a better grip on the ground.

☐ **PORTABLE CASE** Find or make a padded carrying case for your tripod. Make sure it has a good strap for transportation to observation sites.

☐ **ACCOMMODATING MOUNT** Pick a tripod that can accept a wide variety of mounts. Cameras, binoculars, and telescope mounts can all be used with the same tripod.

☐ **BUBBLE LEVEL** Make sure the tripod has a built-in bubble level.

170 SET UP YOUR GEAR FOR SAFETY

Tripping over your equipment in the dark is a bad way to end a night of stargazing. Protect your setup with the following tips.

ORGANIZE TOOLS Forcing yourself to put away anything not in use is good for many reasons: You probably won't lose anything, your tools will all be in the right spot, and you're less likely to drop stuff. Many skywatchers have accidentally crashed into their equipment while fumbling around looking for a lost eyepiece. Don't be one of them! Wear a vest or cargo pants with lots of pockets so you can store gear, have designated cases for all your pieces (and use them!), and group like parts with like parts so related items are stored together.

ILLUMINATE WISELY For visibility, keep a few dim red lights lit around your setup. Marking your equipment's boundaries in the night shows you and visitors clearly where you're camped—and where not to tread. Attaching red lights to your observing table and around your tripod legs is a good way of letting everyone know where your equipment is in the dark (see #173) without making your eyes any less adapted. The less time you spend fiddling around in the pitch black, the more time you can spend seeing.

171 TAKE A (COMFORTABLE) SEAT FOR SKYGAZING

Picking the right kind of chair for your observing is important. It's ultimately a personal choice—what fits best for your observing habits, equipment, and budget? For many binocular and naked-eye observers, a reclining lawn chair or beach chair is ideal. Sometimes, even a blanket and pillow to lie down on can do the trick.

If you are going to be spending a lot of time looking through your telescope or high-powered binoculars, you will want to upgrade your chair to something with more skygazing specific functions. Many observing chairs actually look like adjustable stools. Qualities to look for in an observing chair are sturdy construction, an adjustable height, and a padded and comfortable waterproof seat. Your chair should also be easily transportable and fairly lightweight, and have stable legs. Nobody wants to take a tumble in the dark.

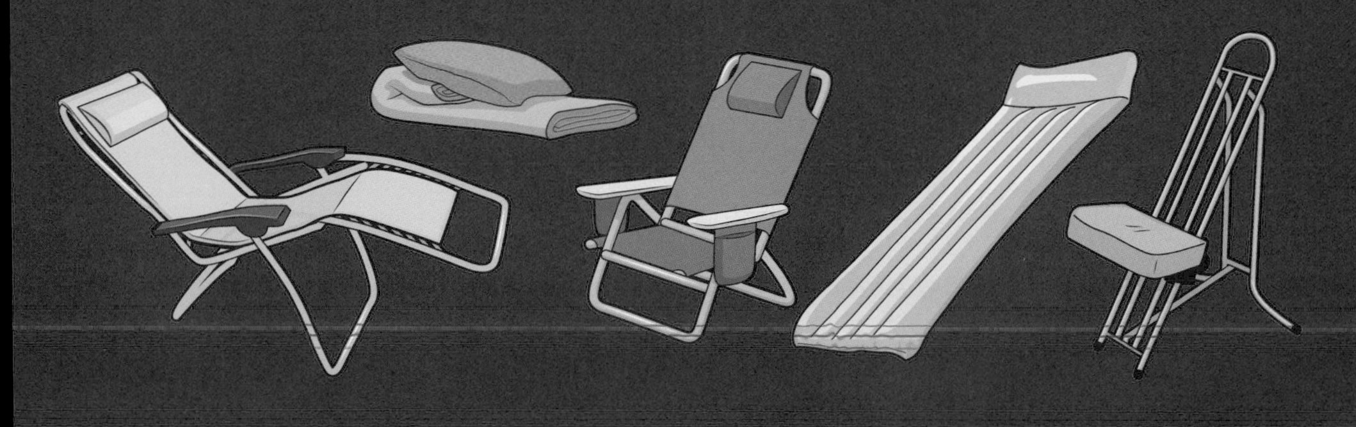

172 COOL YOUR TELESCOPE TUBE

Attaching a cooling fan to your telescope can help chill your tube down quickly, giving you the best possible seeing in no time. This is especially useful if you have a Dobsonian telescope with a large mirror.

STEP ONE Pick up a small computer fan or something similar. You don't want the fan to be too big, as larger fans cause more vibration. You can find computer fans at an electronics shop or online. While you're at it, procure a portable power supply and a DC power adapter (cigarette-lighter-style adapter) with free leads, which will allow you to attach them to the fan's wires.

STEP TWO Clip the plug off the end of the computer fan's wires. Using screw-on wire connectors or splicers, splice the fan's wires to the wires on the DC power adapter, matching the positive and negative wires. Seal with electrical tape.

STEP THREE Cut Velcro into 1-inch (2.5-cm) strips. Run them along the bottom cell of your telescope. Then cut out a circle of foam, plastic, or Neoprene and attach it to your scope's bottom using adhesive strips. (You may need to cut holes in the disc for collimation knobs.)

STEP FOUR Remove the foam disc and place the fan on top of it, trace its shape around the foam, and cut out the tracing with a hobby knife.

STEP FIVE Mount the fan to the disc with glue or more Velcro. Place the fan facing toward the cell of your telescope's mirror, allowing cool air to blow in.

STEP SIX Plug your adapter into the power tank to make the fan quietly spin. If you notice vibrations, readjust the mount until they stop. You can unplug the assembly once your telescope has cooled down.

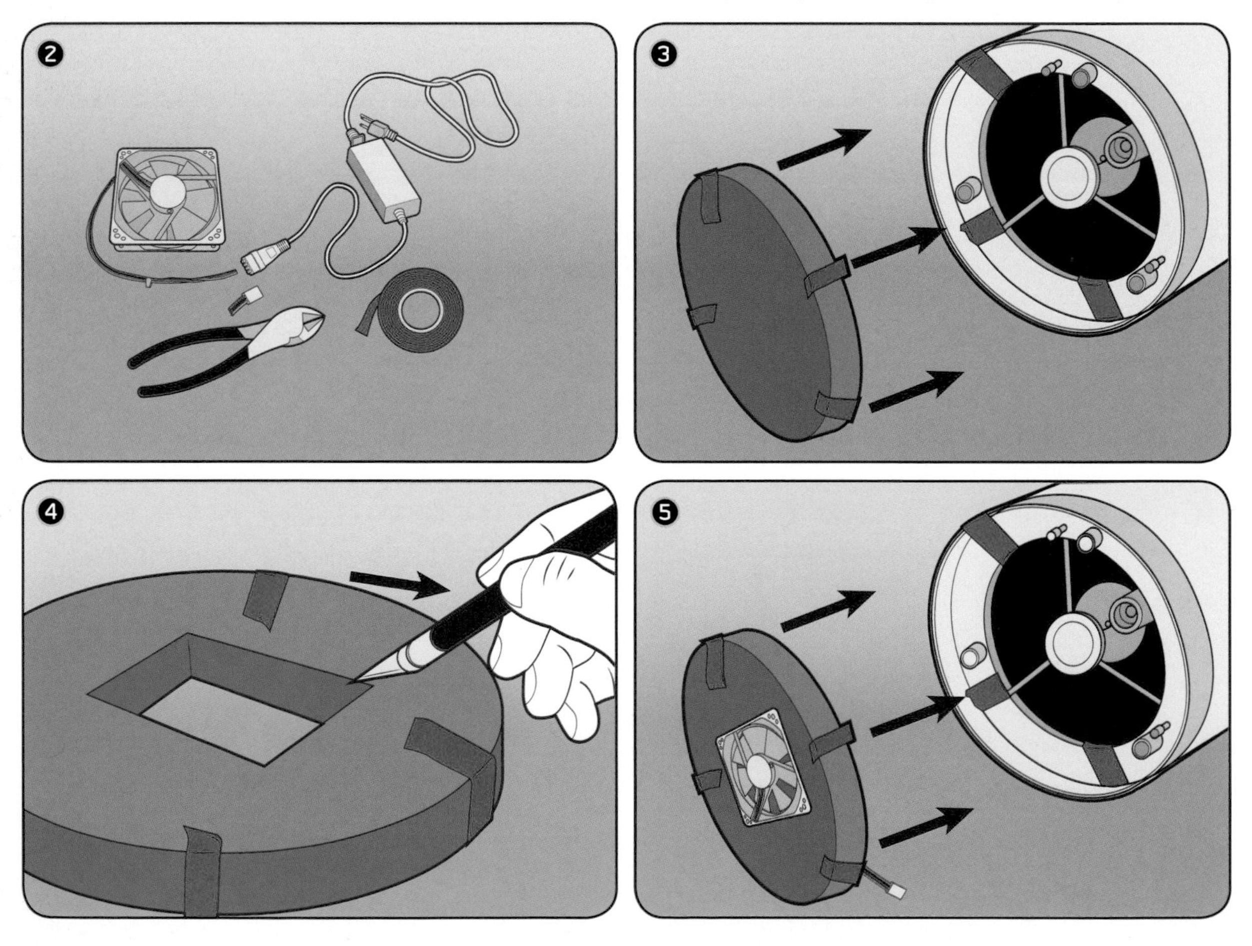

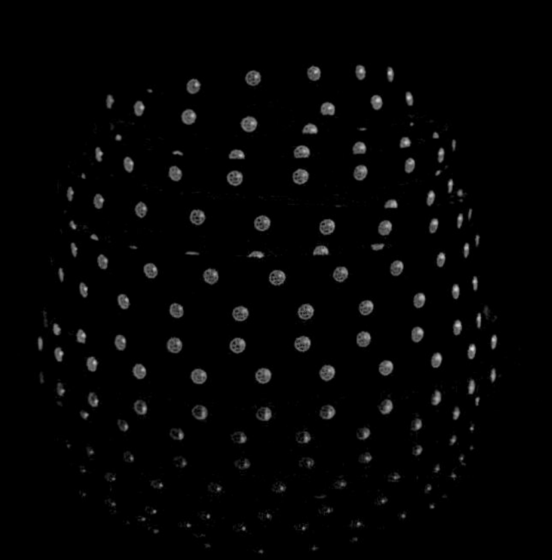

173 LIGHT UP YOUR TRIPOD

Red LEDs can prevent your fellow star-watchers—and yourself—from colliding with your telescope. In a pinch, battery-powered red LED Christmas lights wrapped around your telescope's tripod can work, but they may be a bit bright. Here's a more night-vision-friendly setup.

STEP ONE Grab a red LED light strip and some adhesive Velcro dots. Attach some dots to the back of the light strip.

STEP TWO Wrap the LED strip around your telescope's tripod and mount. Experiment with where the lights hang before really adhering them. If you have multiple strips, arrange them so their exposed lead wires or clips are close. (For example, you may want the LED strip leads for three LED strips to meet near the top of your three tripod legs.) Once you're satisfied, peel off the adhesive backing on the Velcro dots and affix the LED strips to your equipment.

STEP THREE Get a DC adapter and an external 12-volt power supply. Splice the leads of the two wires running to your power supply and twist them with the strips' exposed leads, negative to negative and positive to positive. The lights should illuminate! For more control, you can thread an on/off switch between your power source and the lights.

174 PIGGYBACK YOUR CAMERA

You can get gorgeous, wide-field shots of the sky by piggybacking your camera to your telescope, using your scope's tracking feature to produce beautiful long exposures. While mounts are cheap, making one is cheaper.

STEP ONE Drill a small hole through the center of a small, flat block of wood. Insert a ½-inch (1.25-cm) screw that's flat on both the top and bottom. Since your camera's threaded bottom adapter will attach to this screw, make sure the screw pokes out at least ¼ inch (6 mm) for a secure fit. If the screw is too long, cut it with a rotary tool.

STEP TWO Attach two smaller wooden blocks along the length of the block of wood to serve as a resting point on your telescope. For a better fit, cut the blocks into two triangular wedges and attach Velcro strips. Secure the Velcro under the wedges and attach them to the outside of the scope.

STEP THREE Use two flat-head screws to attach a hose clamp to your wooden camera mount. Make sure the screws aren't so long that they scratch your scope!

STEP FOUR Now slide the hose clamp down your telescope and clamp it together with butterfly screws (or screws and washers, in a pinch).

STEP FIVE Gently screw your camera body onto the flat thread protruding from your new mount.

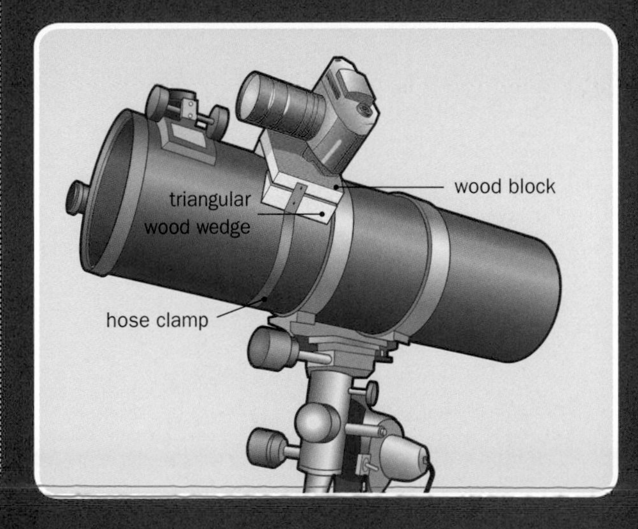

175

TOP FIVE
TREAT YOUR SCOPE TO A TRIP

Before you try lugging large and sensitive telescopes on a plane, consider contacting amateur astronomy clubs in or around your destination; they may loan or rent gear—or might even join you. Another option is to just bring a good pair of binoculars. That said, if you must fly with a telescope, keep these tips in mind.

☐ **SMALLER IS BETTER** Make sure your optics can fit in a carry-on case. Stick to smaller scopes, usually under 6 inches (15 cm). Try out telescopes designed for travel, if possible. Read the reviews before purchasing anything, ever.

☐ **PLAN AHEAD** Give yourself plenty of time at security. Telescopes are rare and can cause some extra scrutiny with the authorities.

☐ **CHECK YOUR LASERS** Pack your laser pointer, collimator, and finder scope in your checked luggage with the batteries removed. While these may be just fine, it's much better to check them in your bag than have to throw them away before passing through security.

☐ **BRING A TRIPOD** A sturdy, collapsible tripod can also travel with your checked luggage. Any tools or knives are best in checked luggage, too.

☐ **HAVE POWER** Rechargeable batteries are ideal; they can last an evening of observing. Remember to bring a charger as well as a converter for it, if you'll be traveling overseas.

176 CAMP UNDER THE STARS

If you live in a populated area, camping is a great way to get under some really dark skies with few distractions. As far as where to go, the International Dark-Sky Association (IDA) has a classification program for parks around the world that "possess exceptionally starry skies." (See #284 for more on the IDA.) But look around—plenty of unregistered sites have great views, too. Besides standard seeing preparation—like planning around the Moon and weather to ensure the darkest and clearest skies, choosing a flat area at high altitude, and coming prepared with chairs, blankets, friends, and provisions—there are a handful of other things to keep in mind when skywatching in the wild.

KNOW YOUR SURROUNDINGS Have an idea of what animals may be sharing your seeing spot with you. Be aware that nocturnal animals may not be expecting you out late at night. Know how to behave if anything more dangerous than a deer approaches.

CHOOSE APPROPRIATE GEAR Most experienced campers will recommend a smaller scope and some good binoculars. Hauling a large or heavy telescope over bumpy terrain can be both difficult and dangerous.

CAMP IN THE SHADE Since you'll likely be up oohing and aahing all night, pick a shady spot for your tent so the Sun doesn't bake you during the day. A sleep mask is ideal, too.

GET SOME REST Being well rested, avoiding caffeine and alcohol, and eating a little something sweet can all add to the enjoyment of the evening by improving your eyesight and energy levels.

177 BEHAVE AT STAR PARTIES

Large, multiple-night *star parties* (gatherings of amateur astronomers who pool gear and resources for making observations as a group—see #286) are a great way to find others with a passion for dark skies. They're also not bad for working on personal observing goals and sharing knowledge. Before you load up, remember to:

SIGN UP EARLY These events are often booked to capacity, so do your research and register and finalize your travel well in advance.

GET SITE DETAILS Check for RV hookups and camping amenities. Large events often have both and will certainly include restrooms of some variety. Note what times you are allowed to drive in. Often there are restrictions after dark to minimize stray light on the telescopes. Find out if the observation site is near the camping site and plan appropriately.

POWER OFF Cover all car lights with red tape—or just turn them off. You'll get a few choice words if you open your car door and spill bright white light at 2 AM.

LEAVE YOUR PETS AT HOME Often dogs are banned because of their possible interference with equipment.

WEAR HEADPHONES Music and other loud noises are generally not welcome. Use headphones if you like to rock out to the stars.

KEEP A CHARGE Charge batteries during the day so there's no need for generators at night. They are often banned at observation sites because of noise restrictions.

MAKE FRIENDS We all learn something at star parties. That's the joy of observing with others—whether you're a newbie or know the sky like the back of your hand.

178 TRANSPORT GEAR SAFELY

Proper protection can help keep your equipment safe from damage resulting from drops or falls, or being crunched from overpacking. It can also protect against cosmetic blemishes such as dings and scratches.

Haul your gear safely and ergonomically with a carrying case and padding. Buy one custom-made for your telescope, customize an old trunk or case, or make one from scratch. Whatever you decide, make sure the case is padded with good foam. Ensure a snug fit by using a foam cutter or use pluck foam—which you can customize to fit any shape—to make sure your equipment is secure. Also, padding isn't just for telescopes: Tripods, mounts, and eyepieces should all be protected during transit.

Last, having a good strong carrying strap attached to your case can really help you transport your gear. Better yet, a strong folding dolly with some bungee straps makes transport a snap without breaking your back. It can also reduce the chances that you will drop anything on your way to your observation site.

179 COVER UP, RAIN OR SHINE

A cover for your telescope is an excellent investment in preventative maintenance—otherwise, you risk degrading your optics' performance over time. Covering your telescope when in storage or during transport will keep out both dust and dirt, as well as prevent bugs from finding their way inside. And it only takes a few minutes after each stargazing session to make sure it's wrapped up snug and secure.

Covers are also great for multiday star parties, keeping your equipment safe and protected during the day. In a pinch, a blanket, tarp, or plastic wrap—with a little help from tape or bungee cords—can work. If ever there is a sudden sprinkle or rainstorm while you are away from your equipment, you'll be glad you had it covered—some observation areas have weather so variable it can change in minutes!

If your telescope will be outside in the daytime—say, while camping at a star party—a reflective cover may also prove to be a wise investment. These covers will help keep your telescope cool on a sunny day by deflecting the Sun's rays—similar to a Sun shield for your car windshield. Bonus: This reduced temperature also means a quicker cool-down time for your mirror when the Sun sets and stargazing begins.

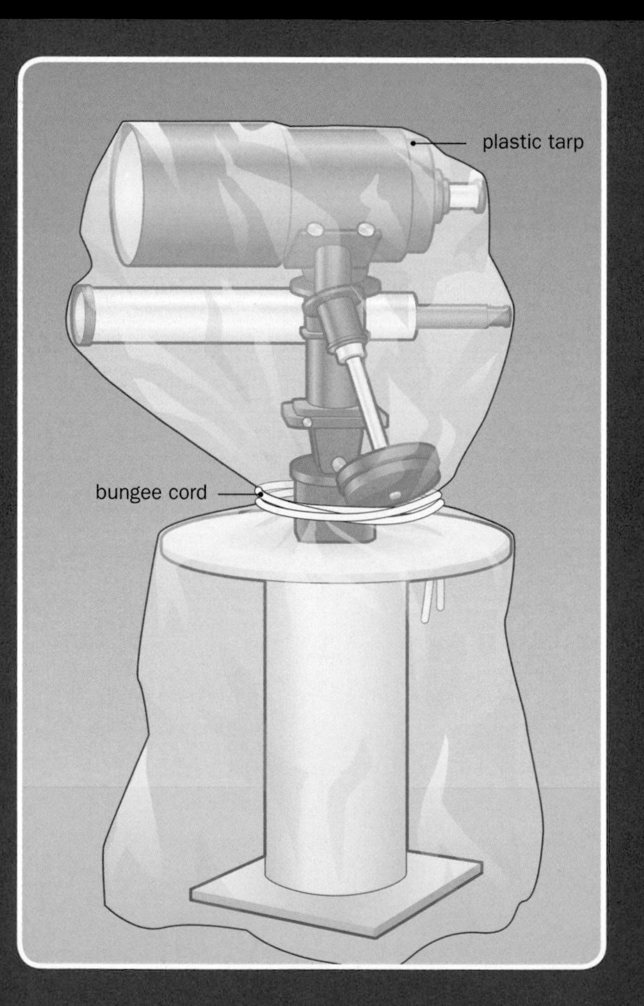

180 SHIELD YOUR SCOPE FROM DEW

Prevent dew from forming on your telescope's optics with a dew shield. Fighting dew is especially important for telescopes like Schmidt-Cassegrains and refractors, whose optics are exposed at the end of the tube.

A DIY YOUR DEW SHIELD You can buy shields for your telescope, or you can easily make your own by wrapping sturdy yet flexible foam or thin flexible plastic around your scope's end and securing with tape or Velcro straps. (Aluminum foil works in a pinch, too.)

B BLAST IT WITH AIR You can also use a small battery-powered hair dryer on a low setting if dew has started to form. But don't fight a losing battle: If the humidity is 90 percent or above, it is probably too wet to observe anyway. Whatever you do, refrain from using antifogging sprays or wipes on your equipment—they can potentially strip the coatings from your optics.

C ZAP UNWANTED DEW Dew zappers are the next level of dew fighting. Small heating elements that strap to the outside of your optics, zappers allow you to heat your gear to remove dew. While they tend to be rather pricey, they can be a worthy investment if you observe in a humid area. Besides, they're much cheaper than replacing your dew-ruined gear.

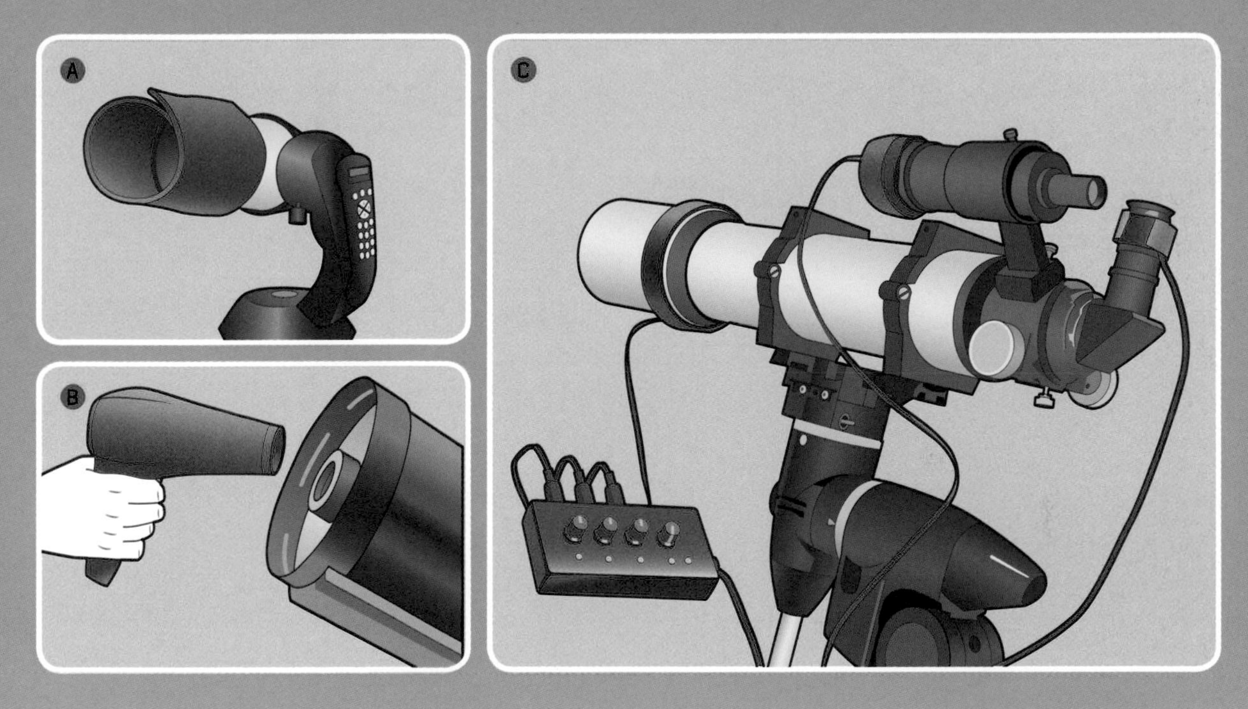

181 CAP YOUR OPTICS

Always keep your caps over your lenses or mirrors—whatever you do, don't lose them! Keeping your equipment capped until you're ready to observe with it prevents dust, fingerprints, and other muck from accumulating. It helps prevent moisture from building up as the evening progresses—and it does its part to keep bugs and other uninvited visitors out of your stuff.

182 PEEP THE SUN SAFELY THROUGH SOLAR FILTERS

While staring at the Sun is very unsafe with the naked eye, it becomes exponentially more dangerous when the power of the Sun is magnified through a telescope. Unfortunately, people commonly use a number of materials claiming to be solar protection that don't actually protect your eyes as much as they should—from Mylar to welder's glass. Avoid these. If you want to look directly at the Sun, keep your eyesight by using one of these filters.

A METALLIZED GLASS Arguably the best choice when it comes to solar filters, this filter is made by coating polished glass with a mixture of metals. Since they effectively reduce the Sun's rays to 0.00001 percent of full intensity when mounted to the top of your telescope tube, you can feel free to explore the surface of the Sun without worrying about scorching your retinas.

B SOLAR SAFETY FILM While not nearly as sturdy as a metallized glass filter, solar safety film is inexpensive and can work in a pinch. Be sure to check for any damage to the filter itself; it only takes one tiny hole or rip for a beam of magnified sunlight to do a serious number on your ability to see.

C HYDROGEN-ALPHA FILTER Available at a premium as stand-alone filters, hydrogen-alpha filters are generally built into specialized scopes. They block out all light except for a very narrow bandwidth on the visible spectrum—the deep red color of light that hydrogen atoms emit, called the H-alpha wavelength, which makes them ideal for viewing solar flares, solar prominences, and a number of other solar phenomena. If you do happen to grab one, make sure your telescope is compatible and won't overheat.

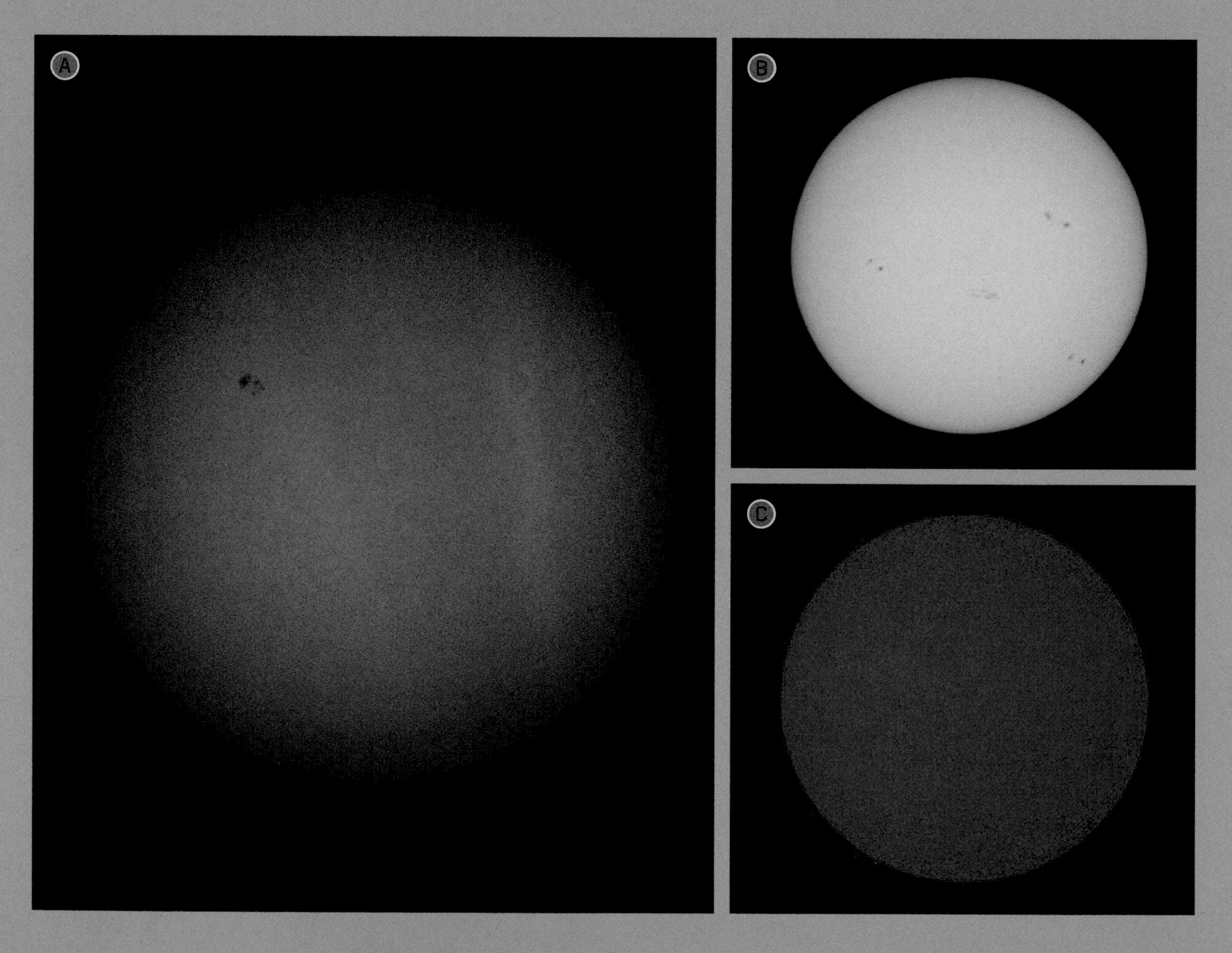

183 ENJOY LUNAR AND PLANETARY FILTERS

Colorful Moon and planetary filters help tease out details from equally colorful lunar and planetary surfaces, rather than cutting brightness like solar models or wavelengths like deep-space ones. Filters are also designed to be used together, so try stacking multiples to create your own cool views.

POLARIZED Perhaps the most important, these are great all-purpose filters. They cut glare and boost planetary and lunar contrast.

#25 RED Good for daylight, red filters can cut glare when viewing Venus as well as enhance surface details on Mars.

#30 MAGENTA Good for darkening green features while enhancing blue and red ones.

#15 DEEP YELLOW Deep yellow is best suited to exploring the atmospheres of Venus, Jupiter, and especially Saturn.

#58 GREEN Great for Venus observations, the red spot on Jupiter, certain features on Saturn, and any naturally white details, like ice on Mars and ice storms on Saturn.

#8 LIGHT YELLOW Great for increasing the surface detail of Mars, it can also make viewing lunar features easier.

#56 LIGHT GREEN Good for viewing surface fogs, polar projections, and frost patches. Note: Many filter kits will throw in greenish "Moon filters." They are crummy and no substitute for polarized filters, and just make the Moon look like Limburger cheese.

#21 ORANGE Orange reveals contrast between dark and light areas, while also helping you see through clouds and haze on Mars.

#82A LIGHT BLUE Most useful for increasing lunar contrast, light-blue filters also make it possible to split some binary stars.

#80A MEDIUM BLUE Good for gas giant cloud bands, it's really cool to look at comet tails with this filter.

#47 VIOLET Useful for observing comets, violet filters are also good at highlighting any non-red Martian features.

184 PROBE DEEP SPACE WITH NEBULA FILTERS

When trying to view deep space with a telescope here on Earth, half the battle is against light from the local environment. Even in the darkest night skies, nebulae and other deep-space objects can be brought into sharper relief with the use of filters—most of which combat light pollution or bring out the contrast of nebulae clouds.

NARROWBAND/LINE These filters transmit only a very narrow set of wavelengths of light, serving to both block light pollution and enhance the light from hard-to-see nebulae. Oxygen III, hydrogen-beta, and ultra-high contrast (UHC) narrowband filters darken the kind of sky glow that would otherwise obscure nebulae, making some otherwise invisible nebulae visible and sharpening the detail of others. Objects will appear monochrome (as at right), but the detail will be enhanced.

BROADBAND Whereas narrowband filters block out obtrusive lights in the sky, broadband filters try to combat the light pollution produced by your local streetlights. By blocking a large portion of the light emitted from sodium, mercury vapor, and other human-made lightsources, these filters help increase visibility of a number of deep-sky objects.

185 MEET SATURN

Sixth from the Sun and the second-largest planet in our solar system, Saturn is a gas giant that's known for its striking system of icy rings, which have captivated skygazers for centuries. While you can see Saturn with your naked eyes, it's best to observe this planet with a telescope so you can experience its mesmerizing rings.

DIAMETER Approximately 72,000 miles (115,000 km). Nine Earths in a row would span the diameter of Saturn.

MASS 568.3 septillion kg, or the mass of 95 Earths

AVERAGE DISTANCE FROM THE SUN Approximately 886 million miles (1.4 billion km), or nine times the distance from the Sun to Earth

LENGTH OF YEAR 10,756 Earth days, or about 29 Earth years

AVERAGE TEMPERATURE –288°F (–178°C) in the cloud tops

DENSITY Saturn is our least dense planet; it is lighter than water. Skywatchers kid that it would float if there were a bathtub big enough.

AGE 4.5 billion years

MYTHOLOGY Saturn is named after the Roman god of agriculture. Characters from Roman mythology inspire the names of large moons, while irregular moons are named after northern Inuit, Gallic, and Norse characters.

SUCCESSFUL MISSIONS
1979: *Pioneer 11* flyby (United States)
1980: *Voyager 1* flyby (United States)
1981: *Voyager 2* flyby (United States)
2004: *Cassini–Huygens* orbiter (United States)

GRAVITY The effect of gravity in the cloud tops at Saturn's equator is very similar to what you would feel on Earth. Depending on where you measure the "surface," you'd weigh about as much as you do on Earth, if only you could find something solid to stand on!

COMPOSITION What we know of Saturn's interior is still mysterious. But scientists theorize that it likely has an iron-nickel core, with surrounding layers of metallic hydrogen, liquid hydrogen and helium, and a gaseous outer layer. That distinct yellow hue? It's caused by ammonia crystals in the atmosphere.

NORTH POLE Saturn's north pole sports an odd hexagonal cloud formation—it's been there for as long as we know. There's no similar feature in the southern hemisphere, but both poles do experience auroras thanks to Saturn's strong magnetic field.

ICY RINGS Saturn's rings consist of tiny particles of dust and water ice shepherded by tiny *moonlets* (satellites that are between the size of a house and a hill). Mainly made up of water ice, Saturn's rings are currently thought to be pieces of asteroids and comets that broke up before reaching the planet. The rings are surprisingly thin, typically between 32 feet (10 m) and 3,200 feet (975 km). See #199 for information on how to view Saturn's rings.

MIND THE GAPS Look through a scope and you'll see that the rings have some sizable rifts—the largest of which is the 3,000-mile- (5,000-km-) wide Cassini Division between the large A and B rings. This gap is caused by Mimas, a small moon whose gravitational force keeps the ring clear of debris. Other small moons—called *shepherd moons*—create smaller divides, such as the Encke and Keeler gaps. Bonus: One of Saturn's rings is formed by the ice water spewed by tiny but active moon Enceladus!

MANY MANY MOONS Saturn has more than 60 confirmed moons and many, many moonlets in its rings, but Titan—which is larger than the planet Mercury—makes up about 96 percent of the mass of all Saturn's moons combined. Titan may be one of the best places in our solar system to look for extraterrestrial life. Its thick nitrogen and methane atmosphere leads to methane rain, which in turn creates methane lakes and seas. This atmosphere is inhospitable to humans (and it is incredibly cold!), but there is likely a liquid water ocean beneath it.

TINY TILT Saturn is tilted about 26 degrees from the plane of the solar system, just a little more than Earth. That means there are times when Saturn's rings appear to face us more than others, including times when they appear to be head-on and so disappear from our view (about once every 15 years).

186 FIRE UP THE FURNACE

When Nicolas-Louis de Lacaille first plucked these faint stars out of the meandering, river-like constellation Eridanus and named them Fornax Chemica (the Chemical Furnace), he was honoring the famous French chemist Antoine Lavoisier, who was guillotined during the French Revolution in 1794.

While there are no real bright points of interest in Fornax, with a large telescope you can enjoy the Fornax Galaxy Cluster near the Fornax-Eridanus border. With a wide-field eyepiece, you may see up to nine galaxies in a single field of view. The brightest galaxy (at magnitude 9) is Fornax A (NGC 1316).

Another furnace sight is the Fornax System. Spherical in shape, it is a large group of very faint stars and globular clusters. It is too faint to see with an amateur telescope, but one globular cluster, NGC 1049, has a magnitude of 12.9, making it visible in a 10-inch (25-cm) telescope under a good sky. Although this dwarf galaxy—a diminutive member of our Local Group of galaxies—appears to be unusual, galaxies like the Fornax System may be common in the universe.

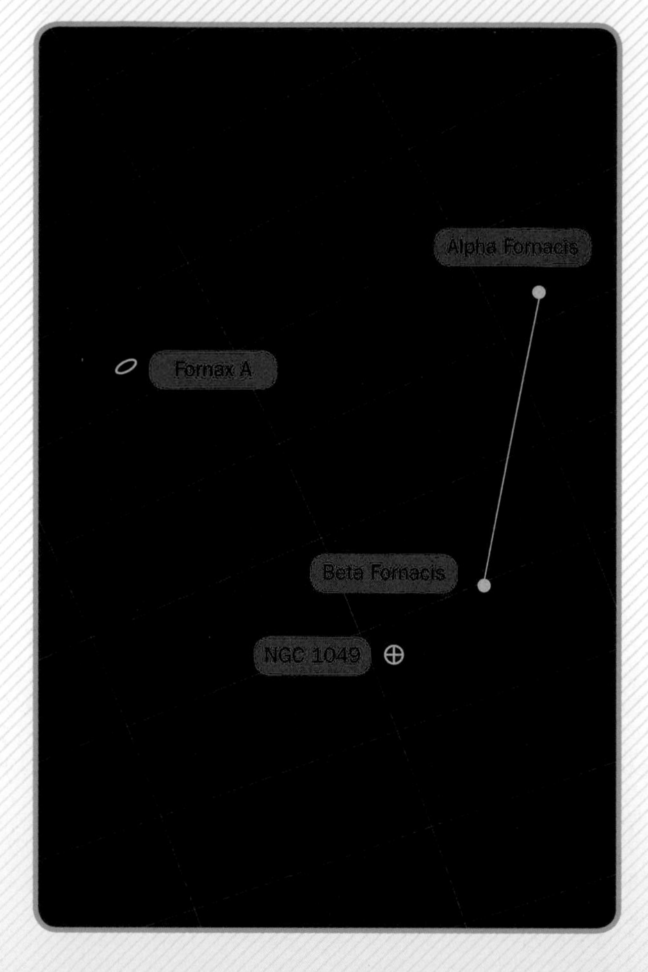

187 GET UP CLOSE WITH GRUS

In his star atlas of 1603, Johann Bayer named this southern constellation Grus the Crane—the bird that served as the symbol of astronomers in ancient Egypt. This group of stars—variously seen as a stork, a flamingo, and a fishing rod—has very little to offer the skywatcher who is using a small telescope, although some faint galaxies provide suitable targets for telescopes of 8-inch (20-cm) aperture or larger.

Grus's three fairly bright stars can illustrate magnitude. Alnair (Alpha Gruis) is a large, blue main-sequence star about 70 times as luminous as the Sun with a magnitude of 1.7. Only 57 light years away, it is the brightest of the three because of its proximity to us. At magnitude 2.3, Beta Gruis is a much larger red giant star, some 800 times as luminous as the Sun, but its 140-light-year distance from us makes it appear fainter than Alnair. Finally, magnitude-3 al Dhanab (Gamma Gruis), a blue giant star more luminous than the others, is faintest since it is 230 light years away.

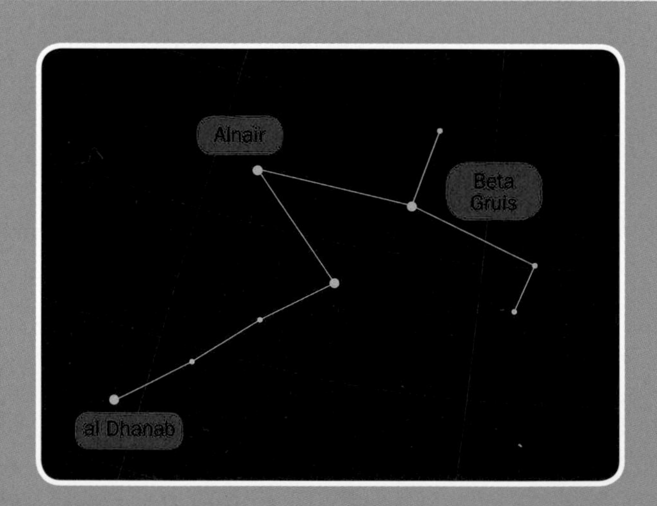

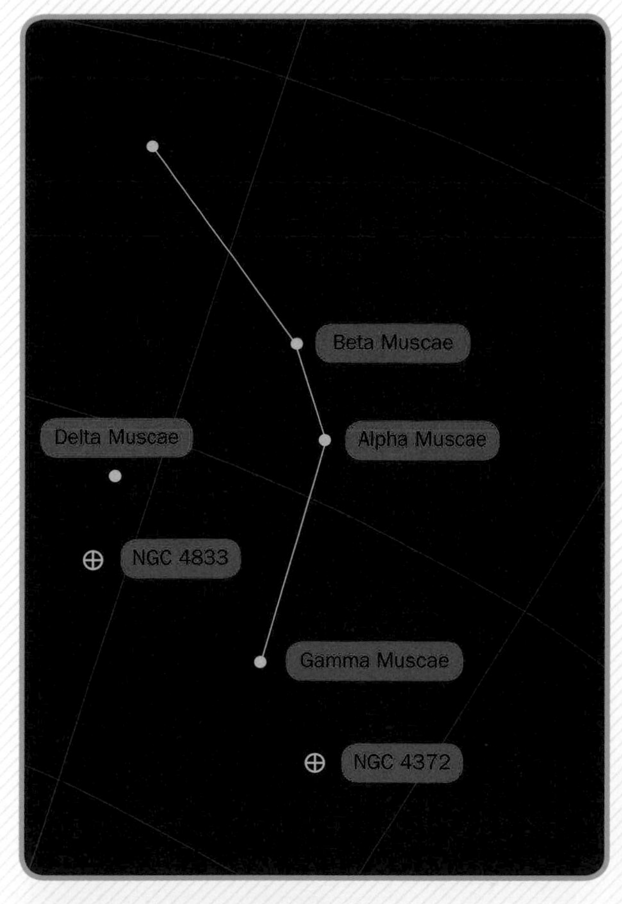

188 NAB THE FLY

Musca is easy to find just to the south of the Southern Cross. Originally described by Johann Bayer in his 1603 star atlas as Apis the Bee, Edmond Halley later called it Musca Apis (the Fly Bee), and then Nicolas-Louis de Lacaille named it Musca Australis (the Southern Fly)—to avoid confusing it with the fly on the back of Aries the Ram. Since the northern fly is no longer a constellation, the Southern Fly is known simply as Musca.

Elegant double star Beta Muscae consists of two magnitude-4 stars revolving around each other in a period spanning several hundred years. The pair is some 520 light years from Earth. They have a very tight separation of 1.6 arc seconds, presenting a challenge for a 4-inch (10-cm) telescope.

Close to Gamma Muscae, globular cluster NGC 4372 has faint stars spread across 18 arc minutes. NGC 4833 is a large, faint globular cluster within 1 degree of Delta Muscae. You need a 4-inch (10-cm) telescope or larger to see individual stars.

189 FIND THE PHOENIX

A classic symbol of rebirth, the Phoenix was a mythical bird that lived for 500 years. It would build a nest of twigs and fragrant leaves that the noontime rays of the Sun would light. The Phoenix would be consumed in the fire, but a small worm would wriggle out from the ashes, bask in the Sun, and become a brand-new Phoenix. While depictions of this miraculous bird have been found in ancient Egyptian art and on Roman coins, the idea of a Phoenix in the sky goes back to the ancient Chinese firebird known as Ho-neaou.

The best example of a dwarf Cepheid variable star, SX Phoenicis changes from magnitude 7.1 to 7.5 and back again in only 79 minutes and 10 seconds! Its range can vary—sometimes it grows as bright as 6.7. Variation probably occurs because the star has two different oscillations occurring at once. Such a small range in brightness can be difficult to monitor, requiring very careful comparison with neighboring stars.

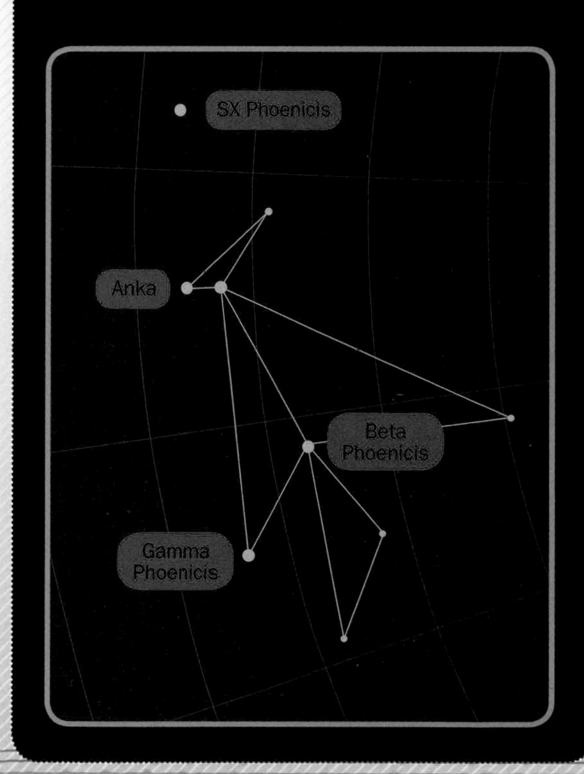

190 ZOOM IN AND OUT OF THE ORION NEBULA

One of the sky's jewels, the famed Orion Nebula (M42/ NGC 1976) rewards observers at magnifications large and small. It is often an amateur's first deep-sky object with a telescope. At a relatively low 7x magnification and a wide field of view (A), the entire complex of swirling gas and glittering stars, as well as nearby deep-sky objects, are a treat. Try an OIII, deep-sky, or light-pollution filter for enhanced views. Zooming in at 25x magnification (B), you'll notice still greater detail in the knots and swirls of the nebula. You'll likely see that the fuzzy bright star in the center of the nebula—the Trapezium—is actually at least four stars. In fact, if you zoom in more—say, at 80x magnification (C)—you will find many more than four stars in this tight cluster.

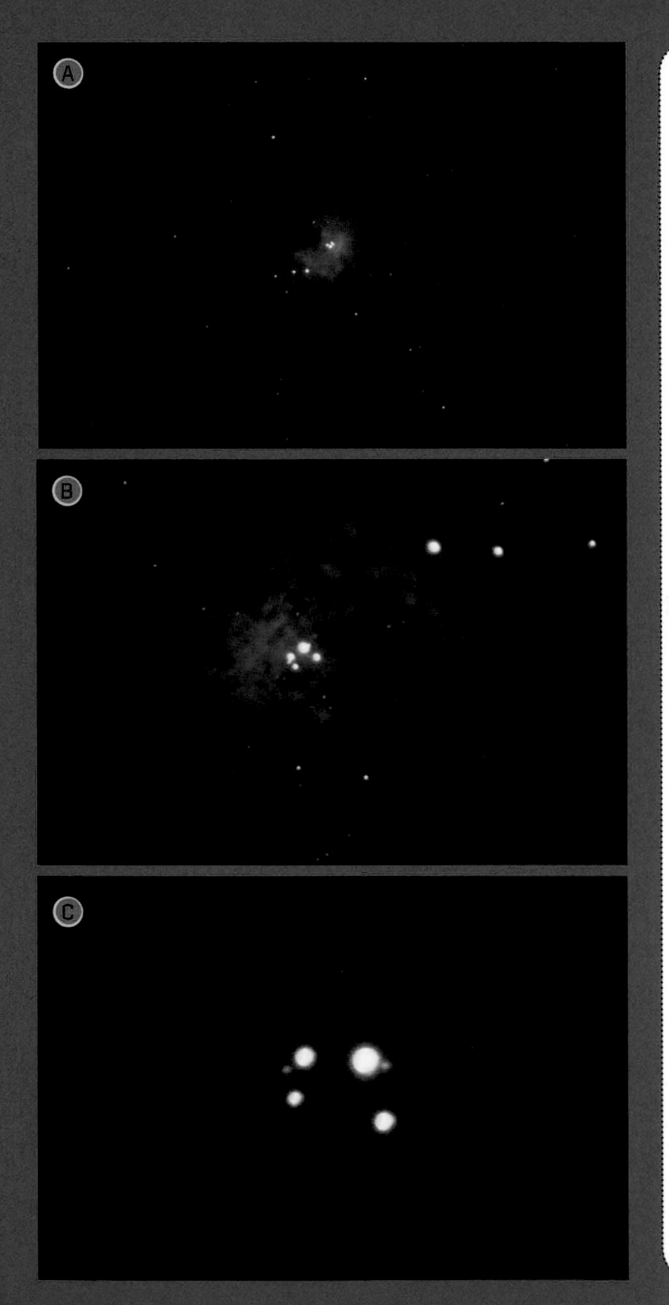

191 GAZE UPON ALBIREO

Many consider Albireo (Beta Cygni) to be the most beautiful double star. To find it, start by locating Cygnus (see #115) and focusing your eyes on the head of the swan—Albireo is the so-called "beak star." Through a small telescope, the star splits into two very colorful and contrasting stars: the golden Albireo A and the slightly dimmer, blue-green Albireo B. Scientists still dispute whether they are a binary pair of stars orbiting each other.

192 FIND THE GREAT CLUSTER IN HERCULES

To find Hercules's Great Cluster (M13/NGC 6205), arm yourself with a wide-field eyepiece and a star map, then:

STEP ONE To find the fairly dim Hercules, find Vega (Alpha Lyrae)—a blue-white beauty in Lyra (see #116)—and Arcturus (Alpha Boötis) in Boötes (see #30).

STEP TWO About one-third of the way between these notable bodies, you'll spot the Keystone asterism, the quadrangle that makes up Hercules's torso.

STEP THREE Look closely at the two stars on the Arcturus side of this shape, called Eta Herculis and Zeta Herculis. Between them is a fuzzy spot of light—M13, one of the most amazing clusters in our galaxy.

STEP FOUR In low-power scopes, the Great Cluster appears as a large smudge. But if you increase your power (both in magnification and scope size), you'll make out distinct stars—of which there are thousands in this globular cluster, orbiting the Milky Way's halo.

193 VIEW VIRGO WITH A TELESCOPE

The largest constellation in the zodiac—and second only to Hydra as the largest constellation in the sky—Virgo is best visible to northern skywatchers in the spring and summer, while southern skywatchers will have better luck seeing it in the fall and winter. Virgo has been associated with a number of goddesses and holy figures—from the goddess Shala's ear of corn in the Babylonian zodiac, to Greek and Roman goddesses of fertility and agriculture, to even the Virgin Mary.

While Virgo is enormous, that doesn't necessarily make it one of the most prominent constellations in the sky. But what it lacks in luster it certainly makes up for in depth. Galaxies (the deepest-sky objects visible in scopes) are abundant in this constellation—there's no better place to view them than in the Virgo Cluster, if you know where to look. In fact, 13 of the 109 Messier objects (see #209) are galaxies in Virgo. A low-magnification (8-inch/20-cm) eyepiece is great for exploring this area with a telescope. Start with Spica (Alpha Virginis), the brightest star in Virgo (see #32)—you're about to witness some real delights.

194 SCOPE OUT THE SOMBRERO GALAXY

Twenty-eight million light years away and 50,000 light years across, the Sombrero Galaxy (M104/ NGC 4594)—so named for its hat-like look—is a galaxy in Virgo. Discovered by Pierre Méchain in 1781, it wasn't on the official list of Messier objects until 1921. One of the most striking of its features is a dust lane that wraps around its rather large central bulge. Some say the galaxy resembles Saturn, with the dark dust lane obscuring stars in the galaxy. (See #209 for a photo.)

After following the Big Dipper's handle, arcing toward Arcturus, and speeding on toward Spica (Alpha Virginis), the next constellation you'll come to is Corvus—a small group of stars that looks more like a sail than a crow (its namesake). Following a small trail of stars from the corner closest to Spica, you'll quickly encounter the edge-on Sombrero Galaxy. (Although the Sombrero Galaxy can be seen in binoculars, a small telescope will best highlight its features.)

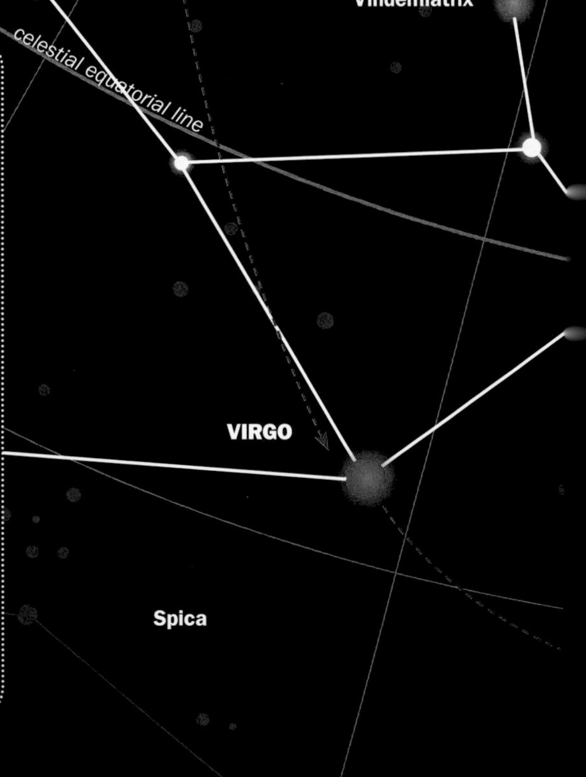

BIG DIPPER

195 CHECK OUT M87

If Markarian's Chain were a bowl, you would find the most prominent member of the Virgo Cluster under its curve. M87 (NGC 4486) is one of the largest galaxies ever observed. It is elliptical in shape and likely merged with many galaxies over time. Those galaxies left more than a thousand globular clusters swarming around M87 (compare that to fewer than 200 globs in our Milky Way), though you won't see them in a telescope. You also won't see the jet of material ejecting from the massive black hole in the center—or the black hole itself. All you will see is a giant smudge in the sky, but keep in mind that you're seeing it from 60 million light years away.

LEO

M86

M84

Denebola

M87

196 SEE M84 AND M86 IN MARKARIAN'S CHAIN

One of amateurs' favorite ways to arrive at one of the more galaxy-concentrated sections of the sky is to begin by making a triangle between the tail of Leo (Denebola, or Beta Leonis), Virgo's brightest star (Spica, or Alpha Virginis), and Arcturus (Alpha Boötis) in Boötes. About halfway between Denebola and Arcturus, shifted a few degrees toward Spica, is the star Vindemiatrix (Epsilon Virginis). You can find most galaxies between Vindemiatrix and Denebola.

If you position your low-powered telescope right between these stars, you'll find two of the brightest galaxies, M84 (NGC 4374) and M86 (NGC 4406). These two are part of a lineup of galaxies called Markarian's Chain—named for the Armenian physicist who discovered that they appear to move in a gently curved band from our vantage point on Earth.

While M84 appears circular, both of these galaxies are *lenticular* (stretched-out discs of old, dim stars) as opposed to *spiral* (younger collections of stars that form into discs with twisting arms, like our own Milky Way Galaxy). M84 is also home to an especially massive black hole that may be up to 1.5 billion times as massive as our Sun.

SOMBRERO GALAXY

CORVUS

197 SPOT THE SHADOWS OF JUPITER'S MOONS

You've caught a glimpse of Jupiter's moons through binoculars (see #135), but did you know that with a medium-size telescope you can also see their shadows as they travel over Jupiter's clouds? All four of Jupiter's large moons regularly cast shadows over the planet's cloud tops as they orbit. While many apps can help you figure out the best times to spot these transits, you'll likely see them on your own with regular observation.

Usually, the moons (and Jupiter itself) are so bright that it's difficult to determine the satellites' sizes, but witnessing the transiting shadows gives you a sense of how large each moon is. You may also see the moons "disappear" and "reappear" from behind Jupiter as well. And once in a very great while, you may happen upon two or even three shadows moving across Jupiter's face. Some skilled amateur imagers have created amazing movies of these transits in exquisite detail.

To bag these shadows, load up your telescope with an eyepiece that has a magnification of at least 150x and a polarized lunar filter.

198 SPLIT DOUBLE STARS INTO THREE

You may remember our ancient seafarers' vision test: splitting what appears to be a single star in the Big Dipper into Mizar and Alcor (Zeta and 80 Ursae Majoris; see #33). But with a telescope, you won't need to strain your eyes at all to resolve this double star. When viewing with a high-powered telescope, you can split Mizar even further. What once appeared to be two stars is now three! (In fact, each of those is a double star, though you won't be able to see them with an amateur telescope.)

199 SPY SATURN'S RINGS

Even with a small telescope, you can see Saturn's rings—two little handles poking out from its golden orb. If you view through a high-powered instrument, you'll get a clearer view of the gap between the rings and the planet, the shadow of the rings on Saturn's clouds, and even gaps in the rings themselves. The most famous gap—and the one easiest to spot—is the Cassini Division. To find this opening, you'll need a 5-inch (13-cm) telescope with 150x magnification and clear skies. You may even be able to spot some of Saturn's satellites; they often seem to dance around the rings. Saturn takes decades to move through the ecliptic, so you'll need to look up its exact location for your observing sessions.

PEEK INTO ORION'S STELLAR NURSERY

Stars, nebulae, and clusters tell the story of how stars are born, live, and die—and it just so happens that Orion makes for fantastic viewing of star birth. If you have excellent eyesight and the sky is dark, you might barely make out a fuzzy patch of light in the sword of Orion. By using binoculars or a small telescope, this seemingly unremarkable smudge of light will reveal itself to be one of the most spectacular (and most photographed) objects in the sky: the Great Nebula of Orion (M42/NGC 1976). This giant cloud is 24 light years across, 1,300 light years away, and the birthplace of hundreds of stars. Research telescopes and other instruments have detected more than 700 baby stars, still wrapped in blankets of gas and dust. The oldest of these stars are about 300,000 years old; the youngest may have been born only 10,000 years ago. About 150 of these may develop orbiting planets. But we will have to wait a few million years for these planets to form.

Betelgeuse

Orion's Belt

201 WITNESS THE DEATH OF BETELGEUSE IN ORION

Turns out Orion not only hosts nascent stars—it is also home to one that is approaching its final days. Peer closely at the bright reddish star that marks his right shoulder—that's Betelgeuse (Alpha Orionis). It began as a ball of gas inside a giant cloud of hydrogen and helium and increased in density and temperature until, millions of years ago, it started nuclear fusion. A star several times larger than our Sun was born. But now Betelgeuse is nearing death.

If you could look deep inside the aging Betelgeuse, you would find the kinds of elements that have been manufactured by nuclear fusion during its life: hydrogen, helium, carbon, oxygen, calcium, and the rest of the elements up to iron. These elements are being stored away and one day, when Betelgeuse explodes as a supernova, it will blast them into space. The gases will form giant gas clouds, like the Orion Nebula, and provide the raw materials for new stars, like those in the Pleiades (M45). When Betelgeuse goes supernova, it could briefly become brighter than the full Moon. This explosion is likely to happen within the next few hundred thousand years, so don't wait up.

202 WATCH THE YOUNG PLEIADES IN TAURUS

When you look at Taurus's lovely Pleiades with your unaided eyes, you might pick out six or seven stars. The ancient Greeks noticed this group and named it "The Seven Sisters," or the Pleiades (M45). A good pair of binoculars will help you see more than a hundred stars in the Pleiades, while very large telescopes have located more than 3,000.

The Pleiades Cluster is actually a huge group of very young stars that were born inside a giant cloud of gas while the dinosaurs roamed Earth. They left their *stellar nursery* (dense molecular clouds where stars form) long ago, but they're still hanging out together. Images of the Pleiades captured by larger telescopes reveal that the young stars are surrounded by a bluish halo of gas, which is caused when the stars' blue light reflects off the gas still enveloping the group. These stars have been together for about 100 million years, but in about 250 million years, all the stars inside the Pleiades will have drifted apart and will mingle with the other stars in our galaxy.

203 SPOT SCORPIUS

Evenings from June through August, you can easily view the constellation Scorpius low in the south for observers in the northern hemisphere, or high in the sky for those down below the equator. Named after the Latin word for scorpion, this large zodiac constellation falls near the Milky Way's center, with Libra to the west and Sagittarius to the east. Unlike many constellations, Scorpius bears a strong resemblance to its namesake, with a head, stinger, and J-shape tail that often dips below the horizon line for viewers in the higher northern latitudes. (Pro tip: You'll never spot Scorpius and its neighbor, Orion, in the same sky. In Greek mythology, Scorpius killed Orion, so they are forever segregated in the heavens.)

Scorpius contains many beautiful stars and star clusters, particularly when viewed through a telescope. For instance, there's Graffias (Beta Scorpii), one of the sky's brightest binary stars that you can see with a small telescope, and in the heart of the scorpion is the bright red star Antares (Alpha Scorpii), whose name means "rival of Mars" in Greek. This huge star will go supernova soon (in astronomical terms, that is—so maybe in a million years, maybe tomorrow). When it does, the remnant of a star will be bright enough to be visible during the day. But for now, Antares is a highly useful star for seeing other grand sights within Scorpius using a telescope.

204 FIND GLOBULAR CLUSTER M4

About 1.3 degrees west of Antares (Alpha Scorpii) in the same binocular view, you'll see the bright M4 (NGC 6121) globular cluster, which was first spotted in 1746 and later cataloged by Charles Messier in 1764 (see #209). One of our nearest globular cluster neighbors, it's 7,200 light years away, just north of the plane of our galaxy, and contains the oldest known white dwarf stars. At first look, this bright beauty appears as a fuzzy ball of light in the sky, but you can resolve its individual stars with a telescope. (If it weren't for all of the interstellar dust between the globular and us, it would be even brighter.) Under perfect conditions, M4 can even be visible to sharp naked eyes. With a telescope, you may be able to see a bar shape of dense stellar objects in the cluster, making it appear flattened.

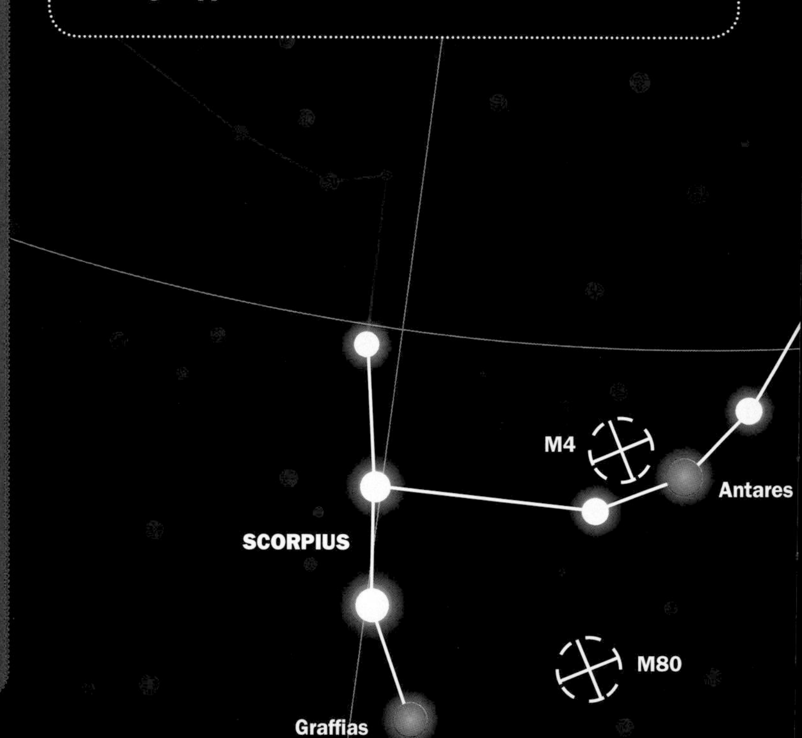

SAGITTARIUS

PTOLEMY'S CLUSTER

BUTTERFLY CLUSTER

206 MAKE OUT M80

Follow the line from Antares (Alpha Scorpii) up to the head of the scorpion. You'll spot Graffias (Beta Scorpii). Between Antares and this binary system is the small and quite compact globular cluster M80 (NGC 6093)—one of the densest in the Milky Way. With hundreds of thousands of stars packed closely together, collisions in this cluster are fairly common. While most globular clusters contain only old red stars, this group features a higher population of *blue stragglers*: larger stars that likely formed when two small stars collided and merged in the packed stellar neighborhood. At 32,600 light years from us, it's more than four times as far away as M4 (NGC 6121).

205 PEEP THE BUTTERFLY AND PTOLEMY CLUSTERS

As opposed to the old globular clusters near Antares (Alpha Scorpii), swinging down the tail of Scorpius brings you to young open clusters with a few dozen bright stars each. Between the curve of the scorpion's tail and the spout of the Sagittarius Teapot, you'll find two bright open clusters that fit in one binocular view.

The brighter of the two and located near Scorpius's stinger, Ptolemy's Cluster (M7/NGC 6475) is about 1,000 light years away. These young stars formed only 200 million years ago, right around the Mesozoic era on Earth. It's so vivid that Ptolemy saw it with an unaided eye in 130 CE.

The nearby Butterfly Cluster (M6/NGC 6405) is perhaps 1,600 light years away and not quite as bright as its neighbor. Most of its stars are hot blue stars, save for one giant orange beauty, BM Scorpii. Probably only about 100 million years old, these stars were born as the first bees began buzzing around newly developing flowers here on Earth. Can you make out the shape of its open butterfly wings?

207 MEET URANUS

Like its sister planet Neptune, Uranus is an icy giant with a cool, pale blue color. It is seventh from the Sun and third largest in diameter in our solar system, and its consistent appearance has earned it a bland rating among amateurs. But there's still a wealth of fascinating facts about Uranus.

DIAMETER 31,763 miles (51,118 km) at its equator

MASS 86.8 septillion kg, or about 14.5 times the mass of Earth

AVERAGE DISTANCE FROM THE SUN 1.78 billion miles (2.87 billion km), or about 19.2 times farther from the Sun than Earth

LENGTH OF YEAR 84 Earth years

LENGTH OF DAY A little more than 17 hours

AVERAGE TEMPERATURE –357°F (–216°C)

THAT BLUE HUE Both Uranus and Neptune get their blue color from the methane in the planets' upper atmospheres. It absorbs the red light and reflects the blue light that we see.

DISCOVERY Uranus was the first planet to be first found with a telescope. The honor goes to William Herschel, who viewed it on March 13, 1781.

NAME Uranus was named after the Greek god of the sky. He was the father of Saturn and the other Titans. Herschel wanted to name it King George's Planet, after the King of England. The last holdouts came around to the name of Uranus in the 1800s. As many jokes as we like to make about the name, scientists place the emphasis on the first syllable, making the pronunciation "YUR-en-us."

SUCCESSFUL MISSIONS
1986: *Voyager 2* flyby (United States)

GRAVITY Uranus's surface gravity is 91 percent that of Earth's. If you weighed 200 pounds (90 kg) on Earth, you would weigh 182 pounds (83 kg) on Uranus—if you could stand on its cloudy surface.

RINGS The rings of Uranus were first discovered by the Perth Observatory and Kuiper Airborne Observatory in 1977. A *Voyager 2* flyby revealed a few more details about these exotic dark rings, and Hubble later observed brighter rings. Thirteen total have been observed, but sadly they are not visible by amateur telescope.

COMPOSITION Uranus has a rocky core surrounded by ices and a gaseous atmosphere.

ICE GIANT Uranus is called an ice giant due to the exotic ices expected to make up much of the planet's interior. Uranus is the coldest planet in the solar system, even though it is closer to the Sun than Neptune! Unlike other giant planets, Uranus generates very little internal heat. This may be due to its formation, or to the event that caused it to tip on its side.

MAGNETIC FIELD Like Neptune, Uranus's magnetic poles are not aligned with its geographic poles. The magnetic poles are inclined 60 degrees from the axis of rotation. And like Neptune, the cause of this odd mismatch is a mystery to scientists. Some speculate that an inner core composed of diamonds might cause this strange magnetism.

WEATHER In comparison to other planets, Uranus enjoys relatively tame weather. In 1986, however, *Voyager 2* captured imagery that shows slight adjustments in the brightening of its polar cap and a narrow band of clouds called the southern collar. So there is some evidence of shifting cloud formations. Under very good conditions, with a very good telescope, you may be able to take images of a strong storm, but don't expect to see any through an eyepiece.

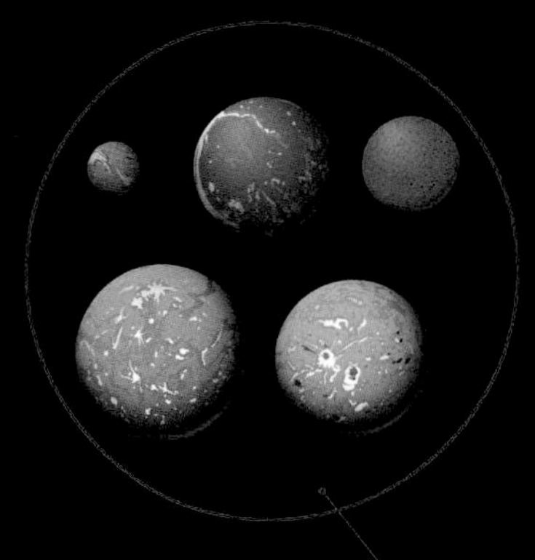

MOONS Uranus has 27 moons that we currently know of, many of which are named after characters from Shakespeare plays. Some notable ones include Oberon and Titania, Uranus's largest moons, discovered by Herschel himself and visible in large amateur scopes. Ariel, Umbriel, and Miranda are the next largest, followed by many smaller moons. Miranda, the innermost moon, has the largest cliffs in the entire known solar system.

TILT Uranus is the only planet tilted almost 90 degrees perpendicular to the Sun. This means that its poles take turns to face the Sun directly as it orbits. Scientists are still not sure what caused this bizarre configuration, but they theorize that a massive impact may have tilted the planet to one side. As a result, each pole gets 42 years of continuous sunlight, followed by 42 years of continuous darkness.

208

GAZE BACK IN TIME WITH A TELESCOPE

When you turn on a lamp, you probably don't think about the time it takes for the light to travel from the bulb to your eye. Light moves at the blistering speed of 186,000 miles per second (299,000 km/s). In the time it takes to blink your eye, a beam of light could orbit Earth three times. Light from the Moon takes less than 2 seconds to reach Earth and light from the Sun takes only 8 minutes. Next time you see Jupiter, keep in mind that its light left 40 or 50 minutes ago. For every planet in our solar system, light gets to us in less than a day.

When you look at an object farther out in space, you aren't seeing it as it is today but as it was when light left it. Space is so vast that it can take a beam of light thousands, millions, and even billions of years to reach us from its faraway starting point. For example, light from anything inside the Milky Way galaxy, but outside our solar system, has traveled anywhere from a few years to tens of thousands of years once it reaches Earth. Light from any other galaxy has been traveling from millions to billions of years to reach you. The farthest objects that the Hubble Space Telescope has seen are galaxies located more than 13 billion light years away.

In this light (no pun intended), you can think of a telescope as a kind of time machine, allowing us to examine how the universe looked back in time. Here's an idea of how far back you're looking when you see light from a number of celestial bodies.

LIGHT TRAVEL TIME

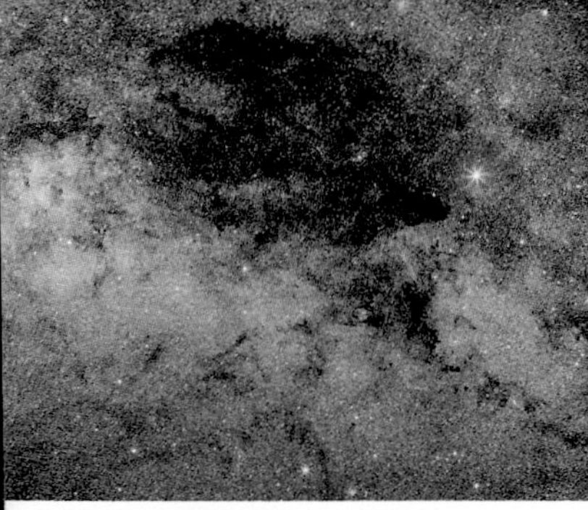

▲ **COALSACK NEBULA** 600 light years

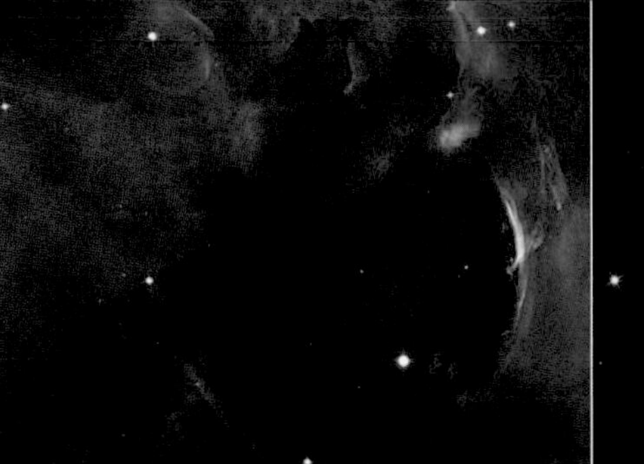

▲ **ORION NEBULA** 1,344 light years

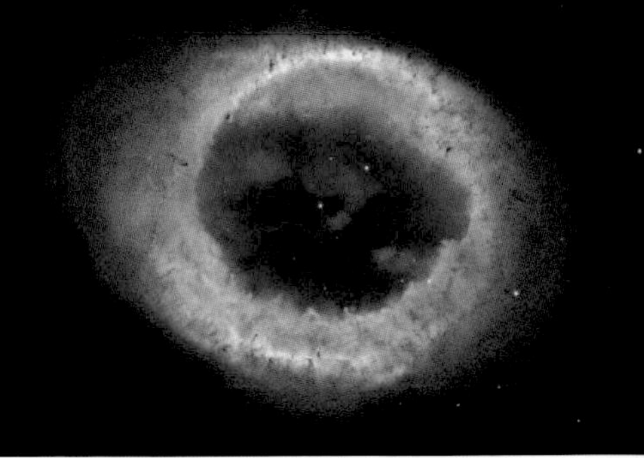

▲ **RING NEBULA** 2,283 light years

▼ **CRAB NEBULA** 6,523 light years

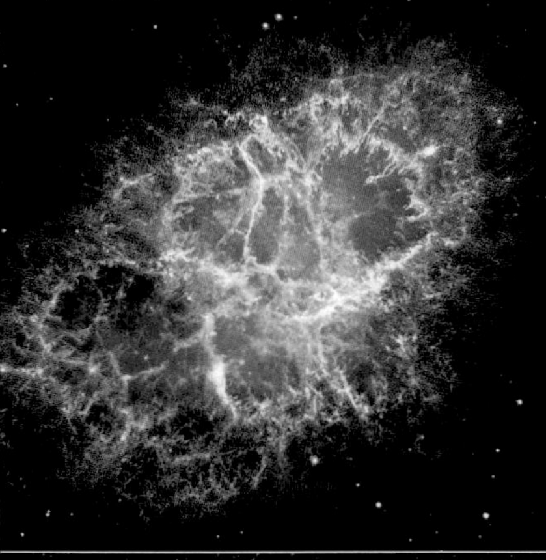

▼ **GALACTIC CENTER** 26,000 light years

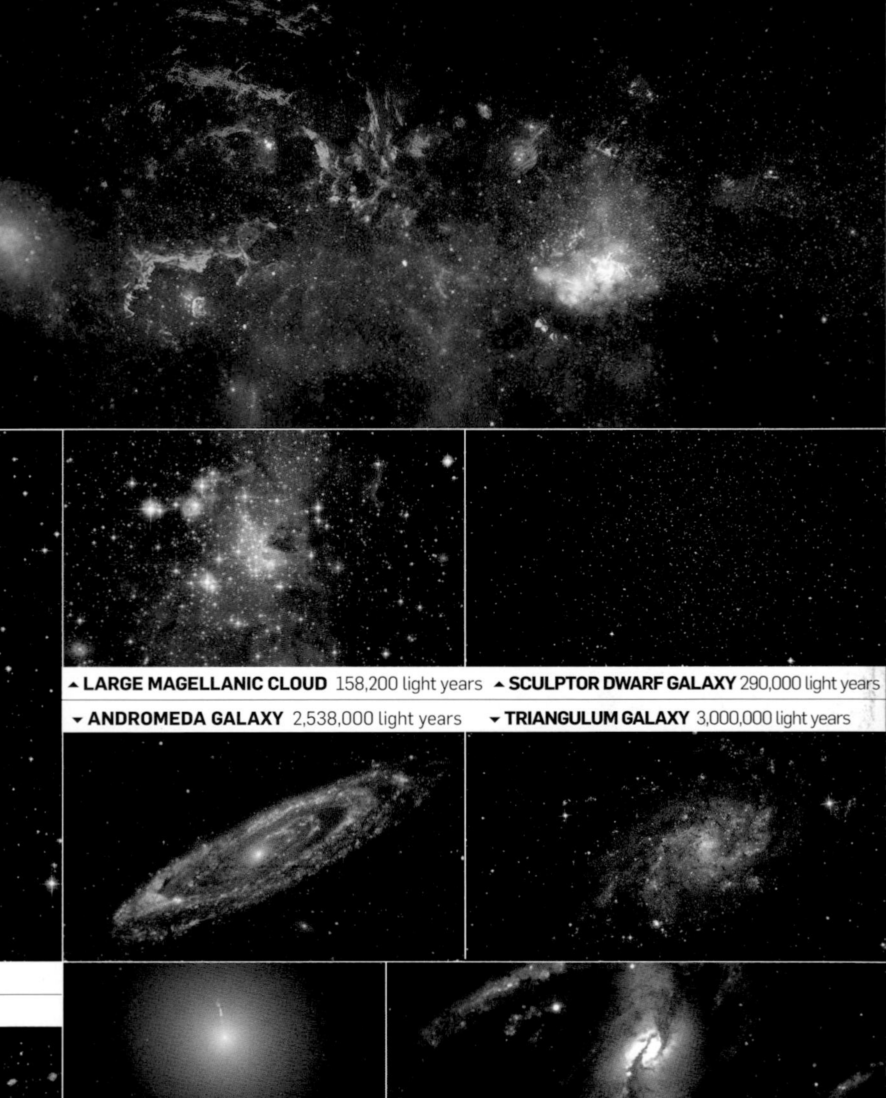

▲ **LARGE MAGELLANIC CLOUD** 158,200 light years

▲ **SCULPTOR DWARF GALAXY** 290,000 light years

▼ **ANDROMEDA GALAXY** 2,538,000 light years

▼ **TRIANGULUM GALAXY** 3,000,000 light years

▲ **SAGITTARIUS DWARF GALAXY** 70,000 light years

▼ **SMALL MAGELLANIC CLOUD** 199,000 light years

▲ **M87** 52,000,000 light years

▲ **GREAT BARRED SPIRAL GALAXY** 62,000,000 light years

▼ **COMA SUPERCLUSTER** 300,000,000 light years

209 RUN THE FAMOUS MESSIER MARATHON

Charles Messier was an 18th-century comet hunter in France. During the course of his life, he used more than a dozen different telescopes to seek out the icy projectiles. Frustrated by the amount of time he wasted peering at comet-like objects that weren't actually comets, he began compiling a list of where not to focus his attention—mostly deep-sky objects like open clusters, galaxies, and nebulae. In a grand stroke of cosmic irony, that catalog of non-comet nuisances (now dubbed Messier objects) would become the bulk of Messier's legacy—well, that and the 13 comets he discovered. Since then, a few other astronomers have taken the liberty of expanding slightly upon Messier's original list, but the namesake remains. (There are other systems for categorizing the galaxies, nebulae, and clusters that Messier noted, such as the New General Catalog, indicated by "NGC.")

So when can you gaze upon these beautifully static pains in the cosmos, such as the Sombrero Galaxy (M104/NGC 4594) here? You can see at least some on any given night, but you can catch them all on the new Moon near the end of March, when some people hold "Messier Marathons"—observing all of the more than 100 Messier objects in one night. While attempts are possible at any northern latitude, the odds of finishing are best at lower northern latitudes. And since Messier was observing from France, not all Messier objects are visible in southern latitudes.

Here's a short list of Messier objects for you to feast your eyes upon. It's not the full marathon, but it will get you started—and hooked on these deep-sky sights.

MESSIER # (COMMON NAME)	Object Type	Distance	Constellation	Magnitude
M1 (CRAB NEBULA)	supernova remnant	4.9–8.1 light years	Taurus	8.4
M8 (LAGOON NEBULA)	nebula with cluster	4.1 light years	Sagittarius	6.0
M31 (ANDROMEDA GALAXY)	spiral galaxy	2,430–2,650 light years	Andromeda	3.4
M45 (PLEIADES)	open cluster	0.39–0.046 light years	Taurus	1.64
M51 (WHIRLPOOL GALAXY)	spiral galaxy	19,000–27,000 light years	Canes Venatici	8.4
M57 (RING NEBULA)	planetary nebula	1.6–3.8 light years	Lyra	8.8
M82 (CIGAR GALAXY)	starburst galaxy	10,700–12,300 light years	Ursa Major	8.4
M87 (OWL NEBULA)	planetary nebula	2.03 light years	Ursa Major	9.9
M101 (PINWHEEL GALAXY)	spiral galaxy	19,100–22,400 light years	Ursa Major	7.9
▲ M104 (SOMBRERO GALAXY)	spiral galaxy	28,700–30,900 light years	Virgo	9.0

Table title: TOP 10 MESSIER OBJECTS

210 BEGIN AT THE LEO TRIPLET

One of the best views through a telescope is the Leo Triplet, consisting of spiral galaxies M65 (NGC 3623), M66 (NGC 3627), and NGC 3628 (aka the Hamburger Galaxy, so named for its slight resemblance to a dark dust-lane patty squished between galactic disc buns). Nestled within the constellation Leo, these three spiral galaxies push and pull on each other like warring celestial neighbors. M65 and M66 were both discovered by Charles Messier in 1780 and are approximately 35 million and 36 million light years from Earth, respectively.

The Leo Triplet's galaxies can fit into the field of view of a single low-powered eyepiece—which gives them the added bonus of being relatively easy to find. If the Big Dipper were instead a big dripper, the drops would fall on Leo the Lion's head. Just under Leo's back haunches, about halfway between Theta and Iota Leonis, you'll quickly spot the spiral discs of M65 and M66 with your telescope under decent skies. The third, NGC 3628, is fainter, requiring a larger aperture to see. From our vantage point here on Earth, the so-called Hamburger Galaxy appears on its side.

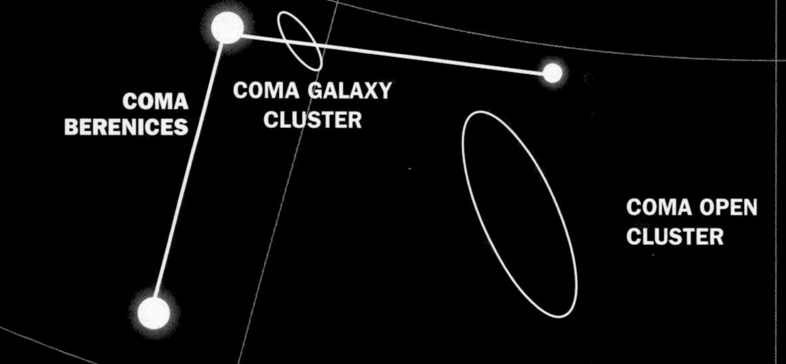

211 BAG THE COMA GALAXY CLUSTER

The famed Coma Galaxy Cluster is between two stars in Coma Berenices. With an 8-inch (20-cm) scope or larger, you can spot a dozen or so of the galaxies inside the cluster, mostly fainter than magnitude 12—they will resemble a faint group of fuzzy stars. A little trivia: It was the Coma Galaxy Cluster that astronomer Fritz Zwicky was studying in the 1930s when he coined the term "dark matter" to explain why galaxies don't fly apart as they speed through the universe. Dark matter clumps around to hold them together with additional gravity. While still a mystery, evidence for dark matter has been found all over the universe (see #278).

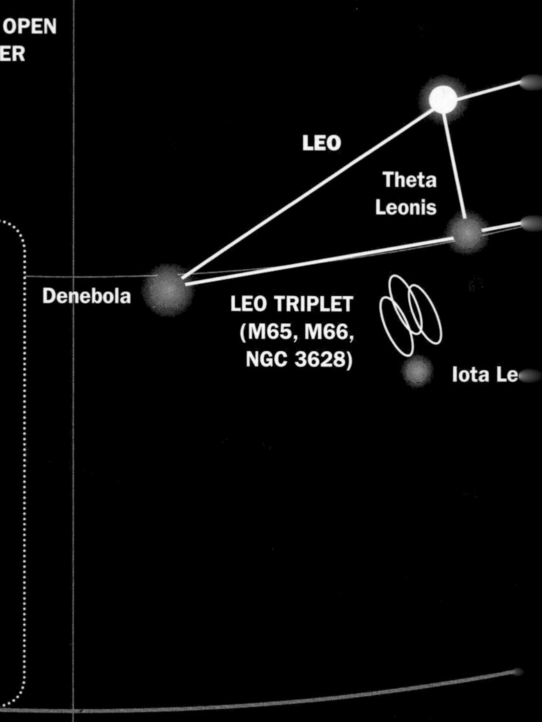

212 CATCH THE COMA OPEN CLUSTER THROUGH A TELESCOPE

Gaze at Leo (see #31) and you'll notice that, just off the tail star Denebola (Beta Leonis), there's a small, faint constellation visible with the naked eye. Coma Berenices, or "Berenice's hair," is made up of three dim stars that together form a right triangle. The constellation's namesake, Egyptian Queen Berenices II, supposedly sacrificed her hair to Aphrodite to keep her husband Ptolemy (the king, not the astronomer!) safe in battle. When the king returned, Berenices cut off her hair and sacrificed it on an altar. The next day, when the king discovered that her hair had been stolen, he threatened the priests with death. Thankfully, no blood was shed: The court astronomer wisely told the king that the gods had put her hair in the sky.

Between Leo and Coma Berenices, 288 light years away, is the large Coma Open Cluster, filled with brilliant new blue stars. About 450 million years old, the Coma Open Cluster is visible in binoculars, or with the naked eye under dark enough conditions.

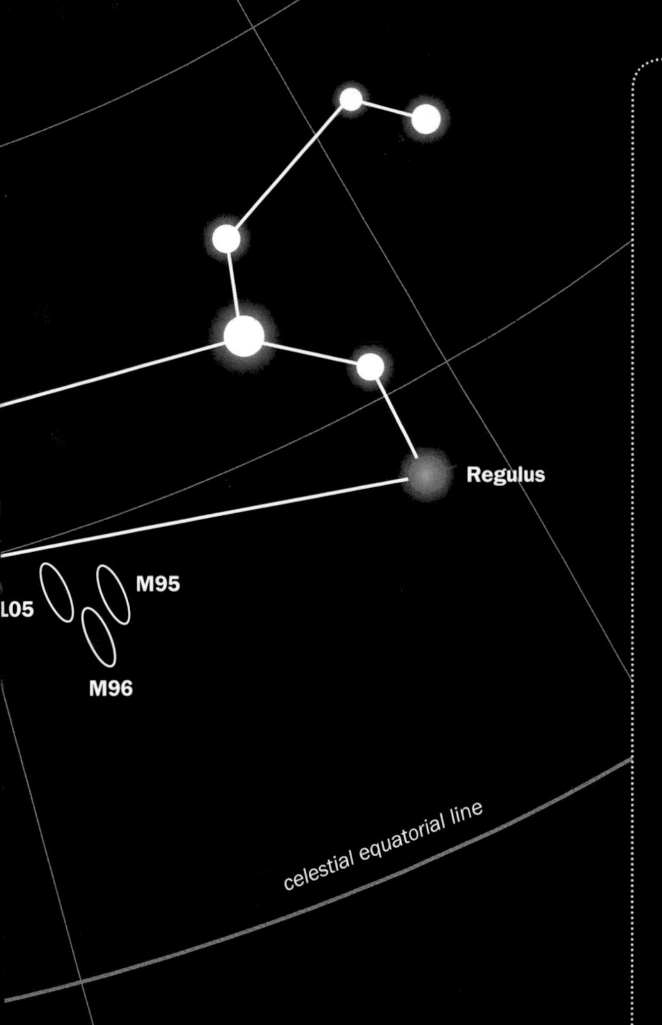

213 TAKE A PEEK AT THE M96 GROUP

Two more beautiful spiral galaxies, M95 and M96 (NGC 3351 and NGC 3368), can be found just below the line between Regulus (Alpha Leonis)—Leo's brightest star (which is actually a group of stars at the lion's heart)—and Theta Leonis. About halfway between the two stars, these galaxies—along with M105 (NGC 3379)—are part of a larger group of galaxies that may contain between 8 and 24 galaxies, called the M96 Group. Astronomer Pierre Méchain found these three galaxies in 1781. While Messier added M95 and M96 to his original list of non-comet, non-star objects, M105 wasn't added until later.

About 38 million light years away, spiral galaxy M95 is visible in most amateur telescopes. In 2012, it was home to a supernova—a star explosion large enough to be visible from Earth even in some 6-inch (15-cm) scopes. M95's neighbor, spiral galaxy M96, is hard to see with binoculars—you're better off with a telescope with at least a 10-inch (25-cm) aperture. These two classic spiral galaxies will look better the bigger your scope is!

The last of the easily visible members of M96 is M105. The brightest elliptical galaxy in the M96 Group, it is about 34 million light years away from us. While you may be able to see faint smudges of M105 with binoculars, you'll need at least an 8-inch (20-cm) aperture to resolve the core and bar structure.

214 WATCH THE WATER BEARER

Aquarius the Water Bearer dates as far back as Babylonian times and is appropriately placed in the "water" section of the night sky—not far from Delphinus the Dolphin, the river Eridanus, Hydra the Sea Serpent, and Pisces the Fish. It's best seen in the northern hemisphere during the fall months and in the southern hemisphere's spring. Of its many mythological associations, it was at times identified with Zeus pouring the waters of life down from the heavens.

One of the largest constellations in the night sky, Aquarius is home to M2 (NGC 7089), a fine globular cluster that appears as a fuzzy spot of light through binoculars and small telescopes. It's possible to see the cluster's mottled appearance through a 4-inch (10-cm) telescope and to resolve it into stars through a 6-inch (15-cm) telescope.

Aquarius also hosts a handful of nebulae, among them the Saturn Nebula (NGC 7009)—so named for its apparent rings—and the Helix Nebula (NGC 7293). At 450 light years away, the Helix Nebula is the nearest planetary nebula to Earth.

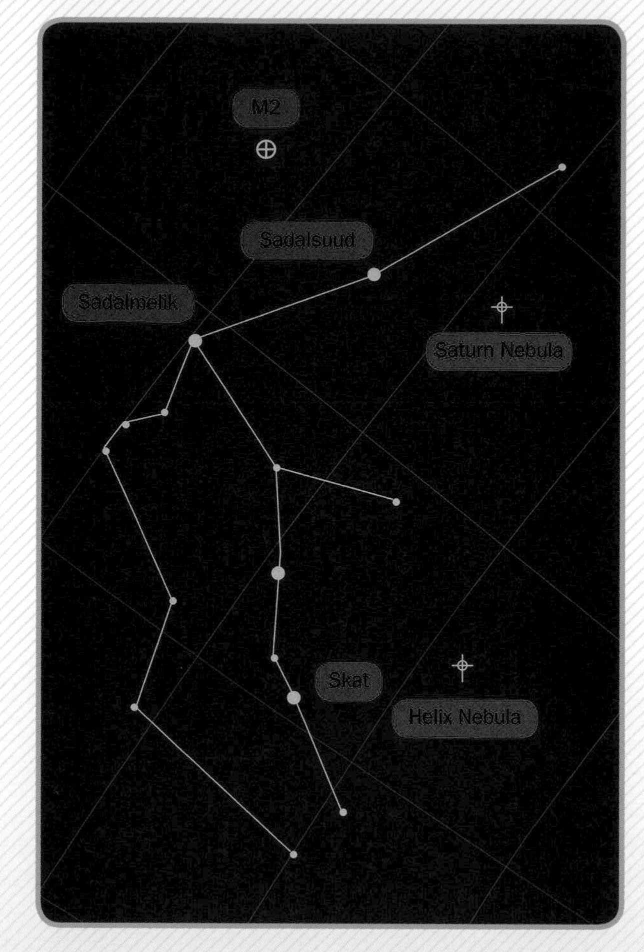

215 NAB THE CRAB

In Greek mythology, Cancer was sent to distract Hercules when he was fighting with the monster Hydra. The crab was crushed by Hercules's foot, but as a reward for its efforts, Hera then placed it among the stars. The zodiacal symbol represents the crab's claws.

Millennia ago, the Sun reached its summer solstice (its northernmost position in the sky—declination 23.5 degrees north) when it was in front of this constellation. As time passed, the Sun shifted overhead to the Tropic of Cancer. As a result of precession (the shifting of the poles caused by gravity and Earth's top-like wobbling), the Sun's most northerly position is now on the border of Gemini and Taurus.

Cancer lies between Gemini and Leo—two of the sky's showpieces. Its only real claims to fame are its membership of the zodiac and the beautiful Beehive Cluster (M44). This feature, also known as Praesepe, is one of the sky's finest open clusters. There are more than 200 stars in the Beehive Cluster. Spread across 1.5 degrees, they are best seen with binoculars.

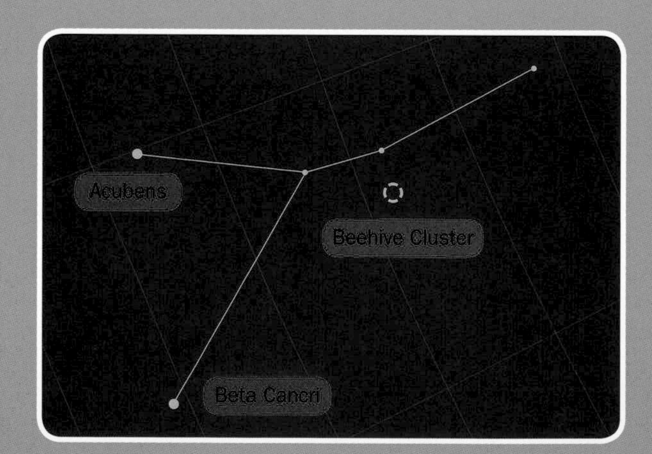

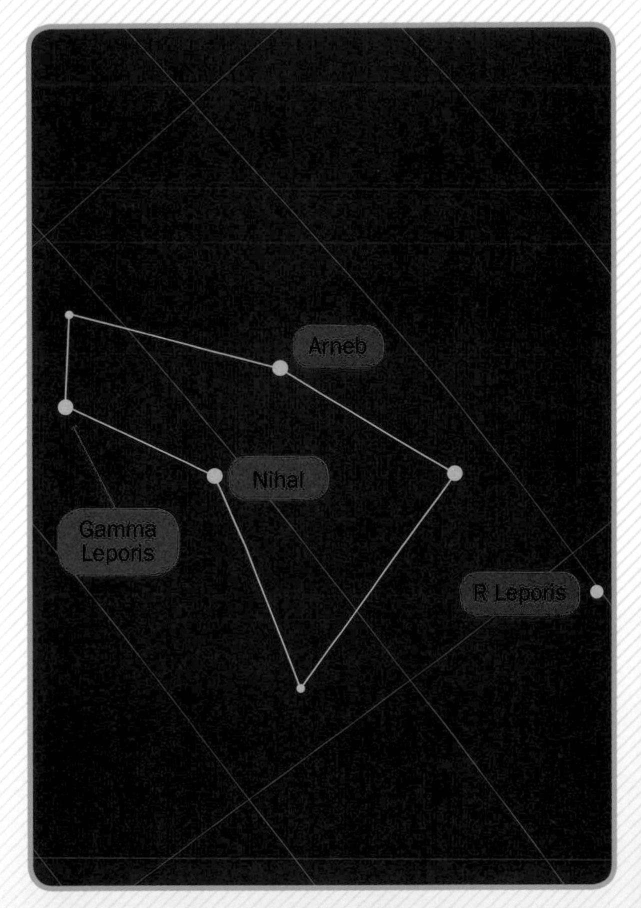

216 LOOK FOR LEPUS

A faint constellation, Lepus is nevertheless easy to find because it is directly south of Orion. In ancient times, it was thought of as Orion's chair. Egyptian observers saw it as the Boat of Osiris. The Greeks and Romans gave it the name Lepus. Since Orion particularly liked hunting hares, it was appropriate to place one below his feet in the sky.

Easy to separate in virtually any telescope, Gamma Leporis is a wide double star with contrasting colors and a separation of 96 arc seconds. It is relatively close to Earth, at a distance of 21 light years. Likened by some observers to a drop of blood in the sky, R Leporis is the variable that 19th-century British astronomer J. Russell Hind called the Crimson Star. During a period of 14 months, the Crimson Star varies in magnitude from 5.5 to 11.7. Its red coloring is at its most striking when the sky is dark and the star is near maximum brightness.

217 DISCOVER THE UNICORN

The German astronomer Jakob Bartsch created this faint unicorn constellation around 1624. Monoceros is the Latin form of a Greek word meaning "one-horned," and it seems that the mythical unicorn may have come into existence as a result of a confused description of a rhinoceros. Anxious to create a winter equivalent to the northern hemisphere's Summer Triangle (see #112), some observers advocate a winter triangle, bounded by Betelgeuse (Alpha Orionis), Sirius (Alpha Canis Majoris), and Procyon (Alpha Canis Minoris). Monoceros the Unicorn and the band of the Milky Way fill the space inside this triangle.

Skywatchers can find a host of celestial vistas in Monoceros. M50 (NGC 2323) is one such beautiful open cluster. Lying slightly more than one-third of the way from Sirius to Procyon, it is fairly easy to find. Some of the cluster's stars are arranged in pretty arcs. Farther up in Monoceros, NGC 2264 is really several different objects—the Cone Nebula and the Christmas Tree Cluster—categorized as one. There's also the beautiful Rosette Nebula, which you can experience via satellite imagery (see #268).

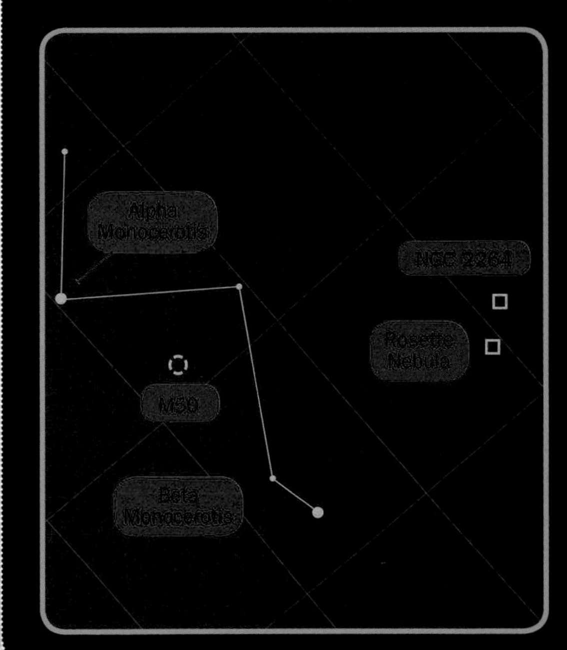

218 LOOK AT THE SUN SAFELY

While there are many safe ways to enjoy an image of the Sun, looking directly at our closest star is definitely not one of them. Luckily, catching a peek with your naked eye is usually painful enough to stop you before you seriously hurt yourself. Still, beware: There are some ways to do irreparable damage to your eyes quickly if you're not careful. Here are some things to avoid, and a few others to try, when peeking at the Sun.

DON'T

- Don't look through an unfiltered telescope or binoculars at the Sun. Ever. As a kid, did you burn leaves with a magnifying glass? Imagine a much more intense heat focused on your eye. Plus, you'll likely damage the instrument, too.

- Don't use a Mylar balloon, welder's glass with a value less than 14, or sunglasses as solar filters, no matter what you're told by strangers on the Internet!

- Don't use a filter at the eyepiece end of a telescope. These can heat, crack, and cause quick thermal damage to your eyes before you even have a chance to look away.

DO

- Use a manufacturer's solar filter on your telescope's light-gathering end.

- Keep the cover on your finder scope when safely observing the Sun. You don't want anyone to accidentally look through it, and you won't be using it during the day, anyway.

219 PROJECT THE SUN WITH A TELESCOPE

While you should never look through binoculars at the Sun, you can safely project an image of the Sun with binoculars or a small telescope. It is important to monitor this setup at all times so that no one accidentally peeps through the eyepiece and hurts his or her eyes.

STEP ONE Place your telescope on its tripod. While it helps to have some sort of mount if you're using binoculars, it isn't necessary.

STEP TWO Make a square collar for your telescope out of cardboard to shade the spot where your projection will be. If you're using binoculars, cover one of the lenses.

STEP THREE Point your telescope or binoculars at a distant object that is not the Sun—like a tree or a house or anything else that won't blind you—and focus it on infinity, using a low-magnification eyepiece on your telescope.

STEP FOUR Set up a white piece of cardboard or other light-colored material to serve as your projection screen.

STEP FIVE Point your telescope or binoculars at the Sun. Do not look through the eyepiece—ever. Adjust your setup until an image of the Sun appears on your projection screen.

220 SEE SUNSPOTS IN TRANSIT

Wait, what's that dark spot on the Sun? It's a *sunspot*: a slightly cooler and therefore darker area that forms on the surface of the Sun as a result of concentrated magnetic fields inside. The first amateur astronomer, Galileo, made extensive drawings of sunspots in 1612, and we've been diligently recording them for the last 400 years, as the Sun is one of the few astronomical objects that you can observe changing in very short time periods. Sunspots tend to appear at the peak of the 11-year solar cycle, the regularly occurring fluctuations in solar magnetic activity.

The Sun rotates every month or so (faster at the equator than the poles, since it is not a solid sphere) and sunspots can last days to weeks, meaning you can track them as they move across the visible part of the Sun. Using any of the projection methods or safety precautions explained here, watch sunspots in transit by observing the Sun around the same time each week for an extended period of time.

221 BOUNCE THE SUN ONTO A WALL WITH A MIRROR

If you don't have binoculars or a telescope, you can still project the Sun for safe observation if you can get your hands on a basic household mirror.

STEP ONE Grab a mirror. Cover all but about 1/5 inch (5 mm) of the mirror with cardboard or masking tape.

STEP TWO Put the mirror on a windowsill, where the uncovered portion can catch some sunlight. Turn off the lights in the room and draw all the blinds to make it as dark as possible.

STEP THREE Adjust the mirror until an image of the Sun reflects onto the wall. You can also tape or tack a sheet of blank paper to the wall to function as a projection screen for clearer viewing.

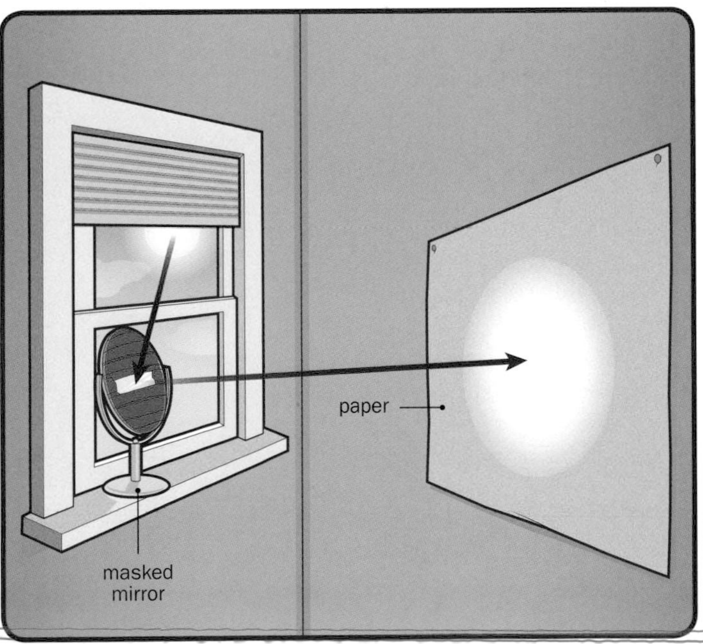

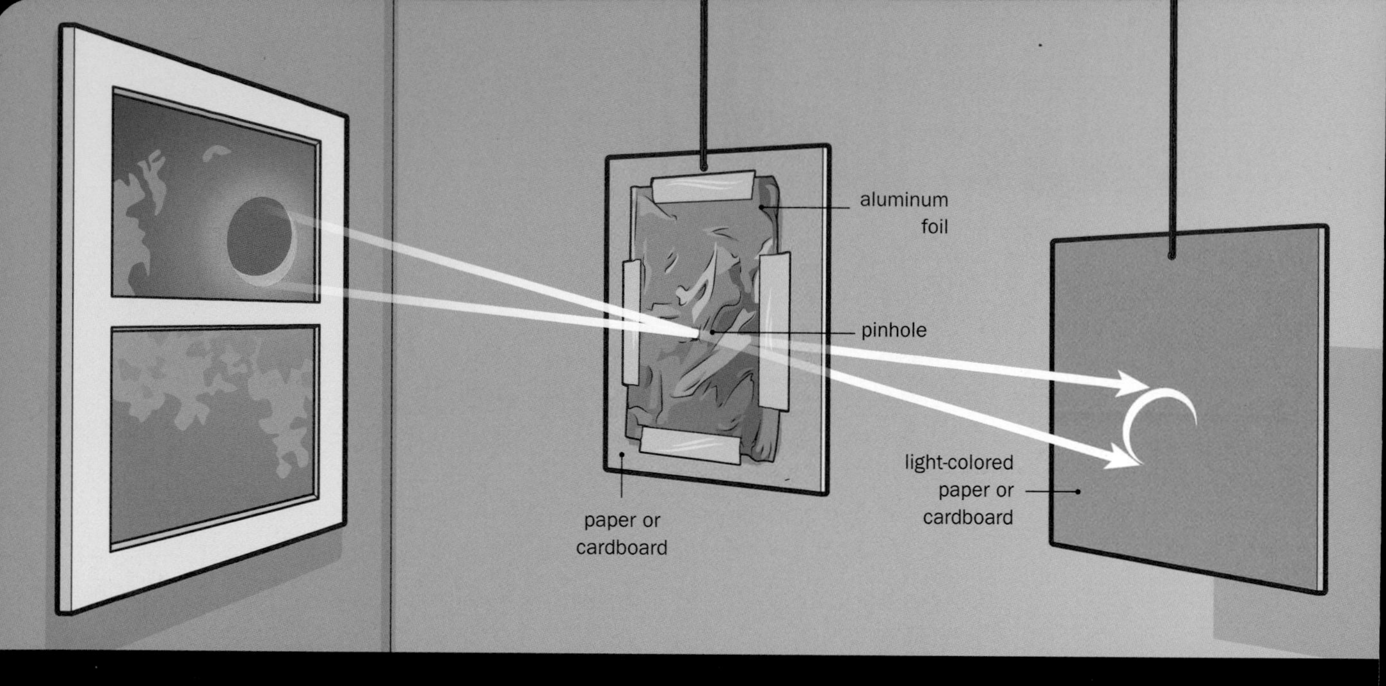

222 WATCH A SOLAR ECLIPSE WITH A PINHOLE VIEWER

Got a solar eclipse coming up on the calendar? Project an image of it with this simple pinhole projector. It's an elementary school classic that lets you protect your eyes and see the eclipse in all its glory.

STEP ONE Grab a sheet of paper or cardboard at least 1 foot (30 cm) square and cut a small hole in the middle.

STEP TWO Tape a piece of aluminum foil over the hole and poke a hole in it with a pin.

STEP THREE Get a second sheet of paper or cardboard, preferably light in color. Hang up or hold the two sheets so that the pinhole's shadow falls on the second sheet.

STEP FOUR That circle of light you're seeing? That's an inverted image of the Sun. Now gaze to your heart's content without fear of ocular damage!

223 SPORT ECLIPSE SHADES

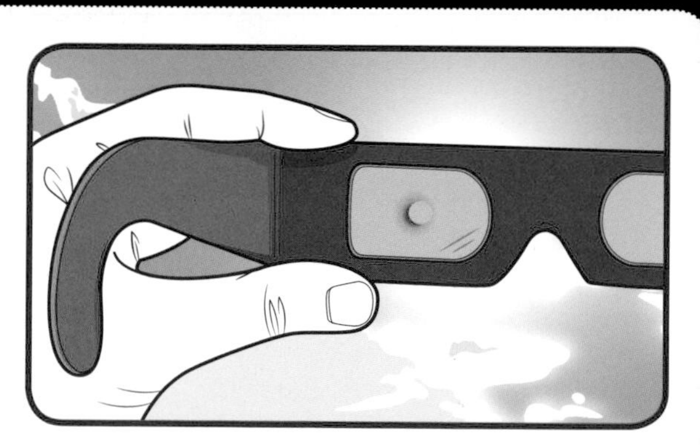

While these nifty little numbers might not win you any fashion awards, eclipse shades will get plenty of wows and keep you from being blinded by the Sun. Unlike observing an eclipse via projection, these sturdy, inexpensive, and easy-to-find-online paper glasses (not to be confused with 3-D glasses) connect you directly to the phenomena: You actually view the Sun, not just a picture of it.

For young children, single-pane glasses are easiest to use. Hold the glasses or card up to your face first; then, slowly look toward the Sun. Since the filter has to be dark enough to block most of the light coming from the Sun, you won't see much of anything around you until you're facing it. When you look up, you should see the Sun with incredible clarity.

224 SEE ECLIPSES THROUGH LEAVES, A COLANDER, OR YOUR FINGERS

Did you know that if you're standing under a leafy tree, feeling the sunlight dapple your face, you're actually covered in dozens of tiny images of the Sun? The leaves of the tree act much in the same way as a pinhole projector.

Next eclipse, stand under a leafy tree and watch the eclipse through the Sun images projected on the ground or a wall (A). Each circle of light will be slowly devoured by the shadow of the Moon as it moves between us and the Sun. Stranded with no trees in sight? Use your hands. Make a crosshatch pattern and look at the light coming through your fingers (B). And if there's a kitchen at your disposal, you might also try a colander (C). Use it as a many-pinhole projector, taking in the array of eclipse images it provides against a background of your choice.

TRAVEL FOR SOLAR ECLIPSES

Eclipses generally happen four times a year: two solar and two lunar. With that kind of frequency, you'll have plenty of viewing opportunities in your lifetime—especially for lunar eclipses, since everyone on the night side of Earth can see those. (See #75 for viewing information.)

Catching a solar eclipse (whether annular or total—see #226) is trickier and much more dramatic. Skywatchers urge you to see one of these in your lifetime, if possible. Folks often organize expeditions around solar eclipses—or you may get lucky and find one nearby on the maps here. If you're going to travel to see one, plan ahead! Hotels are often booked a year in advance.

Each swooping path you see here indicates where a type of solar eclipse will be visible through 2040, along with the date it will appear on.

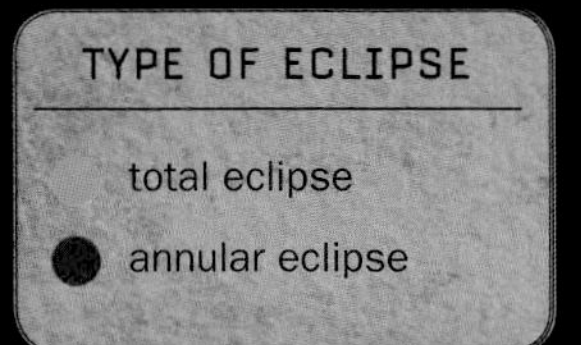

best viewing location

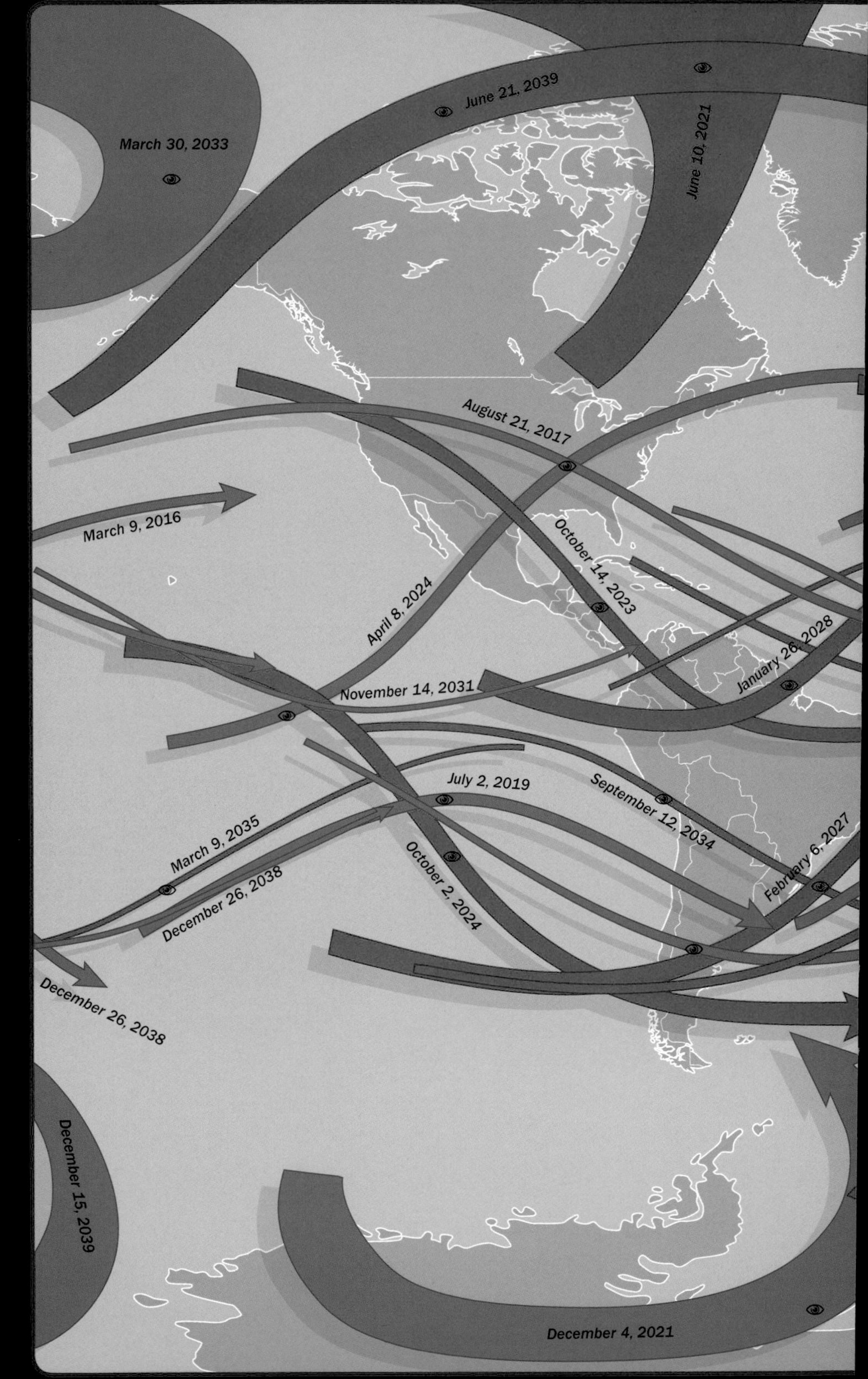

June 10, 2021
August 12, 2026
June 1, 2030
September 2, 2035
August 2, 2027
June 21, 2020
July 2, 2038
March 20, 2034
January 5, 2038
December 26, 2019
April 20, 2023
May 21, 2031
July 22, 2028
September 1, 2016
March 9, 2016
July 13, 2037
February 26, 2017
December 14, 2020
November 25, 2030
December 26, 2038
March 9, 2035
May 9, 2032
February 17, 2026
December 15, 2039

226 IDENTIFY SOLAR-ECLIPSE TYPES

Since solar eclipses usually happen twice a year, you might think they'd be a common sight. However, when they occur, they are only visible on a very small section of Earth. Seeing more than one in a lifetime is a rare (or globetrotting) event. Not all solar eclipses are the same; they're all special in their own ways. Get to know the three main types here.

TOTAL ECLIPSE This is probably the type you think of when you imagine a solar eclipse. During a total eclipse, the Moon's shadow—more specifically, its *umbra*—traces a narrow path along the face of Earth. For a few minutes, the dark Moon occults the Sun's bright disc so that it looks like a hole in the sky.

PARTIAL ECLIPSE When just part of the shadow of the Moon (the *penumbra*), passes over a section of Earth, we see a partial eclipse. This can happen at the same time as a total eclipse, on either side of the path of totality. The Sun will look like a big bite has been taken out of it!

ANNULAR ECLIPSE This eclipse type occurs when the Moon is farther away from Earth in its orbit and the tip of the main shadow doesn't reach Earth. The shadow area that touches our planet during an annular eclipse is called the *antumbra*. Views through filtered telescopes or eclipse glasses can be spectacular—just a ring of orange light where the blazing Sun used to be.

The one thing all solar eclipses have in common? You cannot view them with your naked eyes. Any time you look at the Sun, you must use the correct filters (see #182).

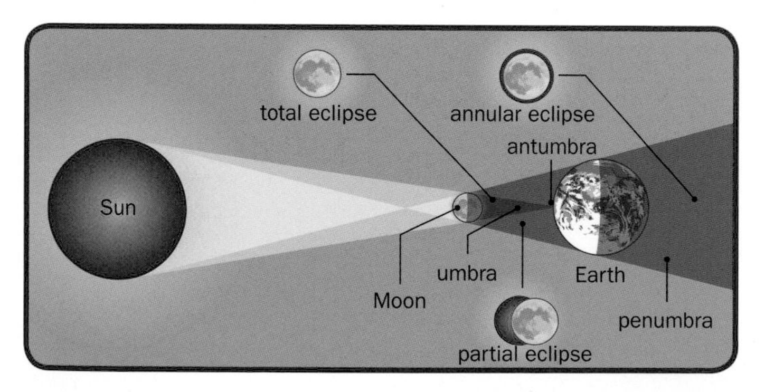

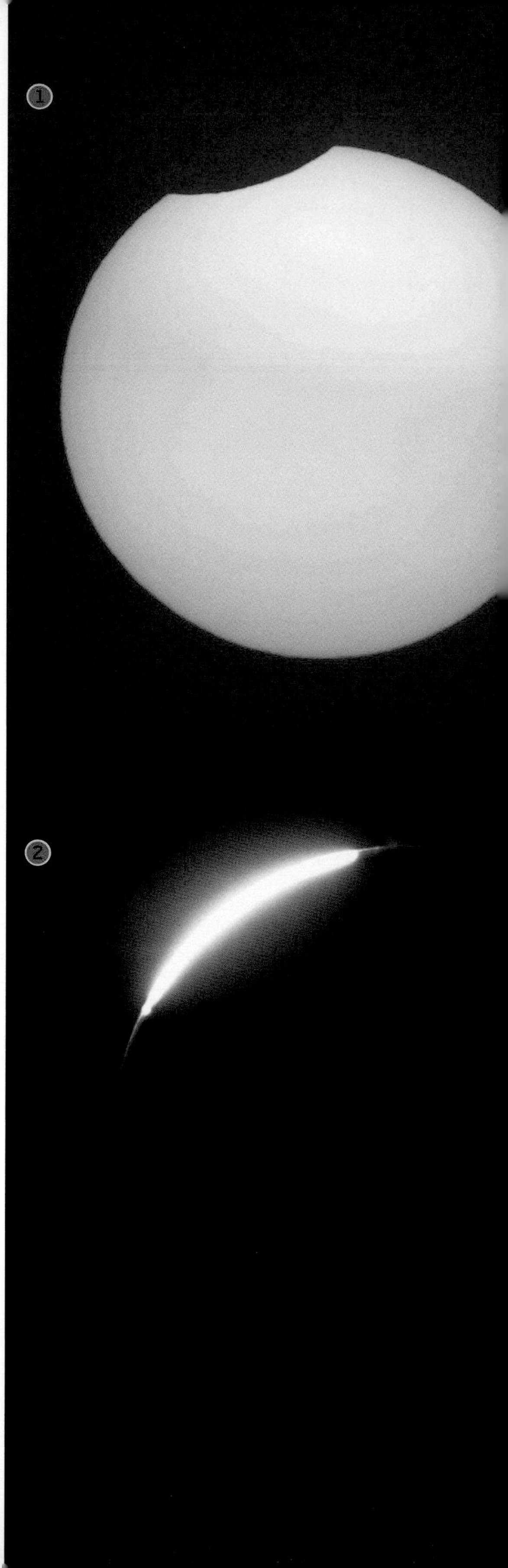

3

4

5

227 TOP FIVE

CATCH A TOTAL SOLAR ECLIPSE'S HIGHLIGHTS

Solar eclipses can last up to 2 hours, but most of the good stuff happens just before or during the period of totality—which typically lasts less than 7 minutes! Here's what to keep your (protected) eyes peeled for.

❶ **FIRST CONTACT** When the Moon initially touches the solar disc, it means the eclipse is officially beginning. For the next half hour, the Moon will steadily occult the Sun, taking a bigger "bite" as it goes.

❷ **BAILEY'S BEADS** As the Moon passes in front of the Sun, its lunar valleys allow the Sun's photosphere to peek through. This creates beautiful dots around the Moon's edge, an effect known as *Bailey's beads*.

❸ **THE DIAMOND RING EFFECT** As the last bit of the Sun's photosphere disappears, you'll be treated to the *diamond ring effect*. The Sun's inner corona will form a white ring around the Moon, and one of Bailey's beads will flare up to resemble a diamond in a ring.

❹ **CHROMOSPHERE PROMINENCES** The Sun's upper atmosphere is a thin, cool layer mainly made of hydrogen gas. It's called the *chromosphere* because it's beautifully colored—but you would never know, unless you checked it out during a solar eclipse! The chromosphere is continuously erupting with giant plumes of glowing gas—called *prominences*—that appear brilliant pink during solar eclipses.

❺ **SOLAR CORONA** When the Sun is completely hidden, the Moon will be ringed with a bright white light—as bright as the full Moon. This is the *solar corona*, the Sun's plasma outer atmosphere. It is so thin and faint that we can't see it from Earth unless the solar disc is blocked.

WITNESS THE ECLIPSE'S EFFECTS ON EARTH

Ancients viewed total solar eclipses as a sign from the gods (usually ominous); some modern observers agree that it's akin to a religious experience. Without a doubt, it is not to be missed if you have the chance to see for yourself. But with your eyes tilted upward to the sky, we often miss eclipses' fascinating effects on the ground.

RAPID CHANGE IN LIGHT About 20 minutes before totality, the light starts to change so it appears to be twilight. Then, when totality begins, the Moon's shadow will rush at you quicker than the speed of sound.

CREATURE CONFUSION Often during the twilight stage before an eclipse, you will begin to hear the sounds of night creatures—crickets and frogs that typically sing in the evening. When the totality is over and the light returns, you may hear a chorus of morning birds, confused about the short spell of darkness.

TEMPERATURE DROPS You'll feel a little chilly during the eclipse, as Earth drops in temperature by about 5°F (15°C) as it passes into the Moon's shadow. Some people also report changes in the force and direction of the wind, too.

SHADOW BANDS Keep an eye out for *shadow bands*—light and dark wavy lines moving across the ground, caused by odd refraction of the crescent sunlight and atmospheric winds. To see this, hold a piece of white paper on the ground—the bands will look like ripples on the bottom of a swimming pool!

HORIZON AGLOW During the totality, spin around to see Earth's horizon. It will be ringed all the way around in orange light! This effect occurs because you're looking at areas of Earth that are so far away they do not fall in the shadow of the Moon.

229 MEET NEPTUNE

Named after the Roman god of the sea, Neptune is a bold, blue, and icy gas giant—of all the gas giants, it is the most dense. As the farthest planet from the Sun, located way far out near the Kuiper Belt, it is invisible to the naked eye—but no less intriguing.

DIAMETER 30,000 miles (48,000 km), or about 6.6 times the size of Earth

MASS 102.4 septillion, or 17 times Earth's mass

AVERAGE DISTANCE FROM THE SUN 2.8 billion miles (4.5 billion km), or about 30 times farther from the Sun than Earth. At times Neptune is even more distant from the Sun than Pluto, due to Pluto's erratic elliptical orbit.

LENGTH OF YEAR Almost 165 Earth years

LENGTH OF DAY About 16 hours

DISCOVERY Galileo first observed Neptune near Jupiter in 1613, but he did not initially identify it as a planet. Neptune wasn't officially discovered until 1846, when astronomers fiercely competed to detect its presence using math. In fact, Neptune was the first planet discovered from calculations. Given Uranus's funky orbit, mathematicians predicted that another large planet must be tugging at Uranus from farther out in the solar system, altering its orbit. That planet turned out to be Neptune.

GRAVITY On the surface, Neptune's gravity is about 1.14 times that on Earth. In other words, if you weighed 100 pounds (45 kg) on Earth, you would weigh 114 pounds (52 kg) on Neptune—if there was a surface to stand on, that is.

SUCCESSFUL MISSIONS
1989: *Voyager 2* flyby (United States)

MAGNETIC FIELD Neptune's magnetic field is 27 times stronger than that of Earth, largely because its magnetic poles are inclined almost 50 degrees from its rotational axis. This provides Neptune's core with the extra motion needed to create a highly powerful magnetic field.

COMPOSITION While we haven't explored the interior of any giant planets, models suggest Neptune may have a rocky core surrounded by ice and a gaseous atmosphere of mainly hydrogen and helium.

RINGS Astronomers long suspected that Neptune had rings, and the *Voyager 2* probe in 1989 confirmed it. Thin and faint, the rings cannot be seen in amateur telescopes—or even most professional observatories. However, someday Neptune will get a much more visible ring, when its moon Triton is pulled and torn apart.

FIERCE WINDS This planet enjoys the fastest winds in the solar system—they can reach up to 1,300 miles per hour (2,100 km/h). The winds near storms (such as the Great Dark Spot) move in the same direction as the planet itself, while the winds in the upper atmosphere move in a retrograde manner.

MOONS Neptune has 13 confirmed moons, all named after gods of the sea and sea nymphs. William Lassell first spotted its largest moon, Triton, mere days after the planet's discovery. Triton is the only large moon in the solar system to orbit its host planet in a retrograde fashion, meaning it orbits against the spin of the planet. Like Pluto, Triton was a Kuiper Belt object (see #274) until captured by Neptune. Triton has one of the coldest surfaces in the solar system—at times hitting a frigid −688°F (−400°C)—and it often erupts with icy geysers, as seen by *Voyager 2*. Other moons of interest: Proteus, a surprisingly irregular-shape, large moon, and the wildly orbiting Nereid.

DIAMOND RAIN Diamonds may rain in Neptune's frigid, dense atmosphere, as on Uranus. This fascinating rainfall occurs when lightning strikes methane, breaking it into carbon clumps and hydrogen. When these clumps fall toward Neptune, atmospheric pressure increases dramatically, possibly forming diamonds!

SEASONAL BANDS Due to the planet's extreme tilt, Neptune enjoys seasons—kind of like we do on Earth. One crucial difference? Neptune's seasons can last for more than 40 years. We discovered Neptune's seasons by noticing that a band of clouds in the planet's southern hemisphere have gotten brighter.

THE GREAT DARK SPOT The Great Dark Spot was a large storm observed by *Voyager 2* as it flew by the planet. Observations by Hubble since then seem to indicate that dark spots will emerge and then vanish every few years. A small grouping of white clouds, nicknamed Scooter, float south of the Great Dark Spot, and a Small Dark Spot rages even farther south.

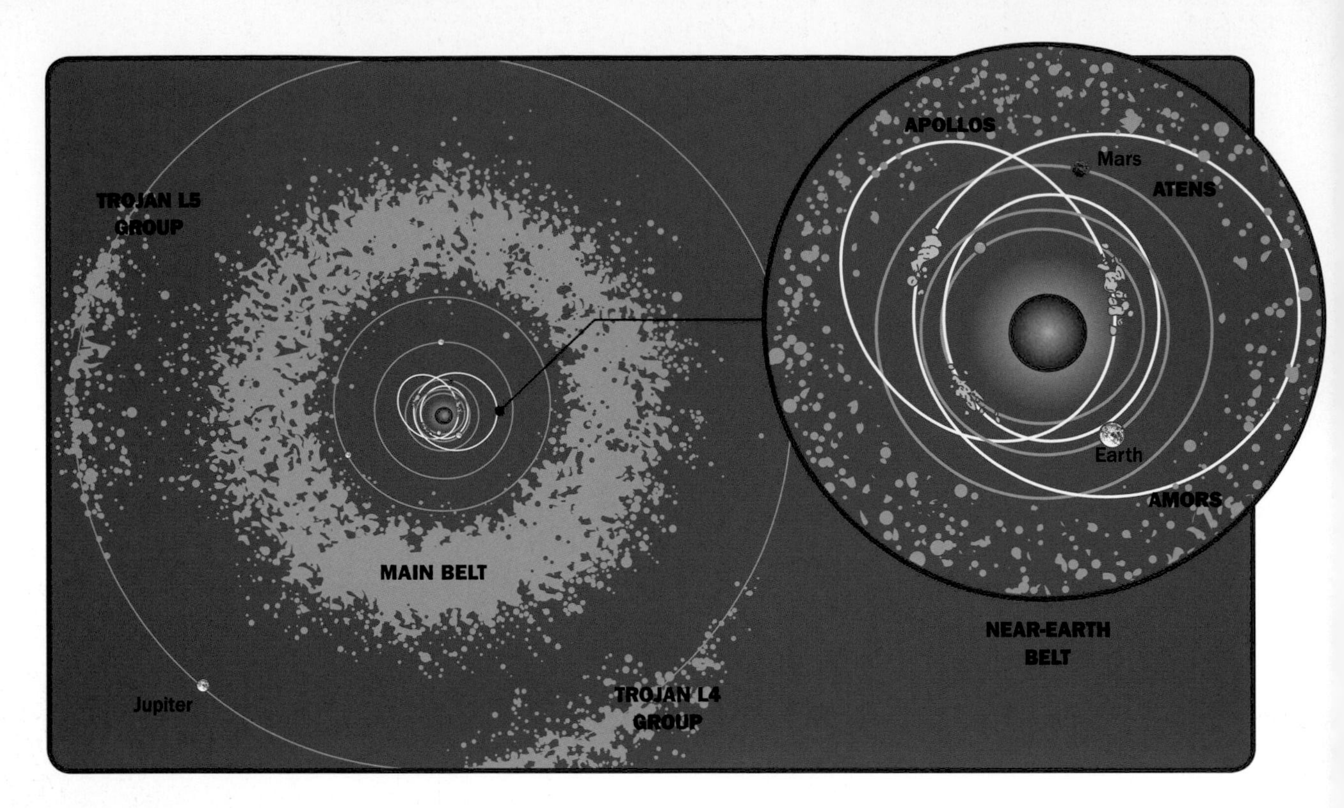

230 STUDY UP ON ASTEROIDS

Chunks of debris left over from the early days of the solar system are called *asteroids*—metallic, rocky bodies that lack an atmosphere, orbit the Sun, and are too small to classify as planets. They range from pebble-size specks to larger chunks. Twenty-six asteroids are known to exceed 125 miles (200 km) in diameter, but many more are out there. Our solar system's asteroids are grouped into these three main categories.

MAIN BELT Most asteroids stretch out in a belt between the orbits of Mars and Jupiter. The main belt contains more than 200 asteroids larger than 60 miles (97 km) in diameter, and scientists estimate that there are up to 2 million larger than 3/5 mile (1 km) across—in addition to billions of smaller ones.

NEAR-EARTH BELT Also known as *near-Earth asteroids* (NEAs), these rocks pass close to Earth's orbit. There are three main groups: the Amor group, whose orbits come close but don't cross our planet's orbit; the Apollo group, whose orbits cross Earth's but are most commonly found outside our orbit; and the Aten group, which not only crosses Earth's orbit but spends most of its time within it. Asteroids that actually cross Earth's orbital path are known as *Earth-crossers* and they've been the inspiration for many a doomsday scenario. We currently know about approximately 11,600 NEAs; the number with a diameter greater than 3/5 mile (1 km) might be as high as 1,000. There are dozens of potentially hazardous asteroids—but luckily, with 30 or 40 years' notice, we would hopefully have time to develop defensive technologies.

TROJANS French mathematician Joseph-Louis Lagrange predicted the existence and location of two groups of small bodies that orbit the Sun on Jupiter's coattails, near two gravitationally stable points referred to as the L4 and L5 regions. You can imagine the positions of these two groups by drawing a straight line between Jupiter and the Sun; the Trojans are found to the left and right of this line within Jupiter's orbit. The leading group, which is positioned ahead of Jupiter's orbit, is point L4. It contains 65 percent of a total 4,800 known Trojan asteroids. The trailing group, which orbits the Sun behind Jupiter's orbit, holds the rest. Other planets have Trojans in their orbits, including Earth.

231 TOP FIVE GET ACQUAINTED WITH THE NEATEST ASTEROIDS

This intriguing bunch of rocks that mostly hang out between Mars and Jupiter is often thought of as a group of *protoplanets*, or early-stage planets that never fully formed because of Jupiter's massive gravitational force. Here are a few fascinating specimens. (See #280 for how you can join the asteroid hunt.)

CERES One of five dwarf planets in the main asteroid belt, Ceres (A) is truly something special. It's 590 miles (950 km) in diameter and makes up one-third of the main asteroid belt's mass, so no surprise that it's the first asteroid we found! (That honor went to Guiseppe Piazzi in 1801.) Ceres is large enough to have been rounded by its own weight, and it has a differentiated interior topped with an icy layer 40 miles (65 km) thick. It often gives off vapor plumes. You can see Ceres as a small, stationary dot with good binoculars or a small telescope a few times a year. Use a planet-tracker to learn when it's visible.

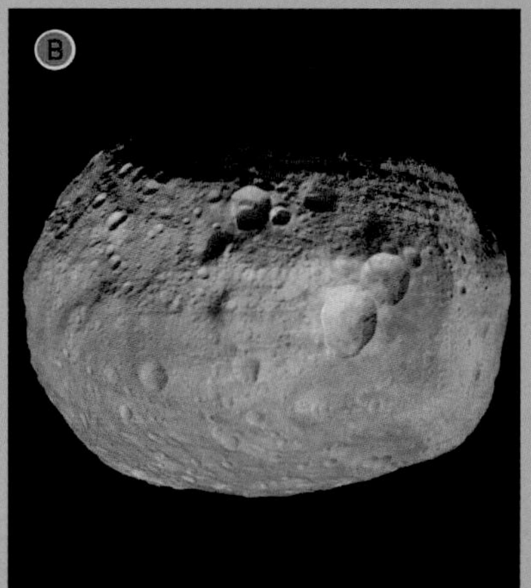

VESTA The brightest asteroid visible from Earth, Vesta (B) is the second-biggest asteroid and, like its larger counterpart Ceres, has a differentiated core. It also has an oblate spherical shape and its surface is cause for much intrigue: There's evidence that lava once flowed and formed a crust over the asteroid's rocky mantel. Perhaps the most prominent surface feature is its large craters, one of which stretches 286 miles (460 km) of Vesta's 326-mile (525-km) total diameter. Debris from the impacts that created these craters has journeyed all the way to Earth; they're called HED (howardite-eucrite-diogenite) meteorites.

EROS This asteroid is one of many firsts: It's the first near-Earth asteroid to be discovered, the first asteroid to be orbited by a spacecraft, and the first on which a spacecraft landed. Eros (C) crosses Mars's orbit but does not quite reach that of Earth, making it part of the Amor asteroid group. Eros was the subject of much observation in 1975 when it was a breezy distance of 14 million miles (22.5 million km) away from Earth. That's about 60 times the distance of our Moon. It is littered with large boulders and experiences shifts in gravity due to its peanut shape.

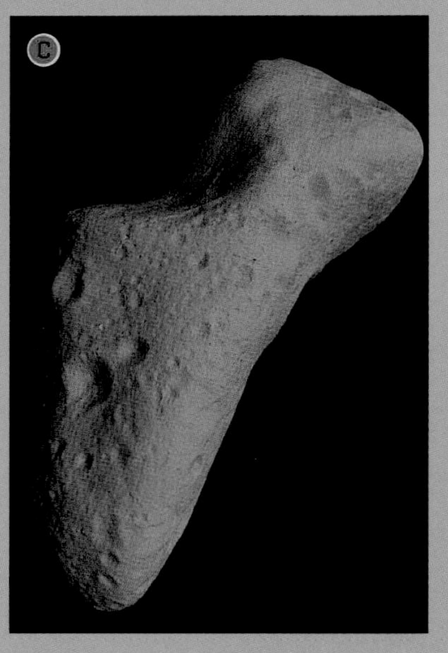

JUNO One of the larger asteroids in the main belt, this stony minor planet has attracted plenty of attention due to its unusually reflective surface. Juno is noted for having a small and irregular shape—excluding it from being classified as a dwarf planet—and it has one of the funkiest orbits of any known body in the solar system, deviating greatly from the shape of a perfect circle.

PALLAS The second asteroid to be discovered (in 1802 by Heinrich Wilhelm Matthäus), Pallas is slightly larger than Vesta in volume but a lot less massive, and it's 22 percent the mass of Ceres. There is evidence that it's partially differentiated, suggesting that it is—like its two larger siblings—a protoplanet. Pallas also has an oddball orbit: It circles at a high tilt of 34.8 degrees.

232 SAIL THROUGH THE SOLAR SYSTEM WITH COMETS

Comets are the celestial family members that astronomers most like to have visit—though not too close! No two comets are the same and they change over time, brightening as they near the Sun and fading as they loop back toward the solar system's cold depths. This happens over weeks or months, so you have time to plan for viewing. While a few comets are visible to large telescopes at any given time, comets bright enough to see with smaller scopes and binoculars swing by every few years on average. The ones we call "great comets"—those that are noticeable even if you aren't looking for them—are visible a few times in a lifetime, if you're lucky.

But what is a comet, exactly? Despite their many differences, comets do have some parts in common:

A NUCLEUS The nucleus of a comet is made of ice, frozen gasses, rock, and dust—not unlike a really dirty snowball. When it sails through the frigid outer solar system, it is very stable.

B COMA As the comet approaches the Sun, it warms up and begins to develop a *coma*—a fuzzy sort of shroud around the nucleus—and releases a trail of gas (ions or plasma, to be exact). Since the gas tail is being blown by the solar wind, it always points away from the Sun. (Think about what your hair does when you put a fan right in your face.)

C DUST TAIL As the comet gets even closer to the Sun, chunks of dust and rock begin to break off and a second tail appears—the *dust tail*. The trail of dust the comet leaves in its wake are like bread crumbs that orbit in the same path. When Earth crosses these dust trails, we see meteor showers (see #89).

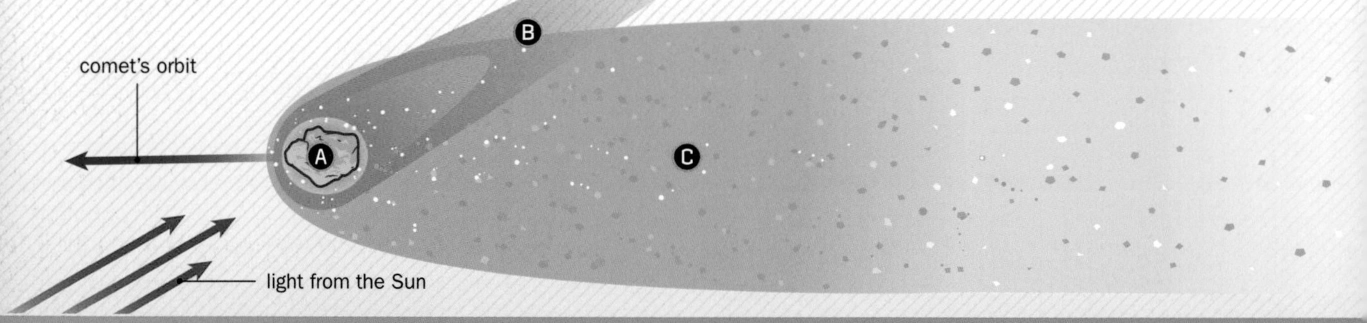

233 NAME THAT COMET

Comet names can seem pretty convoluted, like the famous and fairly recent C/2014 Q2 (Lovejoy). What do all those numbers and letters mean? The letter at the beginning of a comet's name—typically P or C—indicates the comet's *period*. Short-period comets (marked with a P) usually come from somewhere between Jupiter and the Kuiper Belt. They often have orbits in the same plane as our solar system and swing by at least every 200 years. Long-period comets (marked with a C), however, originate in a large region far beyond Pluto called the Oort Cloud. These comets can have eccentric orbits, sweeping near the Sun every few hundred to few million years. Some never come back!

The first number is the year in which the comet was discovered (C/2014 Q2, shown here, was discovered in

2014). The second letter comes from breaking the year into 24 parts and assigning each half month a letter. In this example, "Q" stands for the second half of August. The final number—in this case, "2"—tells you it was the second comet discovered in that part of the month. Last, you'll often see the name of the discoverer added at the end—in this example, Terry Lovejoy.

Robotic telescopes also scour the skies for comets, so some bear the names of robots, such as Pan-STARRS, which found a comet in June 2011.

234 TOP FIVE GET TO KNOW SOME MAJOR COMETS

The sky for the most part is populated by objects that go through predictable patterns, appearing the same throughout a stargazer's lifetime. About once a generation on average, a stunning naked-eye comet will streak through our skies during a period of weeks to months. They mostly show up unannounced from the depths of the icy outer solar system.

☐ **HALLEY'S COMET** Perhaps the best known of all comets, Halley's Comet makes its rounds through the inner solar system and returns near Earth's vicinity every 75 years, making it possible for a human to see it twice in his or her lifetime. Sir Edmond Halley, for whom the comet was named, first predicted the return of the 9-mile- (15-km-) wide comet in 1759. The last time it was visible from Earth was 1986; it is predicted to come back in 2061.

☐ **HYAKUTAKE** Making its first appearance in 1996, this 2-mile- (3-km-) wide comet sports one of the biggest tails ever observed, stretching out more than 100 degrees as seen from Earth. It was discovered by a Japanese amateur astronomer using a pair of binoculars and remained visible to the naked eye for three months. Astronomers have calculated its orbit and say it will not come near the Sun again for another 14,000 years. So don't wait up.

☐ **SWIFT TUTTLE** Six miles (10 km) in diameter, Swift Tuttle was first seen in July 1862 by astronomers Lewis Swift and Horace Tuttle. The comet, which orbits the Sun every 120 years, was rediscovered in 1992. The odds are Swift Tuttle will jet past Earth in 2126 at a safe distance of tens of millions of miles.

☑ **HALE-BOPP** One of the brightest comets ever seen (shown here), Hale-Bopp was discovered outside of Jupiter's orbit by Alan Hale and Thomas Bopp in 1997. Analysis showed that its tremendous brightness was due to its large size. While most comets range from 1 to 2 miles (1.6–3 km) in diameter, Hale-Bopp was estimated to be 25 miles (40 km) across. It was visible through bright city skies, and with the naked eye, for 19 months. It will next appear in 2,400 years.

☐ **LOVEJOY** Comet Lovejoy was discovered in November 2011 by Australian observer Terry Lovejoy. As one of a handful of Sun grazers, this green-hued comet surprised astronomers when it survived its passage just 87,000 miles (140,000 km) above the surface of the Sun (which is one-third of the distance between Earth and the Moon). Lovejoy's gorgeous hues are caused by diatomic carbon (C_2), which is activated by sunlight. The colorful comet won't return to our solar system for another 8,000 years.

235 ROOT OUT RETICULUM

One of the smallest constellations in all the heavens, Reticulum is made up of a number of faint stars halfway between the bright star Achernar (Alpha Eridani) and Canopus (Alpha Carinae). Some of its celestial neighbors include the constellations Hydrus and Horologium.

Reticulum was first dubbed the Rhombus by Isaak Habrecht of Strasbourg. French Astronomer Nicolas-Louis de Lacaille changed its name to Reticulum to honor the reticule—the grid of fine lines in a telescope eyepiece that aids with centering and focusing.

The brightest star in Reticulum is Alpha Reticuli, which is in between a giant and *bright giant* (a term for a star that straddles the boundary between giant and supergiant). About 160 light years from our solar system, Alpha Reticuli shines with magnitude 3.3.

R Reticuli is another sight worth seeing in this constellation. A quite red variable star, R Reticuli shines at about magnitude 7 at maximum light. During a period of nine months, it drops to magnitude 13, then returns to maximum brightness. In a few million years, it will transform into a white dwarf star.

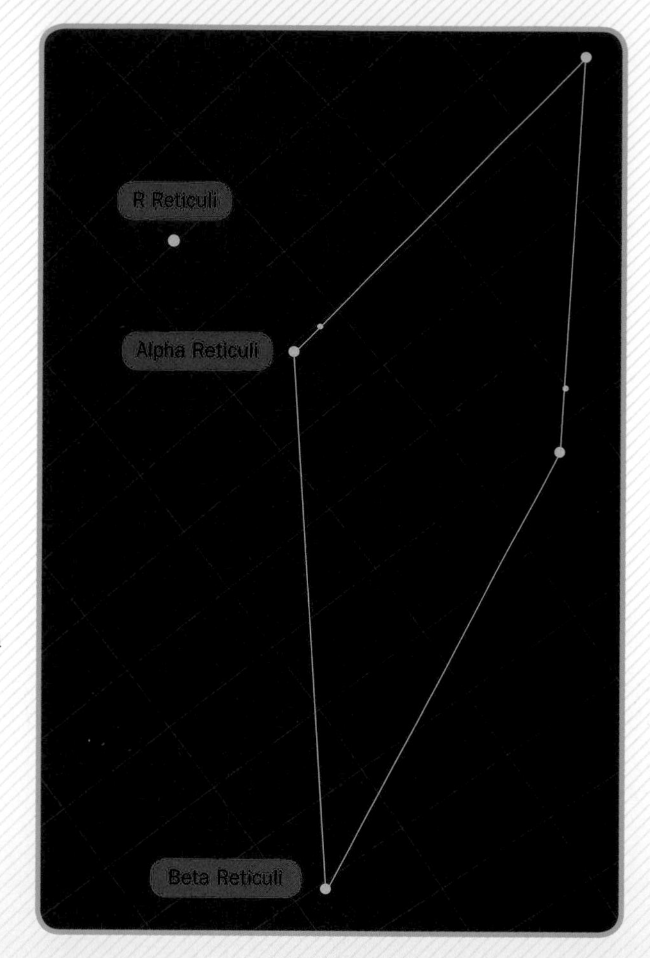

236 SET YOUR EYES ON SAGITTA

Although only a small constellation, Sagitta—not to be confused with the much larger constellation Sagittarius—is easy to find halfway between Altair (Alpha Aquilae) in Aquila and Albireo (Beta Cygni—see #191). It really is true to its name—the ancient Hebrews, Persians, Arabs, Greeks, and Romans all saw this group of stars as an arrow. It has been thought of variously as the arrow that Apollo used to kill the Cyclops; one of the arrows shot by Hercules at the Stymphalian Birds; and as Cupid's dart.

Sagitta can be seen from anywhere on Earth, excluding the Antarctic Circle. Some of its most spectacular sights include giant blue star U Sagittae, an eclipsing binary that drops from magnitude 1.5 to 9.3 every 3.4 days and appears high above Sagitta's main stars. V Sagittae, although faint and varying wildly from magnitude 8.6 to 13.9, is interesting for its almost nightly variation. It might have been a nova a long time ago. South of the midpoint of a line joining Delta and Gamma Sagittae, you'll see faint cluster M71 (NGC 6838).

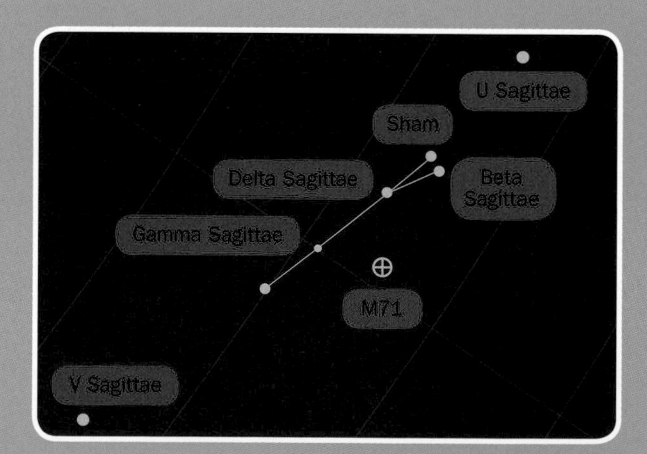

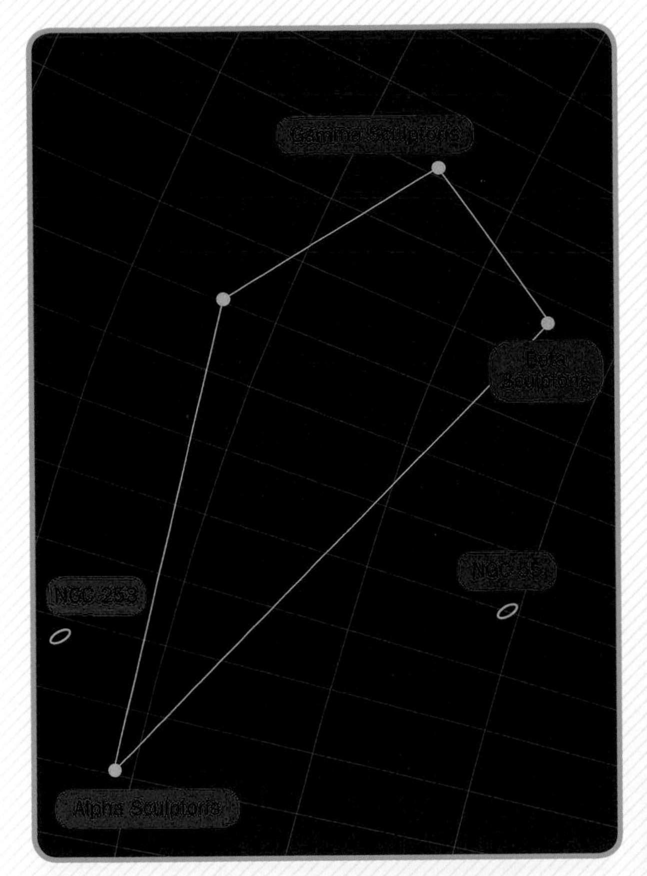

237 GLIMPSE THE SCULPTOR

Originally named L'Atelier du Sculpteur (French for the Sculptor's Woodshop) by Nicolas-Louis de Lacaille, the Sculptor lies south of Aquarius and Cetus. Its most significant feature of interest is a small cluster of nearby spiral galaxies.

For a small-telescope user, the Sculptor Galaxy (NGC 253) is highly satisfying, especially for observers in the southern hemisphere. It is very large and is viewed almost edge-on. It was discovered by Caroline Herschel one night in 1783 while she was searching for comets. It appears as a thick streak in binoculars and begins to show the texture evident in photographs when larger instruments are used.

Another fine edge-on galaxy, NGC 55 is distinctly brighter at one end than the other when seen through an 8-inch (20-cm) telescope. Both NGC 55 and 253 are members of the Sculptor Group, an arrangement of galaxies that might be the nearest neighbor in the cosmos to our Local Group (a group of more than 50 galaxies that includes the Milky Way).

238 SCOPE OUT SCUTUM

Although Scutum is not a large constellation and has no bright stars, it is not difficult to find in a dark sky because it is directly above the teapot of Sagittarius, on the opposite edge of the Milky Way. Johannes Hevelius created the constellation at the end of the 17th century, giving it the name Scutum Sobiescianum (Sobieski's Shield) in honor of King John Sobieski of Poland, after he fought off a Turkish invasion in 1683. Despite not containing many bright stars, Scutum still has some marvelous views—like the Wild Duck Cluster (M11/NGC 6705. This spectacular open cluster is clearly visible in binoculars, rewarding in a small telescope, and stunning in an 8-inch (20-cm) one. One of the most compact of all the open clusters, the presence of a bright star in the foreground adds to its beauty.

Back in 1973, NASA launched *Pioneer 11* to study the asteroid belt and outer solar system. It's outlived its power source, but it's traveling toward Scutum and carrying the Pioneer plaque, which is our species' message in a bottle.

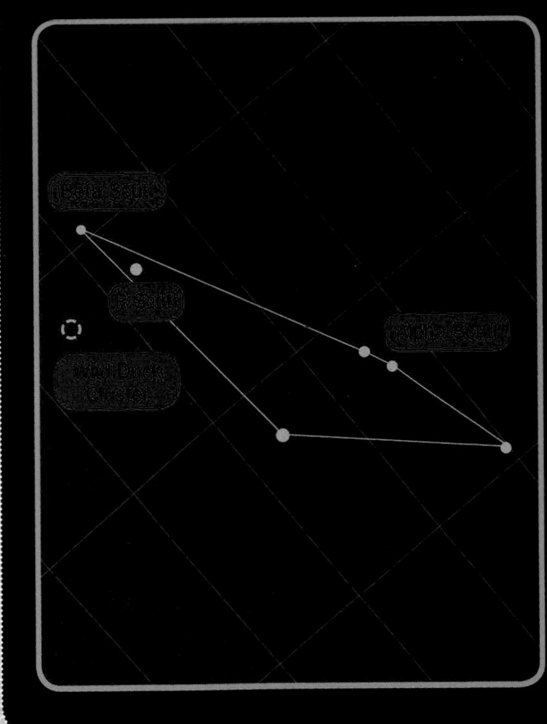

239 PACK A BASIC ASTROPHOTOGRAPHY KIT

Once you start observing the wonders of the night sky, it likely won't be long until you want to document what you see. Here's what gear you need to start capturing all the good stuff going down way up high.

A CAMERA For shooting the night sky, a DSLR (digital single-lens reflex) or ILC (interchangeable lens) camera is best. These cameras allow you greater control over your images: You can swap out lenses and take multiple photographs at preset intervals or long exposures over time. But if you're still working with a compact camera, no worries. Many of today's compacts offer nice control in manual mode.

B LENSES A wide-angle lens (which tends to have a focal length of less than 35mm) is great for sweeping vistas of, say, the Milky Way, while a telephoto lens (with a focal length of 85–300mm) will allow you to zoom in on Taurus's Pleiades (M45) or even Orion's Flame Nebula (NGC 2024). Fisheye lenses (8–10mm focal length) are fun, too, as they let you create hemispherical images of the heavens. Generally speaking, you want a fast lens: one that has a larger aperture (smaller f/number) to maximize the amount of light that gets in.

C TRIPOD Camera shake is astrophotography's greatest enemy. You want a sturdy yet lightweight tripod that you can haul out into the field. Keeping it under 5 pounds (2.25 kg) is ideal.

D REMOTE TRIGGERING TOOLS Astrophotography often requires long exposures or lots and lots of shots taken at specific intervals. Since you likely don't want to stand with your thumb on the shutter all night (which could create image shake, anyway), pick up a *remote release shutter*—a switch that allows you to open the camera's shutter from a distance—or a *remote timer release,* which allows you to open the shutter remotely at various intervals for various exposures for a period of time.

E HEAD LAMP You're going to be fiddling with controls and settings—in the dark. A head lamp outfitted with a red bulb will allow you to keep your hands on your camera. Just turn it off when you begin your exposure.

240 SNAP A MOBILE SHOT THROUGH A TELESCOPE

You can use your cell phone to get great pictures through your scope's eyepiece. This technique is called *eyepiece projection*: The image from the eyepiece projects at your phone's camera. It's easiest to start with the Moon, then move on to other bright objects such as Saturn's rings and Jupiter and its moons.

STEP ONE Aim and center your telescope at the Moon with a low-powered eyepiece. You may also wish to use a Moon filter to cut down on glare—the Moon is very bright through a telescope.

STEP TWO Dim your phone or cover it in red cellophane. Steady it over the eyepiece and find the eyepiece's focus. The Moon will come into view—you should see craters and seas (see #153).

STEP THREE Tap the screen to focus. It helps to pick a crater or other visible feature to fine-tune. Then take the picture!

241 FREEZE STARS WITH THE 500 RULE

Due to Earth's rotation, taking long-exposure images of the night sky with a DSLR can result in dashes and streaks—not fixed stars. If you're after a photograph of a starry, starry night that actually matches what you see, try a little math: Multiply the focal length of your lens by your camera body's sensor size, and then divide 500 by that number to arrive at the longest exposure possible (in seconds) before stars start to streak—aka, the 500 rule. (Those with high-megapixel sensors may want to try dividing by 450 instead.)

242 SHOOT STAR TRAILS

To us Earthlings, it's impossible to detect the slow movement of the night sky rolling by overhead. But star trail images capture the motion of our Earth—and deliver beautiful streaks of stars on the go. You can make a star trail photo by taking one hours-long exposure or by capturing many shots at regular intervals and stacking them in software (see #246).

STEP ONE You'll want the sky to be as dark as possible, so plan your star trails excursion for a Moonless night. (Or, if you'd like a well-lit landscape, plan your session on a quarter-Moon evening.)

STEP TWO Consult the weather and shoot on a still evening. You'll want the night sky to be the only thing on the move in your image, so shoot when clouds won't create trails and wind won't smudge trees.

STEP THREE Head away from cities, the Moon, and artificial light—even faint illumination from a nearby highway can dull your star streaks. Arrive at your location with plenty of time to set up before dusk.

STEP FOUR Place your tripod on level land. Try to set up so there's a terrestrial object in the foreground, such as a tree, rock formation, or man-made structure, to provide context and scale to your image.

STEP FIVE If you're going with the long-exposure method, set the lowest possible ISO (between 400 and 800) and a middle aperture, such as f/5.6 or f/8. For those who plan on stacking multiple 30-second exposures, start with a test shot taken at a large aperture (f/4 or lower) and ISO 1600. Bump the ISO up if no stars are visible in your test image.

STEP SIX Set your wide-angle lens to manual and focus to infinity. (Tack on some duct tape to keep the focus ring from slipping during your long exposure.) Make sure you have extra power sources—no one wants to run out of juice during an hours-long shoot.

STEP SEVEN If you're going with the long-exposure method, put your camera on bulb mode and open the shutter—then nap for 90 minutes. If you plan on stacking photos after capture, use a remote timer release to take a photo every 30 seconds.

243 CREATE RINGS IN THE SKY

Have you ever seen star trails that seem to spiral in concentric circles? Photographers aren't manipulating the heavens—instead, they've trained their camera at either due north or due south. In the northern hemisphere, point your lens at Polaris (#37) to create an array of mesmerizing rings around our North Star. Meanwhile, if you're in the southern hemisphere, locate the South Celestial Pole—#39 (or a compass) should help you out.

244 CAPTURE THREE-WAY TRAILS AT THE EQUATOR

When shooting at the equator, star trails seem to do a unique thing: Straight lines cross your photo on a diagonal, while the trails above that line arc up and the trails below arc down. How come? The streaks in a straight line are made by stars moving along our celestial equator. The ones that stretch up in a semicircle actually form rings around Polaris—you're just so far south that they don't appear to complete the ring in your image. Ditto with the stars below the celestial equator: If you were farther south, you would have a photo of stars streaking in complete circles around the South Celestial Pole.

245 CRAFT A TIME LAPSE OF A LUNAR ECLIPSE

Watching a lunar eclipse can be a breathtaking experience, so no surprise it can result in an equally breathtaking photo. Follow these steps to document the Moon as it goes from a silver dime to a blood-red disc in the sky, and then composite them into one image.

STEP ONE Enlist the proper gear. You can certainly shoot the eclipse with a compact camera or your cell phone set on a self-timer to reduce wiggle. However, a dedicated DSLR with a wide-angle lens—all set up on a sturdy tripod so you can take long-exposure images—is your best bet. A remote trigger release will allow you to fire shots without touching (and shaking!) the camera.

STEP TWO Prepare by testing your settings a night or two before. The Moon rises about 50 minutes later each day, so if the eclipse is happening at 9 PM, check at 8:10 the night before. It's also a good idea to determine the Moon's path across the sky so you can plan composition—try taking one photo of the Moon at the top corner of your viewfinder and see if it drifts into or out of the frame during the next few minutes.

STEP THREE For a wide-angle shot, start with ISO 400; open the lens to the largest aperture (smallest f/number); and play with shutter speeds of up to 40 seconds. Manually set your focus to infinity and take many wide-angle images during the eclipse, keeping your time intervals consistent—say, every 15 minutes.

STEP FOUR During the eclipse, the Moon's brightness will change. You will need to adjust your exposure settings in order to capture the Moon's features as it passes into totality and back again.

STEP FIVE To combine (or composite) your images, download them to your computer and open them up in your chosen software, such as Adobe Photoshop or Lightroom. (You can also use dedicated and free software for astrophotography, like DeepSkyStacker.) Select the Moon from each of your images and copy and paste them onto individual layers in a new document, arranging them in the diagonal line that you saw across the sky. Make any adjustments, then flatten your file into one image.

246 STACK IMAGES IN SOFTWARE

Often when you're shooting the night sky, you may get a lot of *noise*: graininess or discoloration within an image that makes stars and other space objects seem dim. This effect is caused by shooting in low light. Turbulence can also wreak havoc on your images, especially at high magnifications. Luckily, you can fix both problems with image stacking.

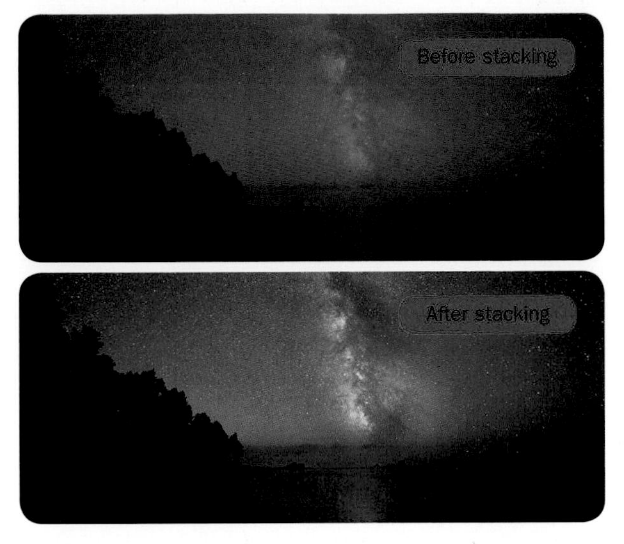

STEP ONE Select a photo-editing software. Adobe Photoshop or Lightroom both work great, and DeepSkyStacker is a fantastic (free!) tool that makes merging photos into one image a snap.

STEP TWO Aim your camera at an interesting swath of sky and take a lot of photos. Crank your ISO as far as you can until you start to get noise and shoot longer exposures—about 10 seconds should do. These might not look supergreat. It's OK—they will soon!

STEP THREE Put the lens cap on your camera and take *dark frames*: black images at the same ISO and shutter speed as the images you made of the sky. Shoot as many dark frames as you did of the sky. (If your camera has digital noise reduction, you can skip this step as long as you have that function clicked on!)

STEP FOUR Time to import your photographs into your computer's image-editing software and open them up. If you're using a dedicated app like DeepSkyStacker, it will automatically map your sensor's noisy spots using the dark frames and subtract those areas from your sky images. Hit "register" and it will align all your images. If you're working in Photoshop or Lightroom, you will need to align the images yourself: Select your base image, then copy another onto a second layer and hit "difference," which will show you when they are aligned. Lower the transparency of the second image so it doesn't completely cover the first, then add more images using the same technique.

STEP FIVE Tweak white balance, sharpening, and other settings until the image is to your liking, then flatten the files into one.

247 CORRECT YOUR SCOPE'S COMA

You can improve both your view and your camera's by attaching a coma corrector to your telescope, which can sharply reduce the warping around the edges of your telescope's field of view. The effect is most noticeable with stars: While they will appear to have tails without coma correction, they will look like bright pinpoints if imaged with the help of a coma corrector.

Made of very high-quality glass, coma correctors are often designed for use on very particular models of telescopes; their optics must be compatible with the optics of the parent scope. Attaching a coma corrector to a telescope not designed for it can actually worsen coma and image quality. Some high-end coma correctors are actually tunable; in other words, they adjust to many varieties of telescope. The most startling jump in image quality comes from using a coma corrector on a "fast Newtonian" style of telescope: a Newtonian reflector with a low f-stop (f/5 or lower).

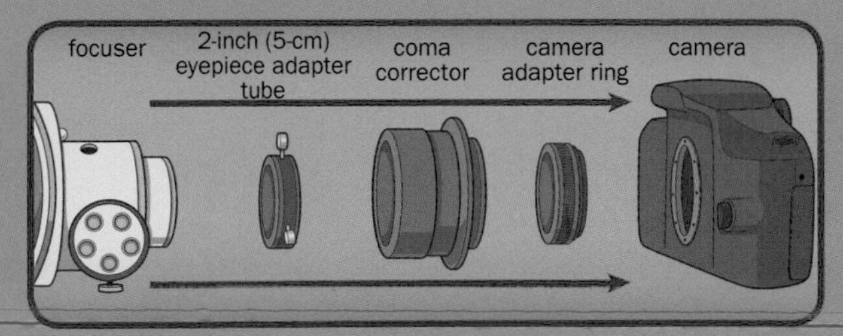

248 MERGE YOUR TELESCOPE WITH A DSLR

Did you know you can take pictures right through your telescope with a DSLR camera? You can basically treat your scope like a giant telephoto lens—because, well, it is one! Shooting through a telescope allows you to take amazingly close, crisp images of nebulae and planets. Using your DSLR in place of the eyepiece to take pictures is called *prime focus astrophotography* and here's how to do it.

"T" UP To secure your camera to your telescope, you'll need two things: a T-adapter for your telescope model that screws into the focuser and a T-ring mount for your camera model, which connects the camera to the T-adapter. T-rings exist for most major DSLRs and you can get T-adapter tubes for most telescopes using both 1¼-inch- (3-cm-) and 2-inch- (5-cm-) wide eyepieces adapters. Screw it together, and your camera now has one powerful lens.

GET TETHERED You may also want to hook your camera to your laptop for additional fine control over your camera's settings, such as ISO, color balance, and (most importantly) focus. Having fine control over focus via software greatly reduces the time spent adjusting your camera and means you don't touch (aka shake!) it. Just put a red filter on your laptop!

249 FOCUS YOUR CAMERA THROUGH A SCOPE

With your prime focus set up, figuring out the focal point can be a bit tricky. Ideally, you should be able to simply latch-thread your camera with its adapter into your telescope and bring it to focus on your camera's sensor with no problems. Unfortunately, it's not always that easy.

A TOO CLOSE If you find your focus is too far in (as in it seems to be inside your telescope), thread a Barlow lens adapter into your T-adapter. This should help to bring the focus up and into the sensor inside your camera.

B TOO FAR If you find the focus is too far out—seemingly farther away from the telescope than your focuser can reach—you can fix it with a simple extension tube. That's exactly what it sounds like: a tube that screws into the T-adapter and extends your focus length from ½ inch (1.3 cm) and up.

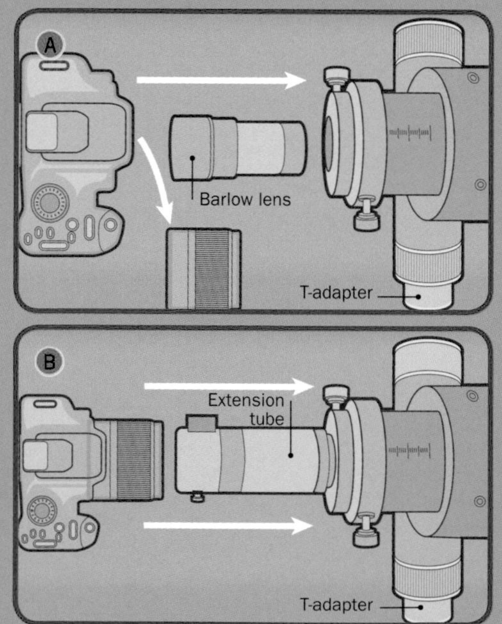

250 SPEED UP FILTER SWAPS

Do you find yourself switching between lunar and planetary observing a lot? Do you like to observe lots of different objects throughout the night—perhaps on a Messier marathon (see #209)? Are you taking advanced astrophotos and processing images of objects taken through several different filters?

A filter wheel or slider can save you a lot of time—and makes swapping filters a breeze. With one click you can switch from a lunar filter to a light-pollution filter—or even no filter at all. Filter wheels also help ensure that the scope stays steady and focused on the object, removing the vibration and hassle of taking off your eyepiece each time you want to swap filters. For hardcore astro-imagers (especially ones using color filters with a black-and-white camera), filter wheels are an essential part of their tool kit, keeping the scope vibration-free and undisturbed during long-exposure sessions.

251 LOCK ON WITH AUTOGUIDERS

If you're getting serious about astrophotography, try taking long-exposure images through your telescope with the help of an *autoguider*. An autoguider's camera and software works like an eye, very finely tracking an object and continually making corrections to the telescope's motors to keep the object in the center of its field of view. Without an autoguider to correct your mount's tracking, stars will smear in your photos after just a few minutes of exposure, even on the best, most well-aligned mounts.

The most common setup is a smaller telescope (a guide scope) mounted next to your main telescope, both calibrated to point at the exact same spot. A guide camera is then attached to the guide scope and linked to both the mount and a computer, which controls the autoguider via software and lets you set your camera functions. If you don't want the added weight or cost of a guide scope, you can incorporate an off-axis adapter to use the same telescope for both your guide camera and main camera. You can even buy a dedicated camera with autoguiding built in.

Some new telescopes come with a built-in autoguider, and many higher-end mounts have a dedicated autoguider port. There are also stand-alone autoguiders if you want to shoot through a telescope without a computer.

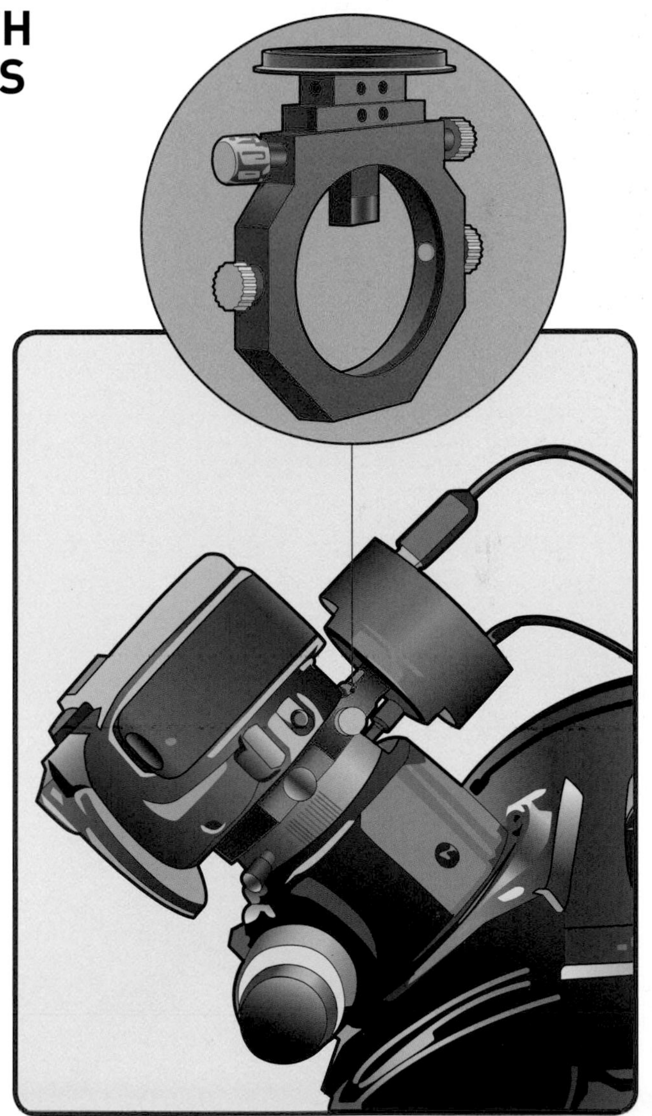

252 CAPTURE THE BEAUTY OF THE AURORAS

Photographing these magnificent, ethereal light displays can be a rewarding challenge. Use these tips to shoot the auroras like a pro.

STEP ONE Know where and when the auroras are active. For best results, plan a trip near one of Earth's poles in the dark season (see #95–97); otherwise, the Sun will outshine any aurora. There are many aurora prediction apps as well as the space weather website of the National Oceanic and Atmospheric Administration.

STEP TWO Once you arrive, get away from city lights and turbulence as much as you can. The best time to catch the auroras is between 10 PM and 3 AM—not exactly the warmest period of the day, especially during colder months. Since you may be in the elements for a long time, grab a heavy coat, warm gloves, sturdy boots with thick soles, and maybe even a couple of chemical hand warmers.

STEP THREE Be prepared. Cold temperatures, long exposures, and large files can sap your batteries and

eat up your camera's memory in a hurry. Keep extra batteries and media cards in your warm pockets.

STEP FOUR Pick your setup spot while it's light out, if possible. While any decent DSLR camera will do, the lower the ISO (64–400), the better. You'll want a good lens with a wide angle and a large aperture—no filters needed for this image. A sturdy tripod is essential for getting clear shots of the auroras. To avoid shaking the camera even more, consider getting a shutter-release mechanism or wireless trigger.

STEP FIVE Turn your camera display to low brightness. Make sure there are no lights nearby save a few distant points of light, which can be useful for focusing and add interest and depth to your images.

STEP SIX Watch for flames of color licking the night sky. Once they get going, rotate your setup so you're trained on the most vivid auroras and arresting composition. Start triggering your shutter (try exposure times from half a minute to 2 minutes). Experiment with your settings, but don't forget to enjoy the show.

253 MEET THE INVENTORS OF SKYWATCHING GADGETRY

While it can still be baffling to consider the number of astronomical discoveries made with the naked eye, we'd be nowhere near where we are today without the gadgets and gizmos that seemingly bring the cosmos nearer to us. Here are just a few of the technological giants upon whose shoulders we gratefully stand.

THE DUTCH (1608)

While no one really knows who first invented the telescope, the three gentlemen associated with its creation all worked in early optics in the Netherlands. Hans Lippershey, a German-born maker of spectacles, attempted to file the first patent for an instrument that magnified faraway objects by holding two lenses apart. Though his patent was denied, he is often credited with the telescope's invention based on this paper trail. Dutch spectacle-maker Zacharias Janssen and lens-grinding specialist Jacob Metius independently developed similar designs around the same period.

SIR ISAAC NEWTON (1668)

Concluding that chromatic aberration could not be prevented using lenses, Newton successfully invented the first reflector telescope. Although his invention was limited by the quality of materials at the time—particularly mirrors—he was successful in both shortening telescope length and paving the way for later reflectors.

JOHANNES KEPLER (1611)

Improving on Galileo's earlier telescope designs, Kepler replaced the concave eyepiece with a convex one, for a wider field of view. While the design created higher levels of magnification, the scope had to be quite long to minimize glitches caused by *chromatic aberration*—a failure to focus all light wavelengths at the same point. But especially long scopes were unusable, since no material was rigid enough to prevent wobbling in the wind.

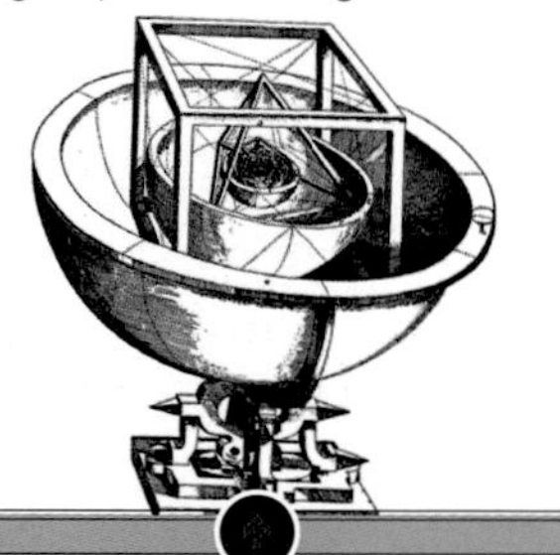

CHRISTIAAN HUYGENS (1684)

To overcome the structural issues that long lengths or refractor telescopes introduced—things like wobbling in the wind, or collapsing entirely under the weight of their lenses—aerial telescopes were born. With the objective lens mounted high on a pole or other structure and attached to an eyepiece with string, a user on the ground would use these "tubeless telescopes" by pulling the string taut, thereby lining up the two optics. While there is some dispute over who really first invented them, Huygens published his designs in 1684 and is often credited. Huygens's other claims to fame? He was the first to propose that Saturn was surrounded by a ring, discovered Saturn's moon Titan, and was able to view the Orion Nebula and see its distinct stars.

GALILEO GALILEE (1609)

Often called the father of observational astronomy, Galileo was the first to point a telescope toward the heavens and record what he observed. His own improvements to the Dutch invention allowed him to observe bodies like Jupiter's moons, Saturn, and Neptune. A proponent of Copernicus's heliocentric model, Galileo was tried by the Catholic Church and placed under house arrest, suspected of heresy.

KARL GUTHE JANSKY (1932)

An engineer with Bell Laboratories, Karl Guthe Janksy became the first person to use a radio antenna to scour the heavens. Charged with the task of determining the sources of potentially problem-causing static, he ultimately identified the first astronomical source of radio waves.

THE HERSCHELS

(1789) Besides a legacy of music, William Herschel made his fair share of contributions to the development of skywatching gadgetry. Of the more than 400 telescopes Herschel is said to have built, his gigantic Newtonian reflector is certainly a highlight. With a primary mirror 49½ inches (1.25 m) in diameter and a 40-foot (12-m) focal length, it was the largest of its time and predecessor to today's giant reflector scopes. William's sister Caroline was also a major force in the astronomy scene: She discovered many comets and nebulae, and published the *British Catalogue of Stars.*

GROTE REBER

(1937) Picking up where Jansky left off, amateur astronomer and radio broadcaster Grote Reber built the first dish radio telescope—which detects radio-frequency radiation from space—in his backyard. While radio telescopes appear all over the globe today, Reber was arguably the only radio astronomer in existence for almost 10 years.

CHESTER MOORE HALL (1729)

By cementing two lenses together, Hall was able to correct the aberration typical of refractor scopes. Called *achromatic lenses*, this invention allowed for the creation of refractor telescopes with greatly reduced chromatic aberration.

SIR BERNARD LOVELL (1957)

An English radio astronomer and physicist, Sir Bernard Lovell spent much of World War II developing aircraft radar systems. Later Lovell built the radio telescope at Jodrell Bank. Measuring 250 feet (75 m) across, it was then the largest steerable dish radio telescope on Earth. In 1987, it was renamed the Lovell Telescope and remains in operation today. Radio telescopes were a breakthrough because they study the longest wavelengths in the electromagnetic spectrum, teaching us things about the universe that we can't detect with our eyes.

254

STOCK UP ON SKY-DRAWING SUPPLIES

Most seasoned astronomical sketchers work on white paper, sketching in gray or black pencil. Of course, what you're looking at is the light parts of the dark sky; if working in negative hurts your brain, there are always white pastels on black paper. Besides a telescope (preferably tracking) and a chair, be sure to pack a box of supplies that includes:

☐ **PENCILS** A variety of pencils from soft (1 or 2) to hard (2H or 3) and a pencil sharpener will keep you drawing for hours.

☐ **PASTELS** While you'll be drawing a lot of bodies in gray scale, pastels are handy for more colorful celestial objects, like Jupiter and comets. You'll need extra black and gray pastels if working on white paper.

☐ **ERASER** A good eraser is important. Many astronomers like to have both the hard type and the soft kneaded type.

☐ **BLENDING STUMP** A stick of rolled paper for smudging can be useful for adding texture to your drawings.

☐ **PAPER** Grab a black- or white-paper sketchbook with a hard backing, or paper and a clipboard.

☐ **SPRAY** If you want to preserve your sketches, using a spray fixative is a simple method.

☐ **LIGHT** Bring a dimmable, red-light headlamp so you can see your work without destroying your night vision.

255 DRAW SPACE OBJECTS LIKE A PRO

Beyond creating a visual record of your observations, sketching astronomical objects ups your skywatching skills and makes you a better observer—and it's how many have documented the beauty of our universe, especially before image-making devices. Here are a few great tips on drawing the night sky.

STEP ONE Pick your paper. Many astronomical artists sketch on white paper, later scanning and inverting the image to create a deep, dark, space-like drawing.

STEP TWO Start with a circle that represents your field of view. While sizes vary, 3 inches (7.5 cm) is a good starting place. Tracing one of your telescope's caps may work well.

STEP THREE Study the object for at least 10 minutes through your telescope before you lay pencil to paper. You will be astounded how much more detail you see after careful observation. To estimate star positions, it helps to imagine the face of a clock on your eyepiece.

STEP FOUR Start by drawing bright stars in the field of view with a dark, soft pencil. Use lighter pencils to fill in faint stars.

STEP FIVE Always record information about the drawing, including date, time, sky conditions, telescope and eyepiece, any filters used, and temperature. If you repeat a sketch, you will be amazed how all of these factors can change your observations.

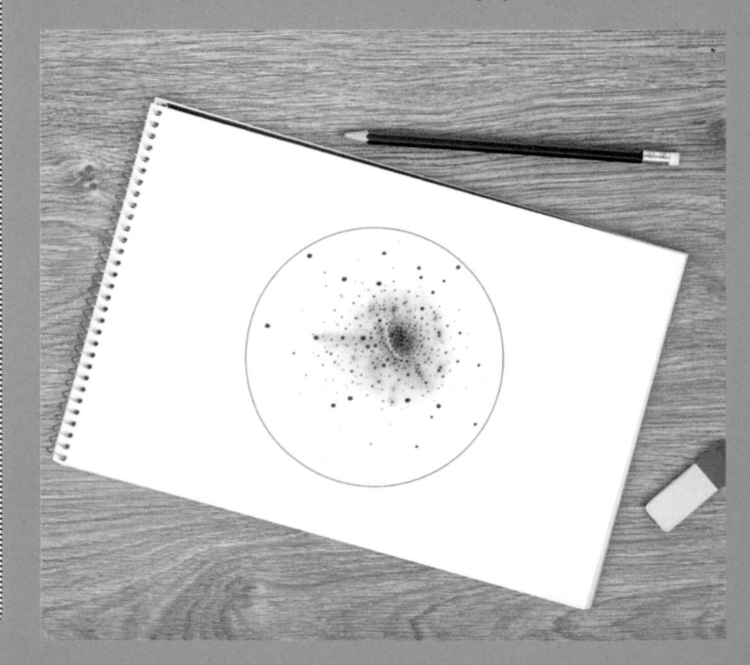

256 TOP FIVE SKETCH OUR UNIVERSE'S HIGHLIGHTS

When choosing celestial muses, the possibilities are as endless as the night sky. Here are a few ideas to get you started.

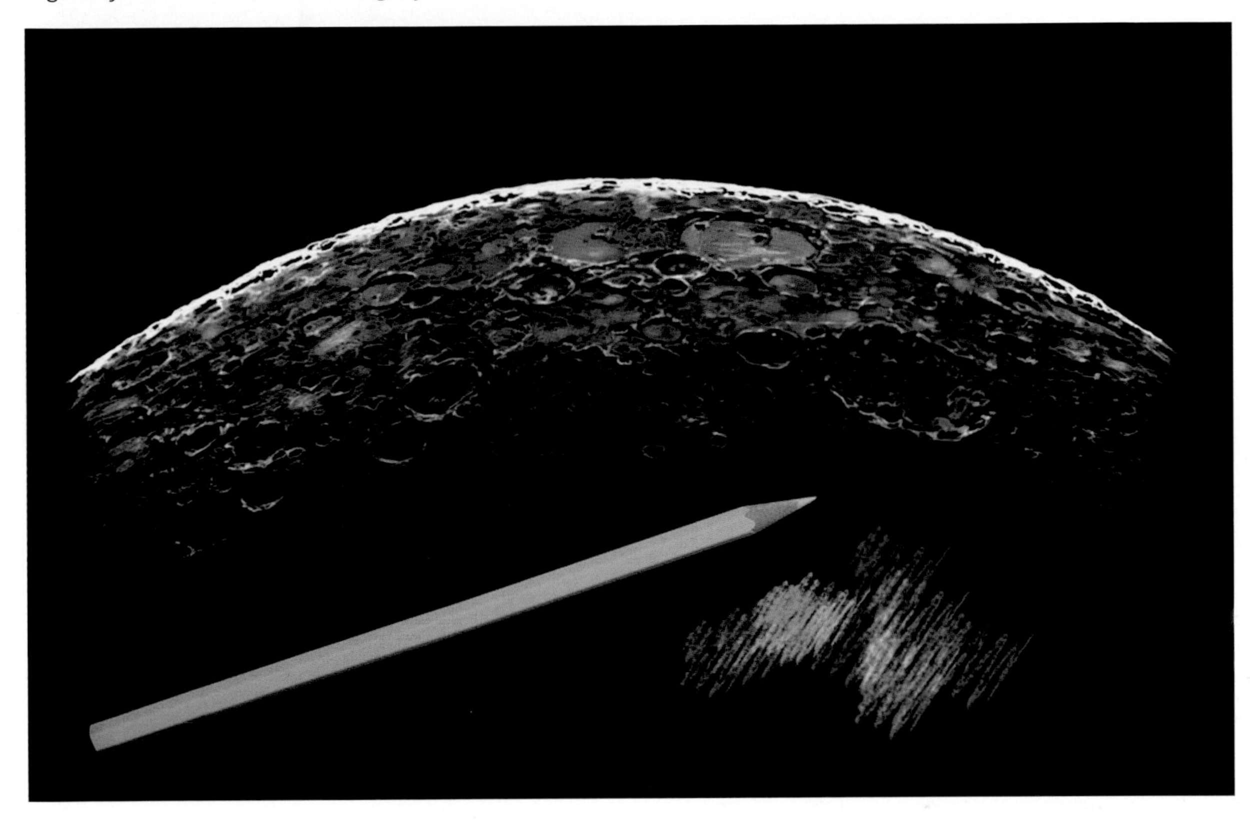

☑ **THE MOON** The Moon is great to practice shading craters, starting with the naked eye and increasing magnification. Observing the craters is best done far from full Moon, when there are shadows creating high contrast. First, sketch a circle and the *terminator* (the division between parts of the Moon that are illuminated and dark), erasing the side you can't see. Outline features and begin shading. Most artists start in the brightest regions, moving to the darkest (or vice versa), so they don't have to keep switching pencils.

☐ **COMETS** First sketch the nearby stars to frame the comet and notice how bright they are compared with the comet. Then draw the comet's coma and use a blending stump to draw out the tail.

☐ **JUPITER** Jupiter as seen through a telescope is a joy to sketch. Try viewing it with a blue or green filter to really bring out the bands and storms. Color pastels can be used with this sketch, as it is one of the few space objects you'll see with color detail. If you're lucky, you may spot the shadow of one or more moons gliding across Jupiter's surface as you sketch (see #197).

☐ **SATURN** Saturn is a great long-term project. Sketching it year after year can give you a great sense of its rings' changing orientation. Getting their size and shape to scale is also a good study. Try making the circle of Saturn with an eyepiece cover, then adding the rings.

☐ **GALAXIES** Large, defined galaxies like the Andromeda Galaxy (M31/NGC 224) or the Sombrero Galaxy (M104/NGC 4594; see #194) are good starts. Making a pool of graphite and "painting" with it using a blending stump will let you shade a little at a time. You can remove a *dust lane*—a band of dense interstellar dust—with a sharp eraser's edge. Since blending makes the shaded area darker, start out light and gradually darken.

257 SEARCH FOR SERPENS

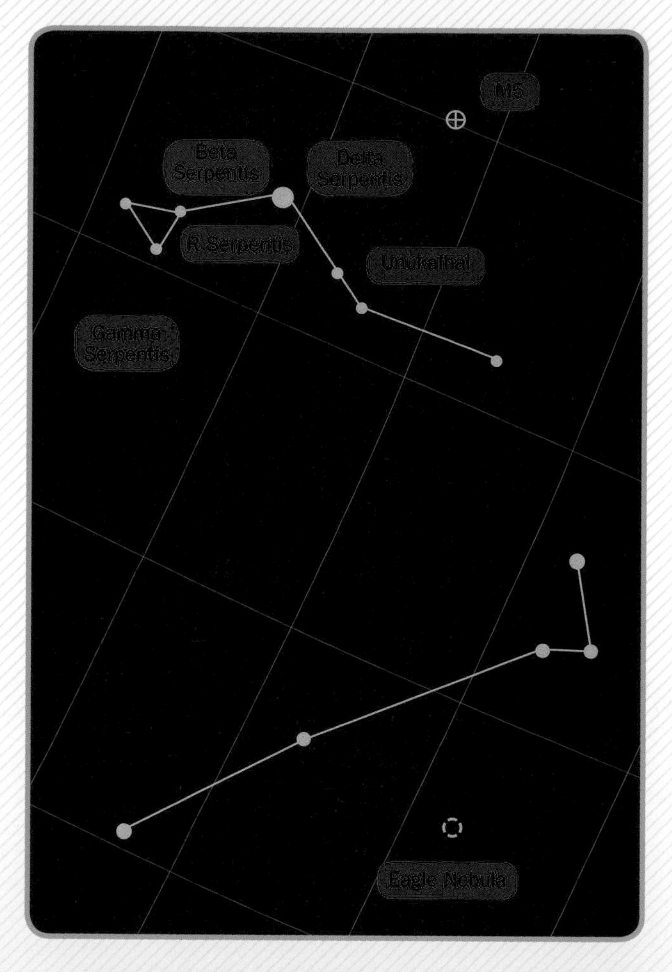

This is the only constellation that is divided into two parts. The head (Serpens Caput) and the tail (Serpens Cauda) are separated by the constellation Ophiuchus the Serpent Bearer. Other neighboring constellations include Boötes, Libra, Virgo, and Corona Borealis. At one time, both Serpens and Ophiuchus formed a single constellation. Serpens was familiar in ancient times to the Hebrews, Arabs, Greeks, and Romans.

Of the more notable sights within Serpens is R Serpentis, a variable star almost midway between Beta and Gamma Serpentis. It has a bright maximum magnitude of 6.9. It fades to about 13.4, although it can sometimes become even fainter. Its period is about one year. Viewable above Serpens Caput, M5 (NGC 5904) is a striking globular cluster about 26,000 light years away. About 13 billion years old, it's one of the older globular clusters in the Milky Way. Down in Serpent Cauda, the Eagle Nebula (M16/NGC 6611) is another worthy sight. Through an 8-inch (20-cm) telescope on a dark night, this combination of nebula and star cluster is quite stunning.

258 RECOVER THE SEXTANT

Sextans Uraniae, now known simply as Sextans, was the creation of Johannes Hevelius. He chose this name for the constellation to commemorate the loss of the sextant—a navigational tool used to measure the angle between objects—he once used to measure the positions of the stars. Along with all his other astronomical instruments, the sextant was destroyed in a fire that took place in September 1679. "Vulcan overcame Urania," Hevelius remarked sadly, commenting on the fire god having defeated astronomy's muse.

Placed between Leo and Hydra, Sextans's brightest star, Alpha Sextants (with a 4.5 magnitude), is best seen with binoculars or a telescope. Despite being barely visible to naked eyes, the ancient Chinese chose faint star Bode's 2306 to represent Tien Seang, the minister of state in heaven. Because we see Sextans's Spindle Galaxy (M102/NGC 3115) almost edge-on, it appears to be shaped like a lens. Unlike many faint galaxies, the Spindle Galaxy is quite satisfying to view with a high-powered instrument. It is located down below the sextant's rotational arm.

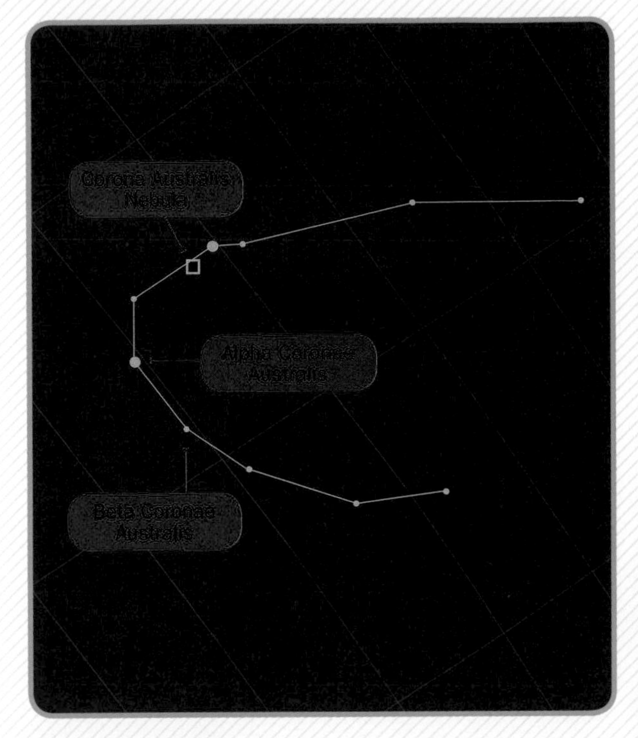

259 SNAG THE SOUTHERN CROWN

One of the 48 original constellations cataloged by Ptolemy in the second century CE, this small semicircular group of faint stars is inconspicuous, especially from the northern hemisphere. Find it bordered by Scorpio, Sagittarius, and Telescopium.

Corona Australis is said to represent a crown of laurel or olive leaves. One story has it that the crown belongs to the centaur Chiron. Another story comes from Ovid's *Metamorphoses*: Juno discovered that her husband, Jupiter, was the lover of Semele, a human. Masquerading as Semele's maid, Juno suggested that Semele ask Jupiter to appear before her in all his glory. Jupiter was appalled at her request but did not refuse it. When she saw him in his splendor she was consumed by fire. Her unborn child was saved, however, to become Bacchus, the god of wine, who honored his mother by placing the crown in the sky.

The Southern Crown's most famous deep-sky object is the Corona Australis Nebula (made up of NGC 6726, NGC 6727, and NGC 6729)—a number of bright, young stars and dust clouds.

260 TRACE THE TRIANGLE

Triangulum is a small, faint constellation extending just south of Andromeda, near Beta and Gamma Andromedae. The ancients knew it and, because of its similarity to the Greek letter delta, it was sometimes called Delta or Deltorum. It has been associated with the delta of the Nile River as well as the island of Sicily, which is shaped like a triangle. The ancient Hebrews gave it the name of a triangular musical instrument.

Located between Triangulum and Pisces, the Triangulum Galaxy (M33/NGC 598) is one of the brightest and biggest members of our Local Group. We have a front row view because it appears face-on. Although it shines at magnitude 5.5, its light is spread out across such a large area that it is notoriously difficult to see. Visible to the naked eye on especially clear nights, the Triangulum Galaxy requires a dark sky and binoculars to see a fuzzy glow larger than the Moon's apparent diameter. You can also use a telescope with a wide field of view to observe this galaxy.

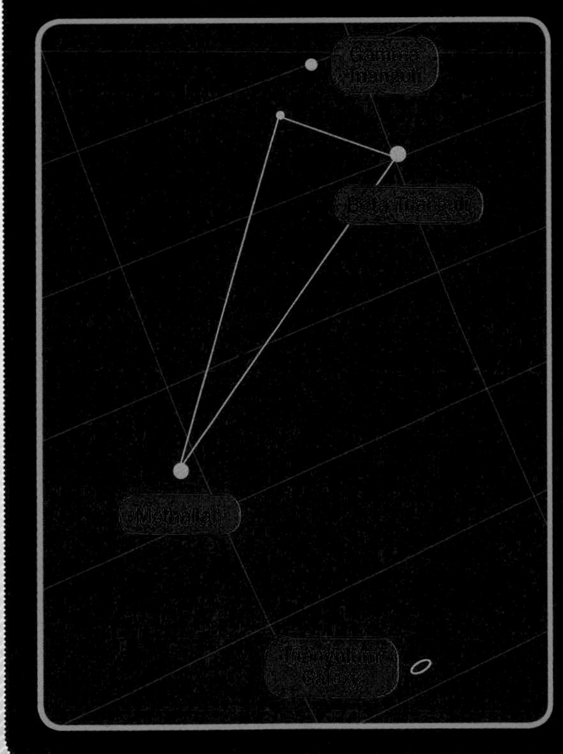

ADVANCED TECHNIQUES

261 TRAVEL TO THE WORLD'S GREATEST OBSERVATORIES

Many great observatories used by astronomers today orbit high above Earth's surface, investigating celestial objects in wavelengths that can't penetrate our atmosphere and reach ground-based telescopes. Still others explore secrets of the early universe from far, far underground, where water and rock block everything but high-energy subatomic particles. But on Earth's surface, astronomers continue to comb the vast expanses above with ever-increasing precision. Here is a list of just a few of the many ground-based eyes on the skies.

ROQUE DE LOS MUCHACHOS OBSERVATORY, CANARY ISLANDS, SPAIN

Situated nearly 8,000 feet (2,400 m) above sea level on the island of La Palma nestled in the Canary Islands, Roque de los Muchachos Observatory is home to many telescopes, including the largest optical reflecting telescope, the Gran Telescopio Canarias, with an aperture size of 34 feet (10 m). The island's landscape enjoys clear night skies about 90 percent of the time during the summer, and its remote location, lack of urban development, and night-protecting laws make it ideal for nighttime observations.

ATACAMA LARGE MILLIMETER/ SUBMILLIMETER ARRAY (ALMA), CHILE

The Atacama Large Millimeter/Submillimeter Array (ALMA) is the largest astronomical project on Earth. Sixty-six precision antennae combine powers to make up one gigantic telescope, situated more than 16,000 feet (5,000 m) above sea level, on the Chajnantor plateau in Chile. From its clear vantage point in the incredibly dry Atacama desert, the ALMA has discovered long-elusive *starburst galaxies*: young galaxies that formed 1 billion years after the Big Bang and where new stars formed at a frenzied pace. Inside these stellar nurseries, ALMA also managed to detect water molecules!

SOUTH AFRICAN ASTRONOMICAL OBSERVATORY, CAPETOWN, SOUTH AFRICA

The South African Astronomical Observatory is home to a number of astronomical instruments, including the aptly named South African Large Telescope (SALT), the single biggest optical telescope in the southern hemisphere. SALT has allowed a number of astronomers to further probe the evolution and makeup of the Milky Way, as well as view objects outside the viewing capacity of telescopes in the northern hemisphere.

BEIJING ANCIENT OBSERVATORY, BEIJING, CHINA

Built in 1442 during the Ming Dynasty, the Beijing Ancient Observatory is one of the oldest in existence, boasting more than 500 years of continuous astronomical observation. Originally called the "Platform for Star Watching," the observatory is home to eight ornately carved bronze instruments, a handful of which visitors can still use to sneak a peak at the heavens. While it's not the most technologically advanced of existing observatories, the Beijing Ancient Observatory is a monument to cross-cultural exchange between East and West.

MAUNA KEA OBSERVATORY (MKO), HAWAII, UNITED STATES

The 13 working telescopes on the summit of Mauna Kea—a dormant volcano on the island of Hawaii—make up the largest astronomical observatory in the world. Besides doing their part to demote Pluto from its former planetary status, the telescopes of Mauna Kea (and the astronomers from 11 countries who operate them) have been instrumental in determining the center of the Milky Way Galaxy, charting universal expansion rates, and exploring exoplanet systems.

AUSTRALIA TELESCOPE NETWORK FACILITY (ATNF), AUSTRALIA

From discovering the first pulsar outside our galaxy to helping transmit television signals from the *Apollo 11* Moon landing, the Parkes Telescope has a storied history and is just part of the Australia Telescope Network Facility in Western Australia. Besides functioning individually, the radio telescopes of ATNF are also used together as the Australian Long Baseline Array.

SOUTH POLE TELESCOPE (SPT), ANTARCTICA

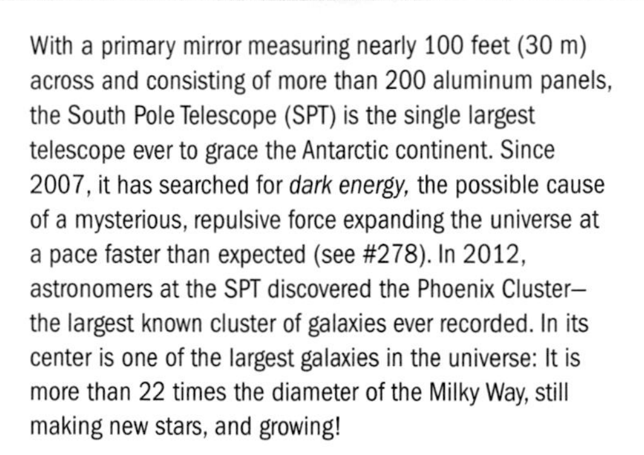

With a primary mirror measuring nearly 100 feet (30 m) across and consisting of more than 200 aluminum panels, the South Pole Telescope (SPT) is the single largest telescope ever to grace the Antarctic continent. Since 2007, it has searched for *dark energy,* the possible cause of a mysterious, repulsive force expanding the universe at a pace faster than expected (see #278). In 2012, astronomers at the SPT discovered the Phoenix Cluster—the largest known cluster of galaxies ever recorded. In its center is one of the largest galaxies in the universe: It is more than 22 times the diameter of the Milky Way, still making new stars, and growing!

262 PEEK AT DEEP SPACE VIA THE HUBBLE SPACE TELESCOPE

Almost everything we know about the universe comes from studying the light that reaches us—not just the visible forms but also parts of the spectrum that your eyes don't detect. Since Earth's atmosphere blocks out gamma rays, X-rays, infrared radiation, long wavelength radio waves, and much ultraviolet light, observing from Earth literally keeps astronomers "in the dark" when it comes to parts of the spectrum.

With access to those unseen wavelengths, and little turbulence or light pollution to speak of, telescopes above our 62-mile- (100-km-) thick atmosphere pull in dramatically sharper views than those on the ground. One exceptional tool is the Hubble Space Telescope. Launched in 1990 and still orbiting more than 200 miles (320 km) above Earth's atmosphere, Hubble studies the universe in visible wavelengths. Unhindered by atmospheric distortion, it has collected hundreds of thousands of images with astonishing clarity, giving astronomers their first highly detailed view of cosmic events that were once just the subject of speculation: the birth of new stars, the formation of planets, the creation and evolution of galaxies, and a peek at what the universe was like 13 billion years ago.

Hubble completes one orbit around Earth in 97 minutes. Moving at a speed of 5 miles per second (8 km/s), its primary mirror collects light, reflects it to a smaller secondary mirror, then sends it through the main mirror to several cameras and instruments. Sadly, the telescope is nearing the end of its usable life. After nearly 25 years of service, the time is coming to retire it. It will slowly lose altitude and reenter Earth's atmosphere with a bright streak across the sky. Luckily, the James Webb Space Telescope will take its place (see #265).

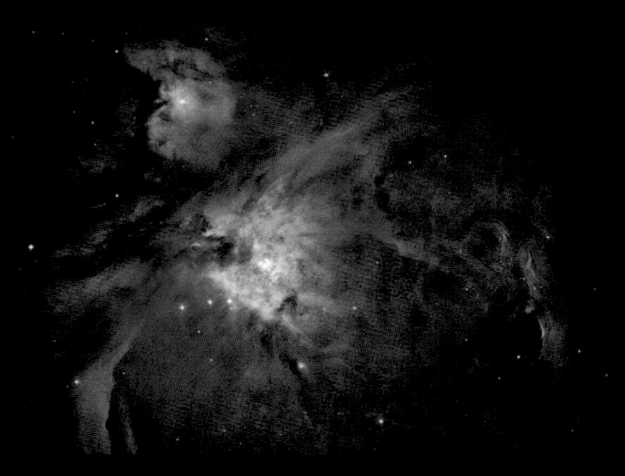

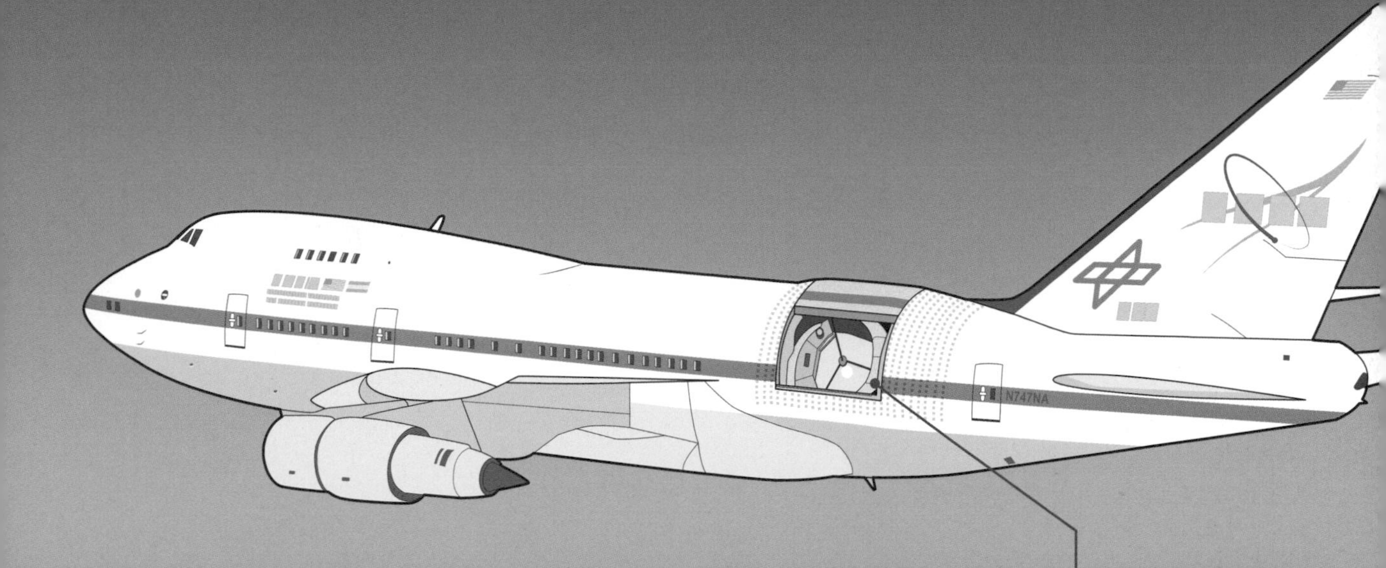

263 SEE STAR BIRTH IN INFRARED WITH SOFIA

Somewhere between our handful of space-based telescopes and myriad land-based observatories flies NASA and the German Aerospace Center's Stratospheric Observatory for Infrared Astronomy (SOFIA). Small compared to ground-based telescopes, this 17-ton (15,000-kg), 8-foot (2.5-m) telescope lives on a modified Boeing 747SP airplane and is primarily intended to observe the infrared universe.

Flying at altitudes somewhere between 37,000 and 45,000 feet (11,275–13,700 m), the SOFIA crew uses its 10-hour, overnight flights to take advantage of being above a lot of the turbulence ground-based scopes have to cut through. Of SOFIA's many targets of exploration, some of the most notable include star birth and death; the formation of new solar systems; the study of planets, comets, and asteroids in our own solar system; and black holes at the center of distant galaxies. The image on the far right below shows SOFIA's bright, colorful view of the infrared emissions coming off Jupiter, compared to the visible-light image of Jupiter on the left.

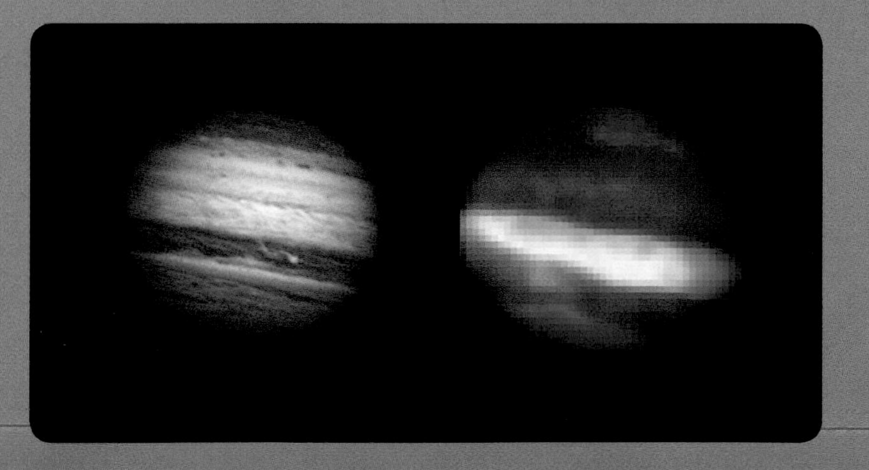

264 PLANET-HUNT WITH KEPLER

Kepler was launched in 2009 with a profound purpose: to examine our region of the Milky Way Galaxy in search of planets capable of supporting life. To do so, Kepler used a transit method, relying on the fact that, when it observed a planet passing in front of a star, a tiny fraction of starlight would be blocked. When transits repeated at regular times, Kepler sometimes discovered a planet—such as Kepler-20e (shown in this artist's conception). Of the more than 1,000 Earth-size planets that Kepler has found, eight orbit at a distance from their sun that could possibly support life. This number will only grow as scientists study Kepler's data.

Kepler is a 3¼-foot- (1-m-) long reflecting telescope with a wide field of view and an extremely sensitive light meter (also known as a photometer), designed to precisely detect changes in brightness. Kepler's wide field of view allows scientists to observe more than 150,000 stars simultaneously.

265 OBSERVE IN INFRARED VIA THE JAMES WEBB SPACE TELESCOPE

The James Webb Space Telescope launches in 2018. The successor to the Hubble Space Telescope, it will "see" infrared light as clearly as Hubble sees visible light. The James Webb's 18 gold-coated hexagonal mirrors, working together in a 21-foot (6.5-m) arrangement, will collect and focus the exceedingly faint infrared light coming from the edge of the observable universe, shortly after the Big Bang, sending us images of the birth of the first stars and galaxies. Inside our galaxy, the Webb will use infrared to see through dust and search for new details about the birth of stars, planet formation, and conditions that lead to life.

While the Hubble is just a few hundred miles from Earth, the James Webb will be almost 930,000 miles (1.5 million km) away—almost three times farther than the Moon. At this spot, the combined gravitational pulls of Earth, the Moon, and the Sun will keep the telescope's solar shields in a position to constantly block sunlight, avoiding the creation of excess heat that would disturb the infrared cameras on the telescope. Keeping cool is important for ensuring that the infrared radiation detected is coming from the edges of the universe—not from inside the telescope itself.

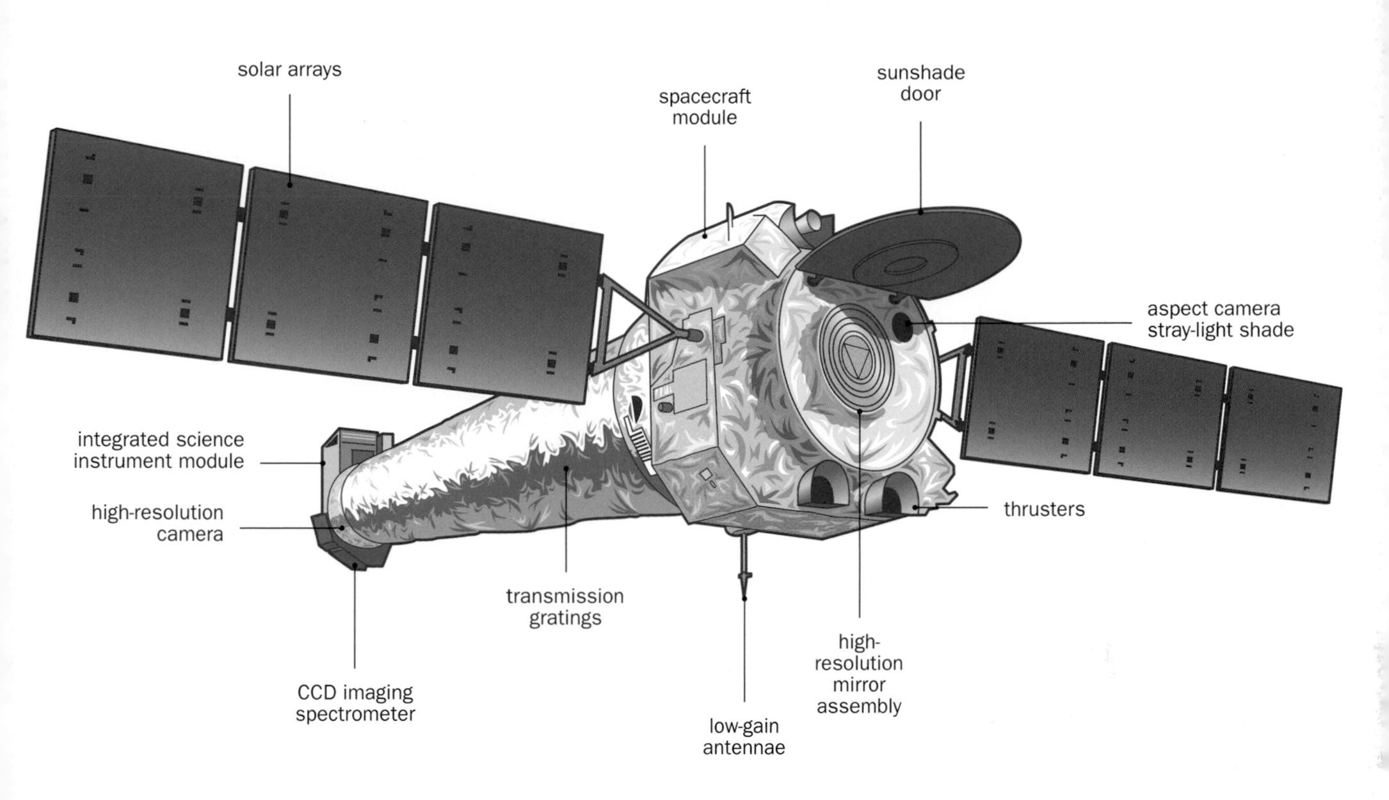

266 LOOK AT X-RAYS OF SPACE WITH CHANDRA

In the dentist's office, X-rays are made inside a vacuum tube by smashing electrons into metal at high speeds. When the electrons slow down, collide, or stop moving, energy is released as X-rays. Cosmic sources of X-rays include supernova explosions, gases orbiting around black holes, stars' hot coronas, and superheated gases surrounding galaxy clusters. Because X-rays don't penetrate our atmosphere, X-ray telescopes need to be sent into space for the best views—like this image of the Cat's Eye Nebula in the constellation Draco.

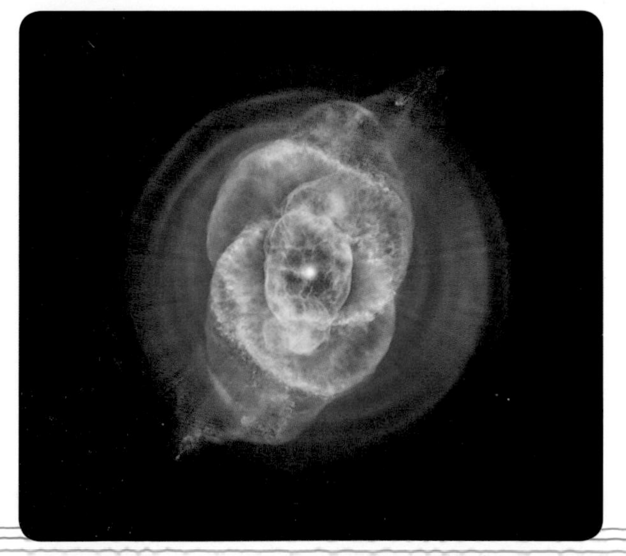

The Chandra X-ray Observatory, launched in 1999, orbits our planet at an altitude of 86,500 miles (139,200 km). Since X-rays hitting a mirror head-on will pass through it the same way they pass through skin, X-ray telescopes are designed so that X-rays graze the mirror's surface and change direction gradually, eventually combining at a focal point.

267 TOUCH THE HAND OF GOD NEBULA

While looking at nebulae through a telescope is amazing (see #190 or #200), looking at satellite imagery is the best way to get superclose views of these clouds of cosmic gas and dust—some of which trick our brains into seeing familiar forms among otherwise random astronomical phenomena! One of the most powerful is the newly imaged Hand of God. While NASA's Chandra had previously captured parts of the nebula (in green and red), NASA's Nuclear Spectroscopic Telescope Array (NuSTAR) imaged higher-energy X-rays, filling out the blue portions of the hand.

But what is the hand? It's a *pulsar wind nebula*: dense material left over from a supernova coupled with the PSR B1509-58 pulsar—a remnant star that spins rapidly, throwing off a wind of charged particles traveling at nearly the speed of light. As the energetic particles interact with magnetic fields surrounding the supernova, they light up the hand with X-ray luminescence.

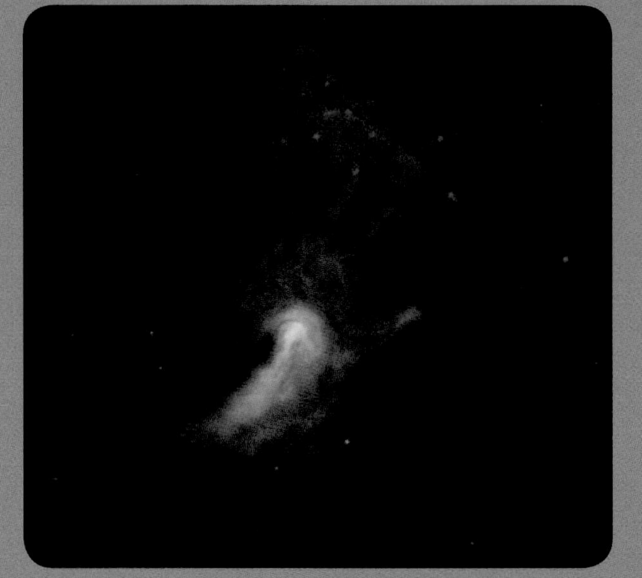

268 STOP AND SMELL THE NEBULA ROSES

NASA's Wide-field Infrared Survey Explorer (WISE) is an infrared space telescope set up with the goal of mapping the sky in great detail. Beyond detecting a number of objects outside light's visible spectrum, the WISE team has released a stunning image of the Rosette Nebula, a composite of four computer-colored images derived from four infrared wavelengths. A cluster of young stars lies in the nebula's heart, while gas, dust, and star-forming regions are represented by the turquoise-green of the "rose leaves." Described by some as a "champagne flow" of outward radiation pressure, we can thank the huge O and B stars (see #11) inside the cluster for providing the emissions that sculpt this heavenly rose.

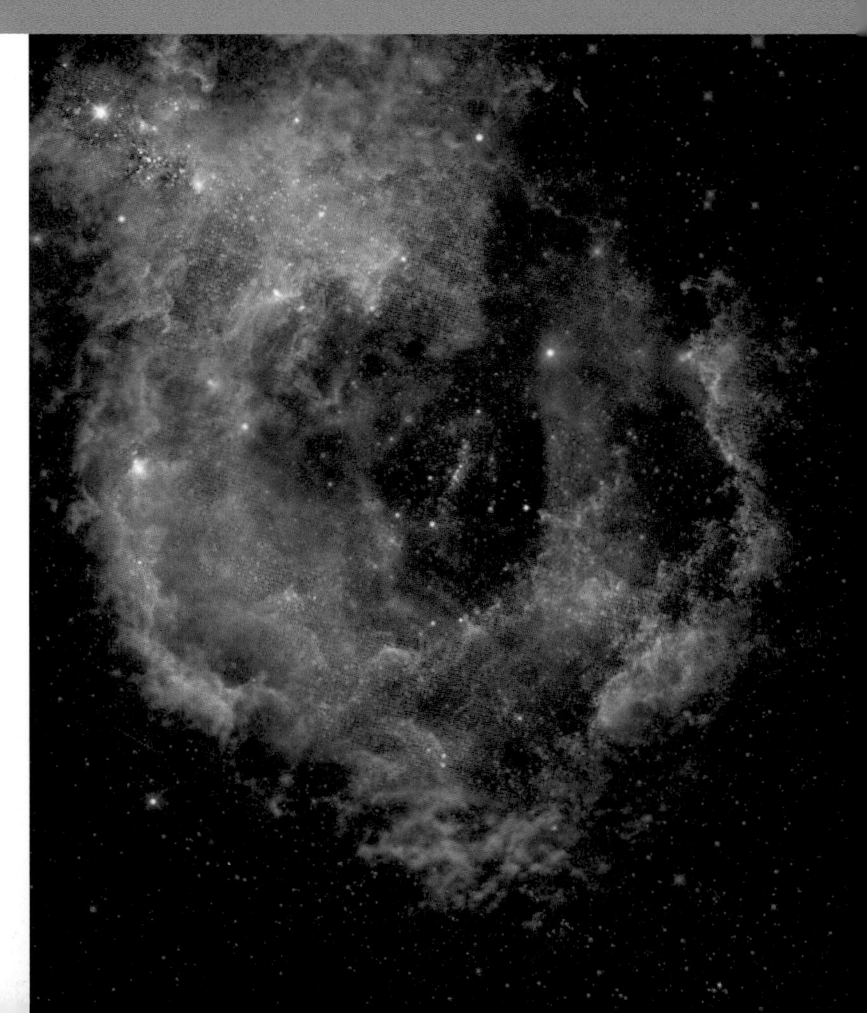

269 PEEP PILLARS OF CREATION

One of the most iconic photos from the Hubble Space Telescope is of the Pillars of Creation in the Eagle Nebula (M16). In 1995, the Hubble team compiled 32 individual images to capture the full magnificence of these large, column-like clouds of interstellar gas and dust. In 2014, Hubble revisited the region and investigated these columns in higher resolution. Astronomers call these spectacular columns the Pillars of Creation because M16 is an active star-forming region. Radiation emitted from the largest young stars in the columns has been eroding the gases away; in fact, evidence suggests the Pillars may have been completely worn away 6,000 years ago. But because the Pillars are 7,000 light years away, and due to the travel time of light, we can still see them—like beautiful ghosts of the heavens. Thankfully for us, the Hubble team captured this amazing image while they still had the chance!

270 SIGHTSEE FROM SPACE THROUGH SATELLITES

There are innumerable awe-inspiring sights to see in the skies above us. But what about back here, at home, on Earth? If you're one of the few lucky enough to have gazed on our planet of origin from afar, you'd have to admit that it's a pretty good-looking rock. And, for when you inevitably finish your astronaut training, here are a few sights to keep an eye on—or look at via satellite imagery here on Earth.

Ⓐ EARTH AND MOON We make a pretty cute pair, bright Earth and tiny Moon. There are only a few images of us together, taken from anywhere but Earth's surface. Passing by the Moon on the way to Jupiter, the *Galileo* spacecraft captured one shot of this match made in heaven, while *Voyager 1* snapped another in 1977.

Ⓑ ALGAE BLOOMS As warmer weather brings deep ocean nutrients to the surface, algae blooms in the sea create beautiful patterns. Earth-orbiting satellites let us glimpse these blooms in their full majesty. Since they're very sensitive to changes, monitoring these blooms can tell scientists about the ocean environment in real time. Algae blooms are a critical part of the food chain, creating meals for the smallest organisms. And these microscopic creatures are also vital to our atmosphere, sucking up carbon dioxide and releasing oxygen.

Ⓒ STORMS While hurricanes and typhoons wreak havoc on Earth, for those lucky few orbiting above in the International Space Station, they appear as majestic swirling disks with incredibly detailed structures; no two are alike. Because these storms form over water—where there are few weather stations—it is hard to take real-time measurements. Observations from space have made huge differences in predicting where these storms will hit land.

Ⓓ VOLCANOES You really don't want to witness a big volcano erupting, unless it's from about 250 miles (400 km) directly above it. In that case, during an eruption, you'll be treated to an explosive sight. Huge plumes of gas and dust collide with clouds to create amazing formations. In contrast to the destruction that these erupting mountains can have on Earth, from

space they take on a fluffy, almost comet-like look as wind blows ash away from its origin. In 2013, an orbiting observatory captured a new volcano erupting from the sea in Japan's section of the Pacific Ring of Fire, forming an island that merged with an existing island. Far too dangerous to set foot on, cameras in space give incredible views of the newest earth on Earth!

E FIRES Satellites can also be life-saving, as in the case of monitoring forest fires and other natural disasters in real time. Helping firefighters from far above the smoky atmosphere, heat-detecting sensors on satellites can pinpoint exactly where the fire is spreading by seeing through the smoke.

F INTERNATIONAL BORDERS Astronauts like to say how much more peaceful and whole our planet looks from above. There are no boundaries as there are on political maps. Still, there are definitely some international borders that can be seen even from space. Perhaps the most noteworthy is the India/Pakistan border, which is lit up from end to end by bright orange floodlights, marking one of the most dangerous places in the world. Another stark contrast is between North Korea and South Korea: The absence of almost all light outside of North Korean capitol Pyongyang is remarkable, while the South has light throughout. Images from space also show the differences in wealth and stewardship of the environment, such as between Haiti and the Dominican Republic (pictured). Poverty and deforestation are pressing problems for the island; these disparities can be seen all the way from space. So, while we spin away on this tiny blue marble, evidence of our differences can be seen, even from space.

G LAND ART Over time, cultures have constructed large-scale structures and land masses that are only discernible from way, way above. From Peru's cryptic, 2,000-year-old Nazca lines (pictured here) to a Coca-Cola advertisement etched into the Colombian desert, humans have been making art for the benefit of satellites or unknown life for a while. Others to check out: the prehistoric Uffington White Horse in the United Kingdom and a giant pink rabbit in the Italian Alps.

271 TRACK JUNK IN ORBIT

Humans produce more than 3.5 million tons (3.2 million kg) of garbage each day. While some is recycled or composted, most of the trash we create accumulates in landfills. But not all of our trash ends up buried—or on the planet at all. We're gathering a lot of garbage in outer space. More than 5,500 tons (5 million kg) of garbage orbits Earth, forming a giant "trash cloud" around the planet. Each piece of junk moves at speeds up to 17,500 miles per hour (28,200 km/h), or 50 times faster than the speed of sound. At these speeds, each piece of garbage has the potential to do enormous damage should it collide with, say, the International Space Station. In fact, the ISS is moved about once a year to avoid collisions with trash. While NASA currently tracks more than 15,000 pieces of trash larger than 4 inches (10 cm) across, scientists estimate that there are many tens of millions of smaller pieces. Although most of the garbage eventually burns up in the atmosphere, larger objects can survive re-entry and make it to the ground, with approximately one piece of space junk falling back to Earth every day.

272 DISSECT SPACE TRASH

While some space garbage is large—about the size of a truck—other pieces are no bigger than a grain of salt. But what is space trash made of? Here are some examples of what's inside Earth's orbiting trash cloud.

DEAD SATELLITES Satellites are very important in our daily lives, providing us with cell-phone service, global positioning information, television and radio broadcasts, and weather data. But since the 1950s, more than 2,500 satellites have been launched into orbit, and today only about 800 are working and active. The oldest piece of orbital junk is the 3-pound (1.4-kg), grapefruit-size 1958 *Vanguard 1* satellite that stopped working in 1964. It has orbited Earth well more than 190,000 times.

ROCKET BOOSTERS Multistaged rockets are used to send spacecraft into orbit. While the first few stages help lift the spacecraft from the ground, the final rocket stage ignites high above Earth's surface and gives the spacecraft the speed it needs to successfully orbit. Once the final rocket has used up its fuel, it and the fuel tank are discarded into space.

A GLOVE Ed White, the very first astronaut to walk in space, let go of a glove during the 1965 *Gemini 4* flight. The glove orbited Earth for a month before ultimately burning up in the atmosphere.

A SPATULA During a space shuttle mission in 2006, astronaut Piers Sellers accidently released the spatula he had been using to spread gel-like repair material on heat shields.

A TOOL BAG Astronaut Heidemarie Stefanyshyn-Piper dropped a 30-pound (14-kg) tool bag while doing a space walk in 2008. Since then, amateur astronomers have observed the bag in orbit. In North America, you can check to see whether the tool bag will be in your telescope's field of view using spaceweather.com's satellite tracker.

TOOTHBRUSH On the 14-day *Gemini 7* flight, Jim Lovell lost his toothbrush. Luckily, crewmate Frank Borman was willing to share his.

273 CELEBRATE OUR MISSIONS AMONG THE STARS

While the history of human space exploration may be but a blip in cosmic time, so far it's been a pretty storied blip. Here are just a few of the high points of our decades of seeking among the stars.

FIRST MAN-MADE OBJECT IN SPACE

(1944) Wernher von Braun and his team in Peenemünde in Nazi Germany launched the first V2 rocket, reaching the boundary of space before crashing down in war-torn London and exploding. After being captured by U.S. and Russian forces, the former Peenemünde team members helped start both countries' space programs while under house arrest or in prison. On October 26, 1946, a V2 took the first pictures of Earth from space, 65 miles (105 km) above the New Mexico desert.

FIRST HUMAN IN SPACE (1961)

On April 12, 1961, cosmonaut Yuri Gagarin became the first person in space, launched into orbit around Earth in the *Vostok 1* space capsule. Once an instant international celebrity and hero of the Soviet people, Gagarin's life was tragically cut short in a plane crash on March 27, 1968. Every year on April 12, an international event called "Yuri's Night" is dedicated to celebrating humanity's achievements in space exploration.

FIRST SATELLITE (1957)

The then-Soviet Union shocked the world—and triggered the space race with the United States—when it launched the first artificial satellite, *Sputnik,* on October 4, 1957. With a body only 23 inches (58 cm) in diameter, it was visible from the ground and orbited Earth for three months before burning up upon reentry.

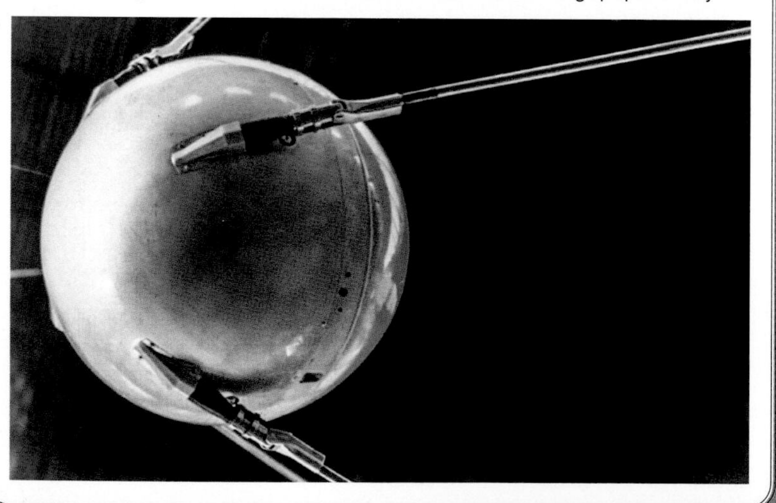

ROBOTS TO THE INNER PLANETS

(1959) Mars, Venus, and our Moon were the first targets of space probes from the United States and U.S.S.R. The Soviet *Luna 3* sent back the first images of the far side of the Moon in 1959. India, Japan, Europe, and China have also sent robots to our nearest neighbor.

HUBBLE TELESCOPE (1990)

Launched by the *Discovery* space shuttle in 1990, the Hubble Space Telescope (HST) has taken some of the most famous images of space and demonstrated the ability of humans to work in space on real engineering challenges. A number of space shuttle missions have allowed astronauts to fix and upgrade the HST, keeping it in service for more than 25 years.

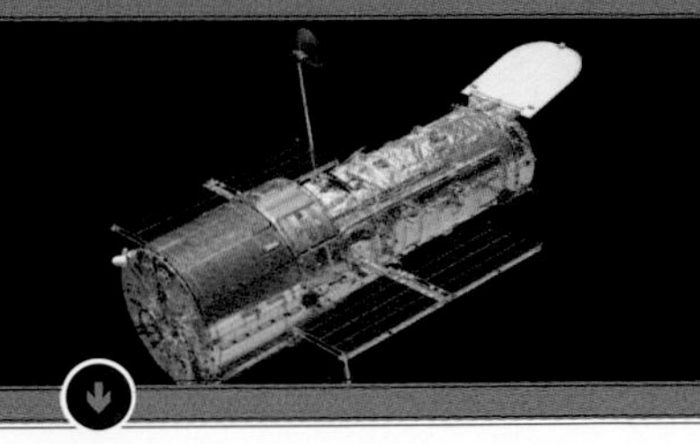

ROBOTS TO THE OUTER PLANETS

(1972) *Pioneer 10* and *11* were the first spacecraft to fly by Jupiter and Saturn. The legendary Voyager probes followed up, flying past Jupiter, Saturn, Uranus, and Neptune. The probes were able to visit all four worlds by utilizing a very rare alignment of the four planets and their gravity to slingshot from one planet to the next.

FIRST HUMAN ON THE MOON (1969)

Neil Armstrong became the first human to step foot on another space object on July 20, 1969. The Apollo program started with a bold speech by U.S. President John F. Kennedy promising that the United States would send a man to the surface of the Moon and return him to Earth by the end of the 1960s. After the tragic *Apollo 1* fire killed three astronauts, NASA created new technologies to land safely on the Moon.

Monday July 21 1969
1969: Man makes his first sp
On the moon after perfect touchdown

INTERNATIONAL SPACE STATION (2000)

The International Space Station is the largest structure in space (larger than a soccer field) and has been continuously inhabited by an international team of astronauts since November 2000. The United States, Russia, the European Space Agency (ESA), Japan, and Canada share responsibility for operation and maintenance, and they supply the station and its crews. Men and women from more than 15 countries have been crew on this orbiting science lab, performing thousands of research projects and experiments.

FIRST HUMAN OBJECT TO LEAVE SOLAR SYSTEM (2012)

In 2012, *Voyager 1* left the *heliosphere*—the Sun-dominated region of space. Now more than 12 billion miles (more than 19 billion km) from Earth, *Voyager 1* is subject to radiation and particles from the interstellar medium, rather than our Sun. Carl Sagan and Ann Druyan placed a "Golden Record" on the spacecraft as a hello from the human race to any potential alien discoverers. After 40 years, *Voyager 1* continues to transmit data back to Earth—though it now takes over a day and half for its signal to reach us!

EXPLORING FURTHER

(2014–2015) Robotic explorers have only just begun visiting dwarf planets, asteroids, and comets. Recently, the European Space Agency's Rosetta mission visited the comet Churyumov-Gerasimenko and sent its *Philae* probe to the surface. Japan's HAYABUSA mission to Asteroid 25143 Itokawa returned samples of its surface. NASA's Dawn mission has orbited Vesta (shown here) and Ceres—the largest asteroids—revealing close-ups of their surfaces. *New Horizons* flew by Pluto and Charon in 2015, imaging the most distant objects yet in our solar system.

274 JOURNEY TO THE SOLAR SYSTEM'S OUTER EDGES

What lurks at the fringes of our solar system, past the icy orb of Neptune? Pluto, shown here in an artist's conception with its moons Charon and Styx. This once-upon-a-time planet is the largest object to date in the Kuiper Belt, a group of objects between 30 and 50 au from the Sun. In addition to Pluto—which was declassified from a planet to a dwarf planet in 2006 due to its small size and irregular orbit—the Kuiper Belt is also home to Eris, the second-largest Kuiper Belt object and the farthest-known object in the solar system; Makemake, the third-largest object in the belt; and Sedna, the reddest object in our solar system after Mars.

Even beyond the Kuiper Belt, however, is the Oort Cloud, an enormous and extremely distant spherical shell of icy planetesimals surrounding the solar system. At a distance between 5,000 and 100,000 au from the Sun, this vast swarm of icy objects is thought to be remnants from the formation of the solar system about 4.6 billion years ago. It marks the outermost boundary of the solar system and is the region where the Sun's gravitational influence is weaker compared to nearby stars. Since the Oort Cloud is so vast, it has not been directly observed, but scientists conjecture that it is the source of all long-period comets that we see streak across the sky.

275 CHECK OUT EXOPLANETS

An *exoplanet* is a planet orbiting a star that isn't our Sun, somewhere outside of our solar system. Scientists estimate that almost every star has at least one exoplanet. Some of these exoplanets are similar in size to Earth and may make excellent homes for our species in the future—making them *super-Earths*. Here are exoplanet systems to watch.

TAU CETI Tau Ceti is a calm star, smaller than our Sun and likely much older. It appears there may be five planets around it. One or two of could be super-Earth-size, and one or even both may be in the star's *habitable zone* (the region surrounding a star with enough pressure to keep liquid water on a planet's surface). At only 12 light years away, the system is pretty close, making it a top target for future searches for life. Sadly, there seems to be a dust disk around it, meaning that bombardments like those that killed the dinosaurs may be common.

GLIESE 667 CC Orbiting a small red dwarf star (the most abundant type of star in our galaxy), scientists have claimed there may be six planets, including two confirmed super-Earths. One of these confirmed exoplanets is likely in the habitable zone. Possibly the best contender for habitability we currently have, Gliese 667 Cc, is only 22 light years away. The small red star at the center of this system isn't alone. In fact, it is the smallest of three stars caught in an orbital dance! You can find a triple-star system with multiple low-mass exoplanets orbiting very close to one star. While you can't see the small red dwarf, you can see the three-star system at the tip of Scorpius's head using a telescope.

KEPLER-186F Heralded as the first true Earth-size planet, Kepler-186f is just 11 percent bigger than Earth, making it unlikely to have a massive atmosphere. It is orbiting very close to its red dwarf star (at about the same distance as Mercury's orbit). Four other planets have been discovered inside this orbit. While the size might be right for this planet, this distance is not. Kepler-186f is almost 500 light years away.

276 SEE THE SKIES FROM OTHER PLANETS

On Earth, the sky is blue because molecules in the atmosphere grab blue light coming from the Sun and scatter it in every direction. But have you ever wondered what space looks like from other vantage points?

FROM MARS The Martian atmosphere is made of tiny iron-oxide particles whipped up by huge dust storms (see #103). When sunlight strikes them, red light scatters in all directions and gives the sky a yellow-brown color, as seen here. As on Earth, the color of the Martian sky glows pink at sunrise and sunset. Earth and the Moon are both visible from Mars, with Earth shining about as brightly as Venus does in our sky. The Moon is much fainter than Earth; you'd need a telescope to see it.

FROM THE MOON Because the Moon lacks an atmosphere, the sky is always black. You'd see the stars at night, but during the day the bright Sun and illuminated Moon would obscure them. Appearing four times larger than the Sun and 50 times brighter than the full Moon, the most spectacular feature in the lunar sky is our planet. Since one side of the Moon permanently faces Earth, only people on that side would see it.

FROM SATURN'S MOONS Saturn's moon Titan has a very thick, brownish-orange atmosphere that would prevent you from seeing the night sky. Even nearby Saturn would be hidden, and Titan's distance from the Sun would make daytime similar to twilight on Earth. Meanwhile, Enceladus's thin atmosphere would make the sky black and the stars visible day and night. The Sun would look tiny—about one-ninth the size of the Sun as seen from Earth. Saturn would be a spectacular sight, with a diameter 60 times larger than the Moon as seen from Earth. Its glorious rings would be seen almost edge-on and be nearly invisible (see #199).

277 SEARCH FOR INTELLIGENT LIFE WITH DRAKE'S EQUATION

Scientists would be thrilled to find evidence of single-celled organisms to show that we are not alone in the universe. However, many of us would like something a little more, well, interactive. Frank Drake, one of the founders of the field of astrobiology (the search for life outside Earth), came up with a number of factors we might need to consider to estimate how many other intelligent civilizations may be in our galaxy right now. It's called the Drake Equation.

$$N = N_{(stars)}\ f_p\ n_e\ f_l\ f_i\ f_c\ f_L$$

While the Drake Equation does not have a solution that we know of, it does help illustrate how different factors may affect the number of intelligent, communicating civilizations in our galaxy. Plug in your own educated guesses to come up with an estimate of how many civilizations may share the Milky Way.

N An estimate of the number of intelligent, communicating civilizations in the Milky Way right now.

$N_{(stars)}$ The number of stars in the Milky Way. This is the factor we know best: We are fairly sure that there are between 100 and 500 billion stars in the Milky Way.

f_p The fraction of those stars with at least one planet. This estimate is in flux, but some astronomers think that as many as 100 percent of stars may have at least one planet.

n_e The average number of worlds per star that are suitable for life. Currently, scientists think liquid water might be the key to life (see #275 for info on inhabitable exoplanets). For our solar system alone, there may be six planets supporting water. Besides Earth, it's probable that Mars once had liquid water. Many moons around Saturn and Jupiter likely have water under layers of ice.

f_l The fraction of suitable planets that ever develop life of any kind. Here is where our assumptions turn into wild guesses because we only have one data point: Earth. What we do know about life on Earth is that it developed very early in the history of our planet, almost as soon as it cooled down enough to have oceans.

f_i The fraction of these planets above where simple life evolves into intelligent life. Of course, we can debate the definition of intelligent. But we only know of one species that has looked out into the galaxy to wonder if it's alone—and it's us.

f_c The fraction of intelligent civilizations who have developed detectable communications technology. While humans have actively sent out a handful of messages into space, we are currently mostly listening. Even if alien civilizations had the technology, is it possible that they might decide not to communicate? We don't know the answer to this and can only guess.

fL The fraction of the planet's lifetime in which an intelligent civilization is able to communicate across interstellar distances. For humans, that would only be this last century. We began sending directed messages into space in the 1970s. That is less than 0.000000001 percent of the lifetime of Earth. How much longer will our civilization survive to send signals into space? That's anyone's guess.

278

DELVE INTO SPACE'S MYSTERIES

We're learning more and more all the time about this crazy universe we live in. But here are some mysteries that still stump scientists.

DARK ENERGY We've noticed that galaxies seem to be moving faster and faster away from each other. One explanation holds that a phenomenon called *dark energy* is countering gravity, pushing space apart at an accelerating rate. Albert Einstein accounted for this energy, including a "cosmological constant" in his formula explaining gravity's role in a static universe. Later, he called this idea his biggest mistake, but a version of the cosmological constant has returned in one of the most popular models of our accelerating universe. Even so, dark energy has never been directly observed and remains very poorly understood.

DARK MATTER This cryptic stuff (which has no relationship to dark energy) seems to compose most matter in the universe. But what is it? Some galaxies don't seem to have enough stars and dust to hold them together, and scientists have proposed that some extra, invisible mass must be present to hold them together. While actual dark matter particles have not yet been detected, the hunt continues.

GRAVITY Sure, this is one of the forces that keeps our planet in orbit around the Sun, and it keeps us humans from floating off into space. But, beyond knowing that it's the force that holds us all together and that it causes a curve in space, we know surprisingly little else about it—gravity waves and particles (so-called *gravitons*) have never been detected, even though theory says they should exist. Several scientists have tried to create a verified, grand unified theory that marries gravity with the other forces of the universe: the *strong force* (the force that binds together small particles, like atoms and quarks), *weak force* (the force that makes neutrons "decay" into protons, then electrons, and so forth), and *electromagnetism* (the force that governs charges within atoms).

279 HELP COUNT THE STARS AS A CITIZEN SCIENTIST

Did you know that you can make real contributions to astronomy—and that all it takes is a computer and some free time? Citizen, civic, or crowdsourced science is actual research done by nonprofessional scientists. Astronomy has a long history of citizen scientists making groundbreaking discoveries—the modern age has allowed even those without telescopes or dark skies to participate. When it comes to crowdsourcing a data set, the more brains that can contribute, the faster discoveries can be made and conclusions drawn.

Case in point: If you can find Orion (see #56), you can contribute to a global map of sky brightness that is growing every year. The GLOBE at Night program is an international citizen-science campaign that raises public awareness of the impact of light pollution. Citizen scientists participate by measuring their night sky's brightness and submitting observations by computer or phone. Comparing trends in light pollution with other data sets (like those on animal behavior) allows us to better understand the effects of light pollution, as well as ways we might mitigate it. Just over 100 years ago, you could look up and see the Milky Way Galaxy, even in large cities. Now, most city dwellers have never seen a truly dark sky, which has a real impact on the lives of humans and other species. (See #283 for more information.)

280 EXPLORE THE ZOONIVERSE

When scientists have a question that seems insurmountable, they often turn to their grad students. When it's even bigger, the public can sometimes be more effective. Enter the Zooniverse—the platform for accessing Citizen Science Alliance projects. Working with many academic partners around the world, scientists use volunteers to climb the mountain of data before them. You can make a real difference on any number of projects, contributing at a variety of levels. Projects come and go, but a few of our favorites are:

GALAXY ZOO The first of the Zooniverse projects asks you to classify galaxies according to their form and structure, or "morphology." This simple task can give scientists a leg up on understanding what's happening in these distant islands of stars. The project has contributed to the discovery of "green peas," a type of tiny star-forming galaxy, as well as many other phenomena.

PLANETHUNTERS Interested in aliens? See if you can find an alien world! Making the Kepler mission's data available in 30-day chunks, Planethunters asks volunteers if they can find a transit of a planet across the disk of a star. While computers are good at this simple task, humans are even better. In the first four years, the project logged more than a million classifications.

CATALINA SKY SURVEY Perhaps the number-one way space is likely to kill you is with a large asteroid impact. It happened to the dinosaurs and it could happen again—unless we find the asteroid and change its path. The Catalina Sky Survey is dedicated to finding all near-Earth asteroids (NEAs—see #230) and it needs your help. Humans have found asteroids in data that computers have already searched. Watch four frames flick by and see if you can spot an asteroid whizzing through. It's addictive—and may save humanity.

281 MONITOR VARIABLE STARS

Variable stars change brightness over periods of seconds to years; the more we learn about them, the better we understand why they behave the way they do. But scientists can't monitor every star over decades—that's where you come in. Despite the name, the American Association of Variable Star Observers has members in more than 50 countries, making more than a million observations of variable stars annually. AAVSO has answered some important questions in the more than 100 years it has been active. Here are some of the ways you can participate.

HELP ON PROJECTS You can participate with your telescope, by analyzing online data, or by helping with other projects.

JOIN THE COMMUNITY A lively community boasts forums, chats, blogs, and meetings—plus a social media presence, of course.

EDUCATE YOURSELF Browse topics ranging from basic questions such as, "What are variable stars?" to things like interacting in a classroom and reporting on discoveries.

GET A MENTOR A mentor system gives newbies an experienced observer to "talk shop" with, introducing techniques and possible projects.

282 IDENTIFY LIGHT POLLUTION

Just a hundred years ago, most humans were able to see dark skies on a regular basis. Today, in most cities and suburbs, we can only see about 3 percent of visible stars due to artificial light leaking into the night. And if you're used to stargazing in a city and then try under truly dark skies, you might as well be on a different planet—the number of new stars can be dizzying.

All through the year, amateur astronomers gather for large star parties under the darkest skies they can reach (see #177), while professional astronomers have even more to worry about: Many once-great observatories are no longer used for research, as surrounding populations have brightened their skies so people can't see clearly.

LIGHT POLLUTION COMES FROM SEVERAL SOURCES:

OVER ILLUMINATION Simply put, we light many places that are nearly unoccupied or we illuminate areas with overly intense light. This results in diminished visibility and also in great expense.

SKYGLOW This soft, luminous effect occurs when lights or parts of lights are directed upward, creating an orangish glow above a city. This can be seen from many miles away and makes the stars in that direction seem to disappear. From outer space, cities across the globe glow clearly, a sign of all the light that is directed upward.

LIGHT TRESPASS Light trespass happens when fixtures are poorly shielded and light annoyingly ends up brightening observatories (or bedroom windows).

LIGHT CLUTTER Several lightsources in one small area create light clutter, or redundant fixtures that flood the space with too much light. Think the Las Vegas strip.

283 STUDY UP ON LIGHT POLLUTION'S EFFECTS

While clearly a bane to skywatchers, light pollution has a whole host of negative effects in several realms of life.

FINANCIAL In North America alone, more than US$1 billion a year is wasted illuminating the sky. If that energy could be redirected downward—where it can serve a useful purpose—the savings would be significant. Calgary in Alberta, Canada, for example, retrofitted its roadway lighting and saved about CAN$1.7 million a year.

CREATURE COMFORTS Light at night can have harmful effects on many animals. Some species of endangered sea turtles are especially at risk: Their seaside hatchlings crawl toward the brightest light, which once was the reflection of the Moon and stars on the water. But when artificial light is nearby, hatchlings become disoriented and never make it to the ocean. For many other species, artificial light creates problems with foraging, mating, and migrating. Birds can be drawn off course, and mammals otherwise hidden by the darkness become easy prey under artificial light.

HEALTH OF HUMANS In humans, exposure to light at night has shown an increased risk of some types of cancers and a prompting of sleep and mood disorders (including depression), as well as obesity and learning difficulties. Bright lights at night also cause *disability glare,* where everything but the light becomes temporarily invisible.

284 COMBAT OVER ILLUMINATION IN YOUR COMMUNITY

Imagine where humans would be if we didn't have the benefit of dark night skies as we evolved. Navigation and our exploration of the universe would all have been nearly impossible without the ability to see deep into the sky, unhindered by artificial light.

So what can be done to reduce the effects of light pollution?

BUY THE RIGHT BULBS Use energy-efficient, low-wattage lights that only give off as much light as is needed. Dimmer switches can help you control the light's intensity, too.

HIT THE OFF SWITCH Personally, only use light when and where it's needed—and no more than you really require.

ANGLE IS EVERYTHING Be sure that the lights around your property are shining down, not up—why illuminate the empty sky? Likewise, fully shielded fixtures will direct light only where it is needed and prevent it from leaking into the sky.

CLOSE THE CURTAINS Simple and effective: When your home's lights are on at night, keep the windows covered to prevent illumination from escaping.

BUILD COMMUNITY Beyond your personal habits, educate your neighbors about the importance of good lighting. The International Dark-Sky Association (IDA) has downloadable brochures and kits to help you learn more and choose improvements wisely. You can even take the issue of lighting to your city council or other government officials. Find a local IDA chapter in your community, see how it is making changes, and join it.

285 TOP FIVE SEEK OUT THE DARKEST SKIES

While it's true that night skies aren't what they used to be, there are spots where the darkness has been protected—and have received official designations as dark-sky reserves. Here are five must-visit sites.

☐ **CHACO CANYON, UNITED STATES** At this ancient spot in northwest New Mexico, you can observe from the ruins of celestial structures, witnessing the sky as the Chacoans once did.

☐ **BRANDENBURG, GERMANY** The Westhavelland International Dark Sky Reserve is just a 2-hour drive from Berlin—making it the closest park near a major city. Visit between May and June for extraordinary views of zodiacal light (see #14).

☐ **KERRY, IRELAND** This haven's position between the Kerry Mountains and Atlantic Ocean acts as a natural buffer for city lights. It's a great spot for viewing meteors and the Milky Way.

☐ **NAMIBRAND, NAMIBIA** Located between the Namib Desert and Nubib Mountains, this 500,000-acre (2,000-sq-km) nature reserve features some of the lowest humidity and clearest skies around—plus gazelle, springbok, leopards, hyenas, and more than 150 bird species.

☐ **AORAKI MACKENZIE, NEW ZEALAND** This 1,600-square-mile (4,300-sq-km) park is nearly free of light pollution. This basin was once home to the Maori, who navigated at sea using the stars.

286 THROW A STAR PARTY

So, you're hooked on skywatching and want to share your newfound love of astronomy. Or maybe your niece's teacher hears that you have a telescope and asks you to throw a star party for her fifth-grade class. Luckily, star parties are a lot of fun—and not so hard to put together.

STEP ONE Pick a date. Dates between new and first-quarter Moon are best for showing off the features along the Moon's *terminator* (the division between light and darkness) without having bright light all night.

STEP TWO Get help. Having more than one telescope or presenter is great. If setting up for a school, be sure there are parents on hand for crowd control.

STEP THREE Plan to view four or five of the brightest objects visible that night. It's neat that light has traveled millions of light years to our eyes, but most people aren't so excited by faint fuzzies at first! Research each object so you can provide context and coordinate among volunteers so there's variety.

STEP FOUR A dry field or parking lot with few bright lights and low horizons is an ideal venue. If possible, set up your gear before dark. Make your party visible but not too bright. Consider marking the legs of your tripod (#173) and all pathways with red lights (#170).

STEP FIVE Enforce a "no white lights" policy, including cell phones and camera flashes. Have binoculars so people who are waiting can preview the general area of the sky that they'll view with the telescope.

STEP SIX Start with the brightest objects (like the Moon) to allow time for visitors' eyes to dark-adapt. Show them where to put their eye and explain *averted vision*—looking slightly to the side of an object to utilize peripheral vision. Let them take their time. If you don't want people to grab the eyepiece (they will want to!), set up a chair and ask them to put both hands on it.

STEP SEVEN Host an afterparty. Print star charts for those who are interested in learning more. Encourage visitors to follow up with unanswered questions and direct them to their local astronomy club. Also, if you're stumped, take it as a chance to learn something new yourself. It's likely you'll be asked that question again!

287

LISTEN TO A METEOR SHOWER OVER THE RADIO

You can listen in to a meteor shower, even if you can't see it with your eyes because of clouds or the Sun (see #89 and #92 for viewing tips). When meteors streak through our atmosphere, they leave a path of superheated gas in their wake, called an *ion trail*. These charged particles briefly reflect radio waves back to Earth, boosting distant signals you normally wouldn't receive.

To listen for meteors with your FM radio, all you need to do is tune the radio to a station that is a bit out of range. Do you hear nothing but static? Good! Now, listen carefully. As meteors streak across the sky, their trails will reflect that distant signal back down to you, "boosting" it for a few seconds. You will suddenly hear a burst of music or talk from that distant station before its signal fades into static once more. Alternately, you can find live feeds from stations tuned to meteor trails over the Internet with just your web browser.

288

WITNESS THE BIG BANG ON YOUR TV

Believe it or not, you can watch the cosmic microwave background by tuning your TV to a blank channel. Some of the static on your television (about 1 percent) is actually the leftover radiation from after the Big Bang that created our universe. While newer digital TVs cannot see this signal, you can tune older analog TVs in to the afterglow of our universe's birth.

289 CONSTRUCT A DIY SCOPE MOUNT

Have you noticed how many different mounts there seem to be for telescopes? People who build their own telescopes especially love to make their own mounts, customized for their skywatching preferences. Since many telescopes ship without a mount, you might as well make your own. Here's how to make a mount for a Dobsonian telescope.

STEP ONE Make the altitude bearing box. Cut four squares of ¾-inch (2-cm) plywood large enough to fit around your telescope. To find the proper size, trace the diameter of your telescope onto the plywood, adding two times the thickness of the plywood to compensate; in this case, 1½ inches (3.8 cm). Pick two sides, mounting a PVC plug (try one that's 5 inches [125 mm] or 6 inches [150 mm] in diameter) on two sides to be used as the bearings. Screw the four squares together, making a cube with two ends open and the plugs on opposite sides.

STEP TWO Make the side boards and mount. Cut two rectangular lengths of plywood, keeping the short end 2 inches (5 cm) wider than the PVC plugs used for the bearings. You will want to make sure that these two boards are tall enough to hold your telescope without its bottom mirror cell hitting the bottom of the mount. A good rule of thumb is to make the box one-third the height of your telescope, but measurements may vary.

STEP THREE Cut a V into the top of the side boards to accommodate the PVC plug bearings. Alternatively, if you have a jig and want to make it fancy, cut half circles into the top in the same diameter as the PVC plugs.

STEP FOUR Cut out a third board from the plywood panel to fit and join the two side boards. This will be the front of your mount. It will need to be much lower than the side boards, as your scope will be swiveling up and down right above this section; you don't want to restrict its freedom of movement.

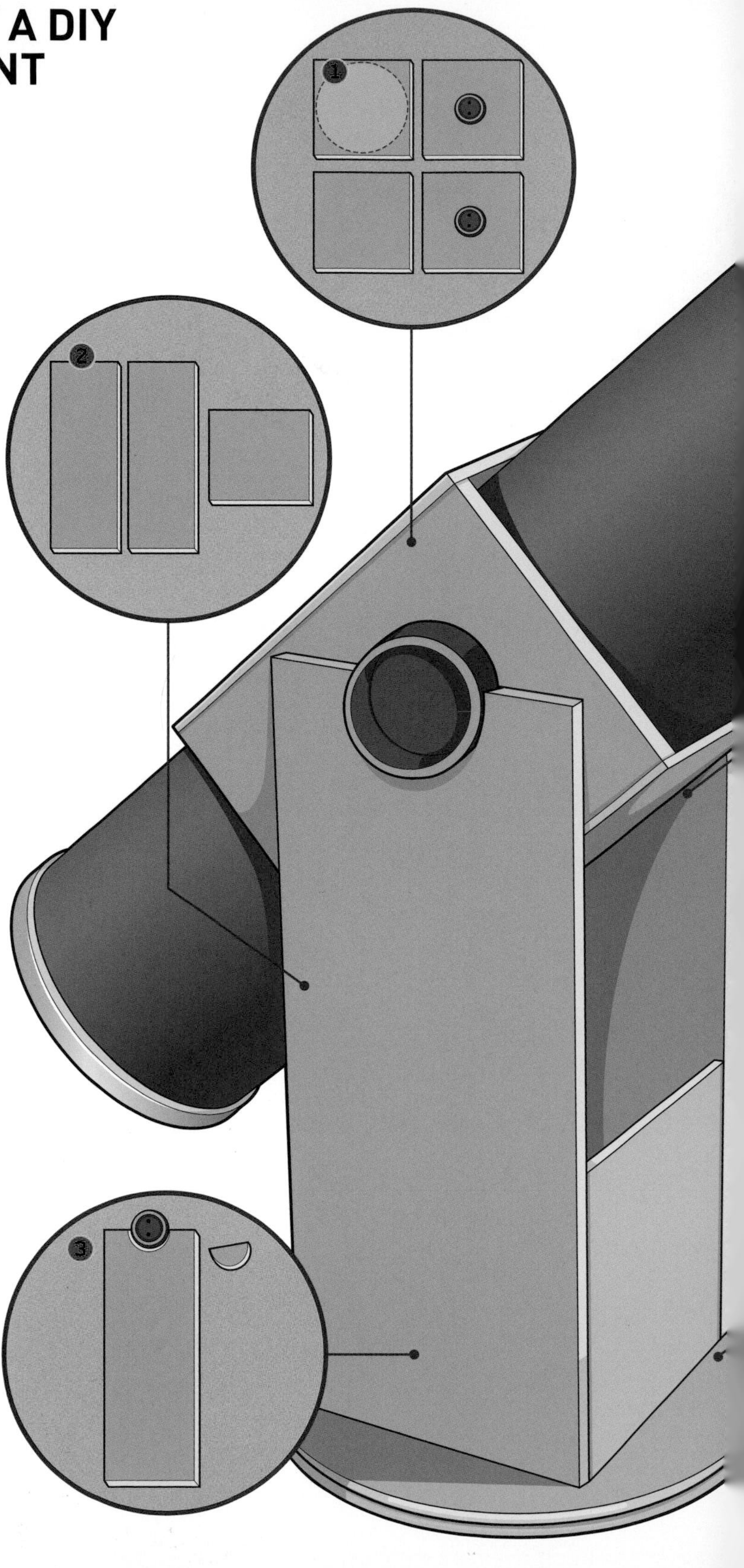

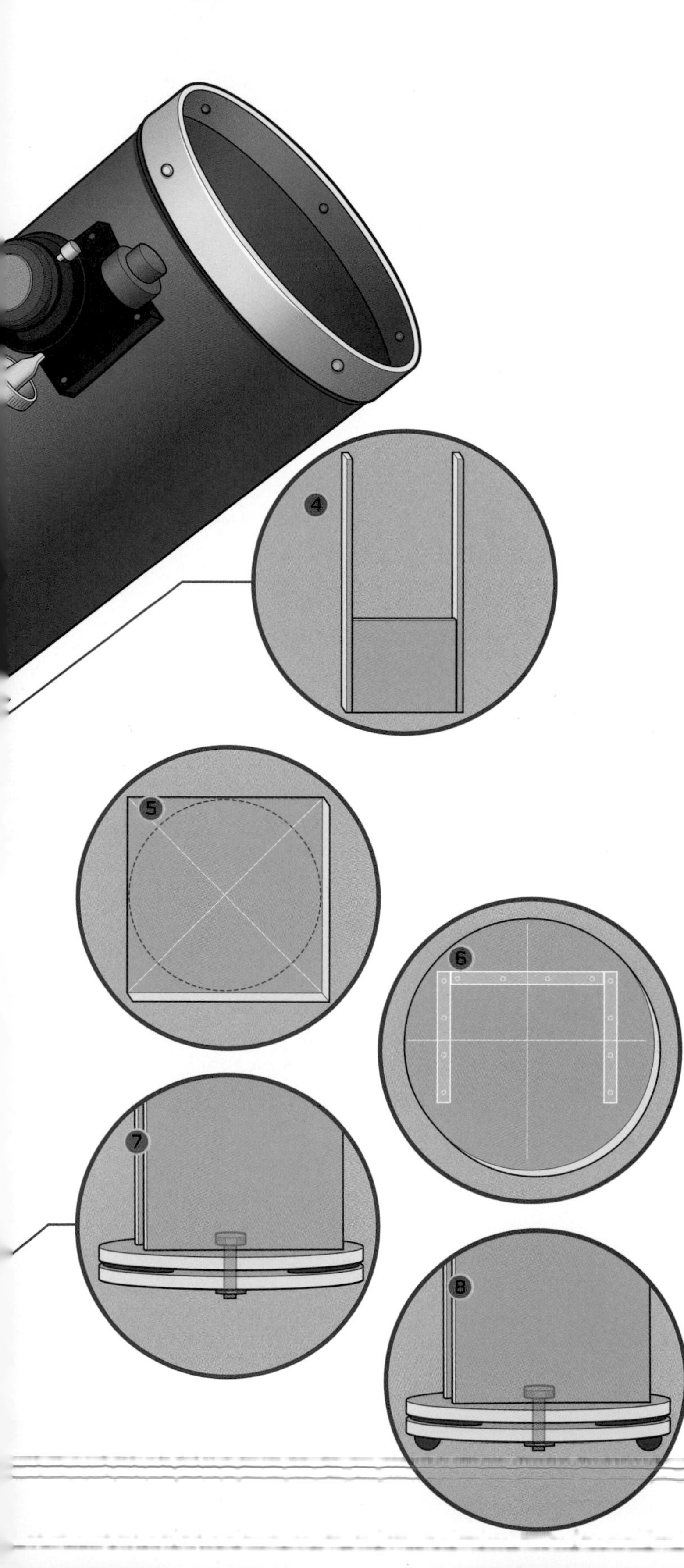

STEP FIVE To make the bearing boards, cut two squares of plywood, making sure they are larger than the width of your mount. On each board, draw two sets of diagonal lines from corner to corner, marking where they join in the middle to find the center. Cut the boards into two large circles.

STEP SIX Set the mount so it's centered on top of one of the circular bearing boards and trace its outline. Mark holes where you'll attach the first bearing board to the mount and drill through them. Attach the mount to the top bearing board with wood screws and glue. Make sure the screws sink in completely; you don't want them to scratch or otherwise interrupt the freedom of movement between the two bearing boards.

STEP SEVEN Bore a hole between the two bearing boards, just the right size to fit a large bolt. Push the bolt up from the bottom, attaching a self-locking nut on the top. Don't connect them too tightly; you want the boards to have the freedom to spin. You can also help the bearing boards move by attaching small felt, Teflon, or nylon pads between the two bearing boards (Teflon will work best).

STEP EIGHT Attach three or four 2-inch (5-cm) rubber pegs—"feet"—to the bottom of the bearing boards with screws and glue. Make sure they are large enough to keep the stand raised above moderately uneven ground, such as gravel or a typical lawn. If you would like, you can add a couple of handles to the mount and bearing box to make transport a bit easier.

STEP NINE You have just made your first Dobsonian mount. To use on the next starry night, simply set your reflector telescope tube inside the bearing box.

290 GET A BIGGER LIGHT BUCKET

Thinking "bigger is better" is a symptom of what we call aperture fever. While bigger isn't always better, the larger your mirrors or lenses are, the more light you can gather. More light gives you more detail and the ability to look farther and farther into the universe. Faint galaxies are much more visible; dim nebulae now stand out with sharply defined wisps of dust lanes; tiny details on the planets become simple to spot; and faint asteroids and comets are revealed with ease.

Still, don't let your aperture fever make you lose your reason! Don't buy a telescope bigger than you can store or transport. Bigger mirrors can be hefty, and you may need to hoist and set up by yourself in the dark. Make sure that you can physically handle larger telescopes and their mounts. You may even need a stepladder to view through your new telescope comfortably. If you can handle those requirements, a huge "light bucket" may be just right for you.

291 SPLURGE ON A HIGH-END REFRACTOR

So you've been stargazing for some time now and feel the itch to upgrade your equipment. Refractors are a great place to start, as they're often the tool of choice for deep-sky imaging and splitting binary stars. (See #142 to revisit refractor basics.) Modern optics allow high-end refractor lenses to produce practically error-free views. Made from fluorite or other special glass, three apochromatic lenses work together inside the telescope tube to eliminate chromatic aberration across the field of view, with none of the distortion near the edges that appears when using reflectors.

Some of the other benefits of a high-end refractor include minimal maintenance and ease of transport. Since the refractor is a sealed tube, your optics are safe inside. The lenses come already very well aligned so there is little need for the kind of collimation required with a reflector. These qualities make refractors easy to transport and set up at dark-sky viewing sites.

Of course, the more precise and well-crafted an instrument you want, the more you will pay. It is normal to pay a couple thousand dollars for a 105mm "apo." A flagship telescope from one company will run US$20,000 for the tube only, after you sit on a waiting list a few years before your number comes up—kind of like buying a high-end supercar.

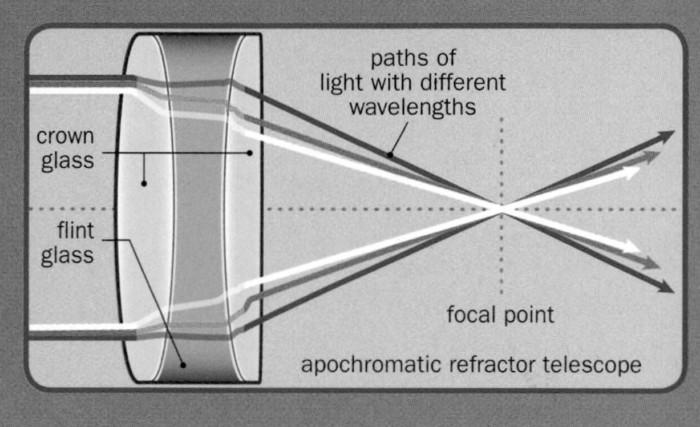

292 GO ALL OUT WITH FANCY EYEPIECES

There's a wealth of eyepieces to explore (see #161), but ponying up for a high-grade one has a huge reward: clarity. It'll seem like you have a whole new instrument. In fact, if you want to dip your toe into the high end of astronomical equipment, eyepieces are the perfect place to start. A set of high-quality eyepieces can be used across multiple telescopes and can last you the rest of your life.

For starters, 2-inch- (5-cm-) diameter eyepieces are big and sturdy, offering much greater eye relief than conventional 1¼-inch (3-cm) units. Also, many telescopes are able to hold both sizes by removing or adding an adapter. High-end eyepieces feature very high-quality glass with multiple coatings to reduce glare and increase contrast. Look for waterproof models with interiors that have been flushed of oxygen (purged with nitrogen). With those two features, the interior of your eyepieces should remain pristine and free from water, dust, and mold, thus vastly reducing the chances of the interior fogging up during a cold observing session.

Large eyepieces can get heavy. Your telescope and its mount must be sturdy enough to support a bigger eyepiece without drooping or losing its tracking.

293 SEND A CAMERA INTO SPACE

Want great pictures from the edge of space? Turns out, you can take them from Earth's stratosphere—heights between 7 and 31 miles (11–50 km)—with inexpensive, easy-to-find tools. Here's how.

STEP ONE You'll want a simple point-and-shoot camera—one that can automatically take pictures at regular intervals—or a video camera. Canon cameras can be easily programmed to shoot remotely at specified times using software called CHDK (Canon Hacker's Development Kit). Try it out before launch to make sure your batteries and memory can last a few hours. (Go with lithium batteries; they perform best in the cold.)

STEP TWO If you want to track your camera rig only after it lands, you can use a prepaid cell phone with Internet and GPS (cell signals rarely reach as high as you need for tracking in the air). If you want to track it during its entire mission, however, you'll either need to make something akin to a "trackduino" (Google is your friend here!) or buy a tracking system.

STEP THREE Pick up some hand-warming packets to keep your equipment from freezing as it enters the upper atmosphere, as well as a container that provides good shock absorption. A few lightweight lights with a long battery life may also help you find the rig if it lands in a tree or somewhere without cell coverage.

STEP FOUR Buy a weather balloon, helium, and a parachute. For starters, try a $1^1/_3$-pound (600-g) balloon—this will get the rig to at least 30,000 feet (9 km), likely much higher. To find helium, try renting a tank and a regulator from a welding-gas supplier, as these tanks tend to have a better concentration of helium than ones from a party store. As for the parachute, you can either purchase one secondhand or reinforce a heavy-duty garbage bag with duct tape. It should be at least 2 feet (60 cm) in diameter.

STEP FIVE Cut a hole in the side or bottom of your container so the camera can take pictures unobstructed. Activate the hand-warming packets and turn your camera on. Use duct tape, zip ties, and anything else that's lightweight to pack the materials in the container. Newspaper is a great insulator and shock absorber, but many lightweight materials will do. Don't use air-filled packing material, which will burst at altitude. Cover your container in aluminum foil so it is easily visible in the air. Once your balloon contraption is complete, attach a note to the outside with your contact info.

STEP SIX Pick a launch site that's flat, far from trees, and at least 100 miles (160 km) from any military base. Let the Federal Aviation Administration or other relevant authorities know about your plans at least a week in

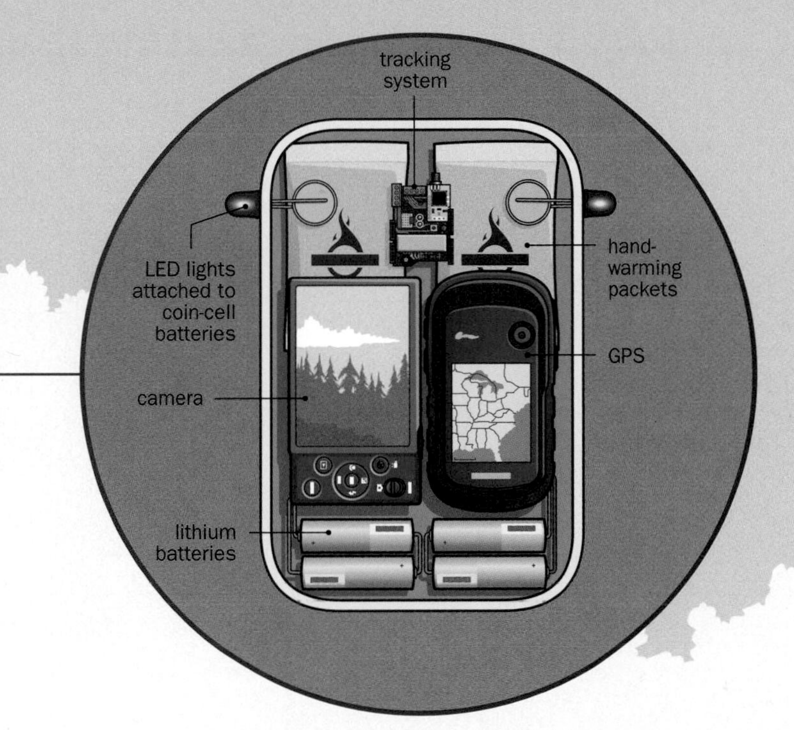

advance. While balloon payloads under 4 pounds (2 kg) are legal, it's a good idea to give them a heads up.

STEP SEVEN For launch, pick a sunny day with little wind and bring some friends. Lay the helium tank on its side and slowly inflate the balloon with the regulator. To estimate the amount you'll need to keep the balloon ascending, multiply your payload by 1.5. For example, with a payload weight of 2 pounds (1 kg), you'll want to hang an additional 1 pound (0.5 kg) of weight while you inflate the balloon. Stop inflating when it can just barely lift the balloon and ballast. When you're ready to launch, simply remove that extra weight.

STEP EIGHT Find it! Weather balloons usually spend 1 to 4 hours ascending before they burst. As a good first guess, you can estimate that your balloon ascended 30,000 feet (9 km). Cross your fingers that it landed in a spot with reception. Check its GPS signal and go on the hunt with lots of friends for extra eyes.

294

MOD YOUR DOBSONIAN WITH A CUSTOM TRACKER

Love your Dobsonian scope but wish you could sometimes fit it with GoTo capability, tracking the night sky and finding obscure objects with computer assistance (see #149)? There is good news! You can actually fit your Dob with a tracker—for a price.

While custom trackers start at thousands of dollars, you can also buy a Dob with GoTo built in. Of course, these scopes will also necessarily be expensive. One trick many folks don't know is that you can actually take the telescope tube from your Dobsonian setup and fit it onto an equatorial tracking mount, as long as the mount is rated to handle your scope's weight. That way, you can use your scope on the fly in Dob mode or track objects in motorized equatorial mode.

295 BOOST YOUR LIGHT WITH A SUPERBRIGHT DIAGONAL

Want to boost the amount of light going through your telescope? One option is to upgrade the diagonal between your eyepiece and your telescope. Look for terms such as "99 percent light transmission" and "dielectric." Diagonals described in this way use special glass and coatings to allow the greatest amount of light possible to pass through. The coatings in dielectric diagonals—made using ultrathin layers of certain oxides—are also much more resistant to corrosion and deterioration than many standard models. Besides, upgrading your scope with one of these high-end diagonals is one of the easiest upgrades you can make.

296 MAKE IT PERMANENT WITH PIERS

Do you have a permanent observing post? If you do, a pier may be your next mount of choice. A pier is a post solidly mounted into the ground, often with a cement foundation. With everything pre-aligned, a pier allows you to strap your telescope onto its post and begin observation quickly. If this sounds like a big commitment, that's because it is. Astronomical piers are often found inside personal observatories.

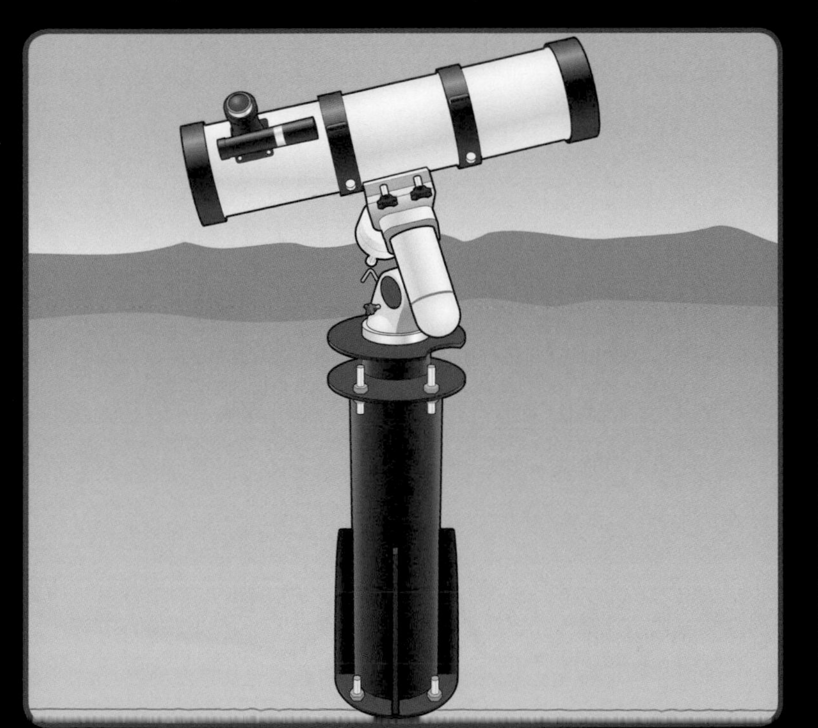

297 SWAP OUT COLLIMATION KNOBS

An odd but very useful trick to help you collimate your telescope's mirrors is to buy an improved set of collimation knobs. Featuring larger grips and smoother play, upgraded collimation knobs can really improve the often tedious but necessary experience of collimation. (See #150 for collimation tips.)

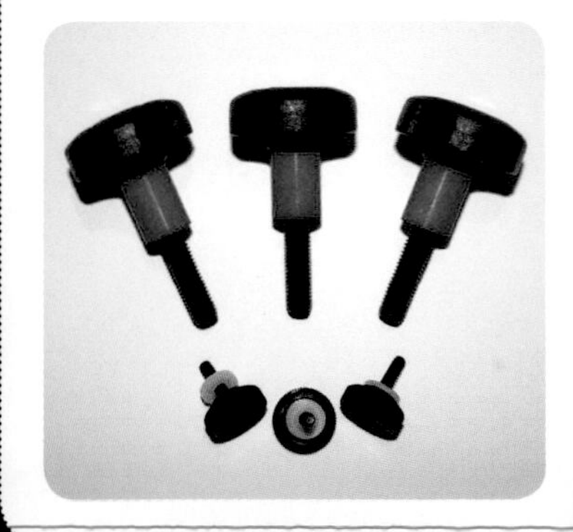

298 BUILD AN AWESOME OBSERVING CHAIR

With just a few pieces of wood, a handful of tools, and some fasteners, you can make your own observing chair with a seat that folds up and down. Plus, the whole chair folds up for easy transport.

STEP ONE Cut a standard 2x4 piece of lumber into four lengths: two 34-inch (86-cm) pieces for front and back uprights, a 24-inch (61-cm) piece for a lower crosspiece, and a 3½-inch (9-cm) block for the seat.

STEP TWO Use a table or circular saw to bevel one end of each of the 34-inch (86-cm) uprights to 22.5 degrees. Center and glue the bottom crosspiece flush with the beveled end of one upright. Secure with two 3/8-by-2½-inch (10-by-64-mm) lag screws and 3/8-inch (10-mm) washers.

STEP THREE Cut two seat supports exactly as shown in the diagram. Drill the 3/8-inch (10-mm) holes.

STEP FOUR To make the seat, cut one 10-by-13-inch (25-by-33-cm) plywood piece. Bevel one long edge to match the angle of the seat supports (about 30 degrees) and round the corners to your liking.

STEP FIVE Center and then glue a 3½-inch (9-cm) block to the bottom of the seat, 1½ inches (38 mm) from the back beveled edge. Fasten from the bottom of the block with four 2-inch (5-cm) wood screws in a square 2 inches (5 cm) apart. Cut and glue a piece of cardboard to one side of the block. This will form a shim that keeps the seat from binding to the front upright section.

STEP SIX Assemble the seat as shown. Align the seat supports and fasten them to the block with four 2-inch

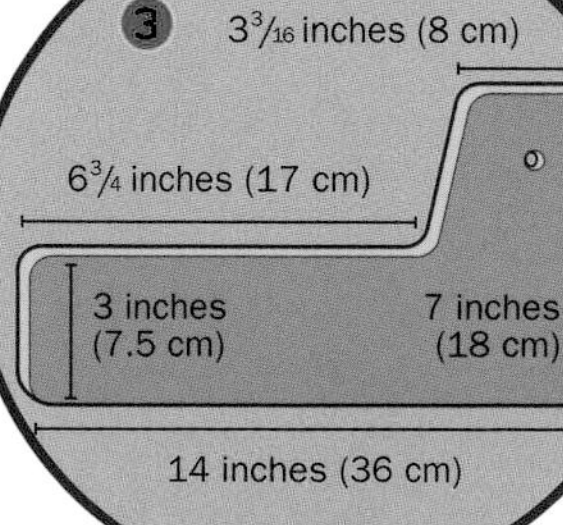

(5-cm) wood screws. Apply several waterproof coatings to the wood. Let it dry.

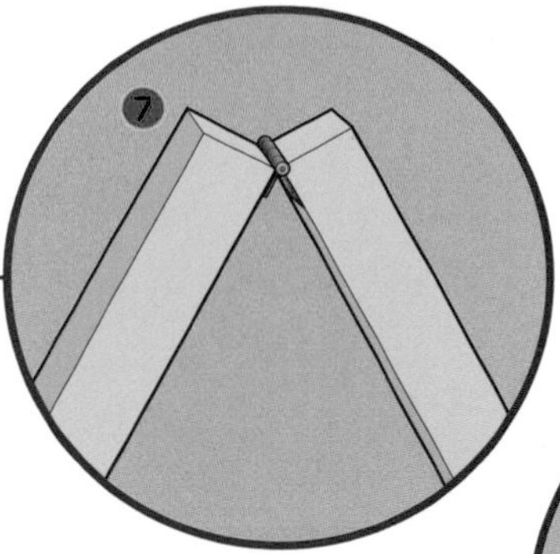

STEP SEVEN Lay out the front and back uprights, butting them at the square ends, with the shorter section formed by the bevels facing up. Install a 3-inch (7.5-cm) door hinge joining the square ends. Close the uprights to a 45-degree angle.

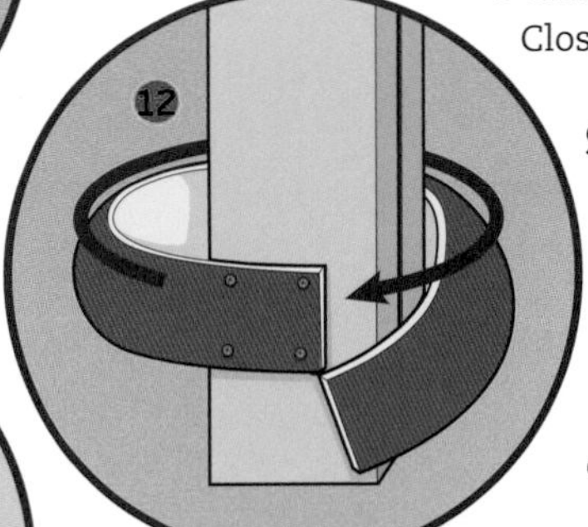

STEP EIGHT Take a 1-by-1/8 inch (25-by-3-mm) aluminum bar, at least 28 inches (71 cm) in length, and drill a 3/8-inch (1-cm) hole about ½ inch (1.25 cm) from one end. Bevel the opposite end to 22.5 degrees. About ¾ inches (2 cm) from the beveled end, drill a 3/8-inch (1-cm) hole and cut a slot from one side.

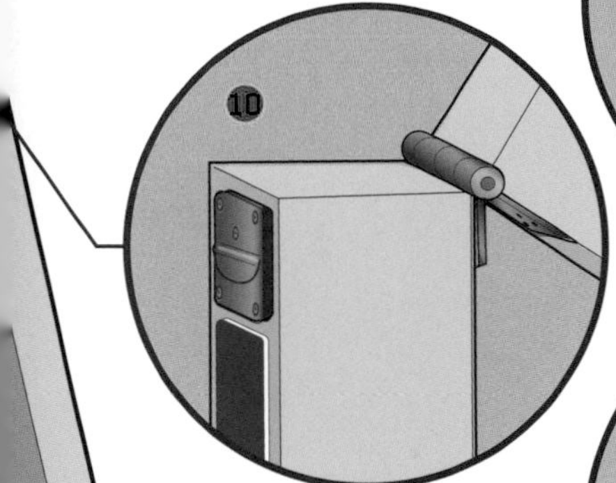

STEP NINE Install the aluminum crossbrace on one side of the rear upright using one 3/8-by-2-inch (1-by-5-cm) lag screw. Install the screw 2 inches (5 cm) from the bottom end of the upright, leaving a bit of slack so the crossbrace freely rotates. Install another 3/8-by-2-inch (1-by-5-cm) lag screw at the corresponding location on the front upright, leaving enough slack so the slotted end of the crossbrace firmly fits.

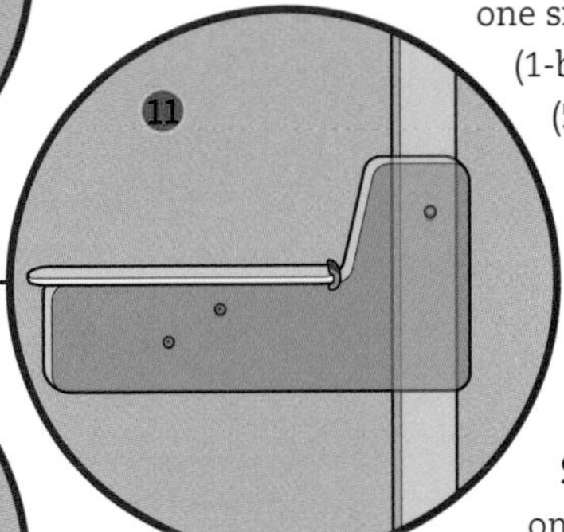

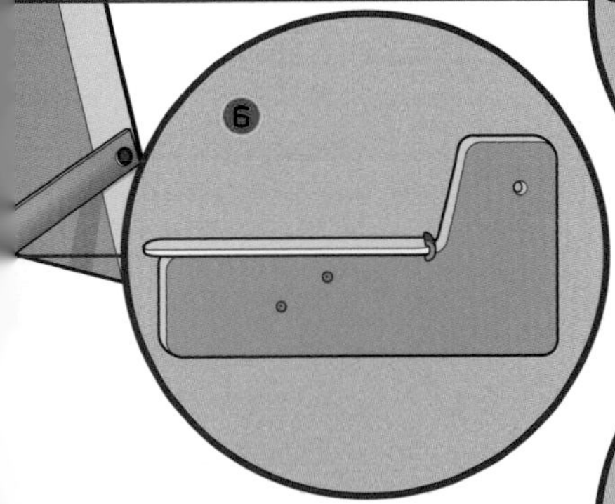

STEP TEN Install a 2¾-inch (7-cm) chest handle on top of the front upright. Trim a 24-inch (61-cm) length of 4-inch (10-cm) safety-tread tape to 3 inches (7.5 cm) wide. Install the safety-tread tape on the front upright, starting about 3 inches (7.5 cm) from the top. Slit a 4-inch (10-cm) length of ¾-inch (2-cm) rubber tube down the middle. Using flathead nails, spread and install the tubing on the rear edge of the seat, between seat supports. Hammer the nails tightly against the rubber to prevent rubbing against the safety tread.

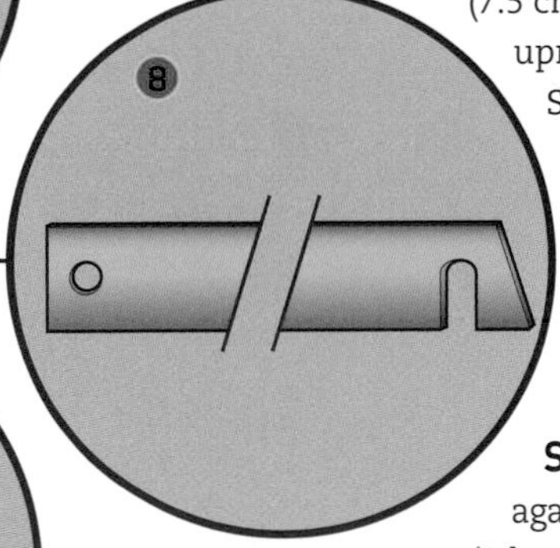

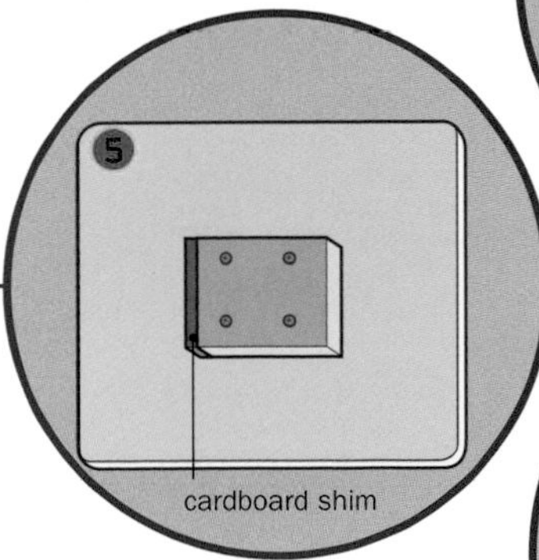

STEP ELEVEN Fit the seat assembly in place against the front upright. Then install 3/8-by-5½ inch (1-by-14-cm) hexbolts through the provided holes on each side; fasten with a 3/8-inch (1-cm) locknut. This holds the seat at the desired height.

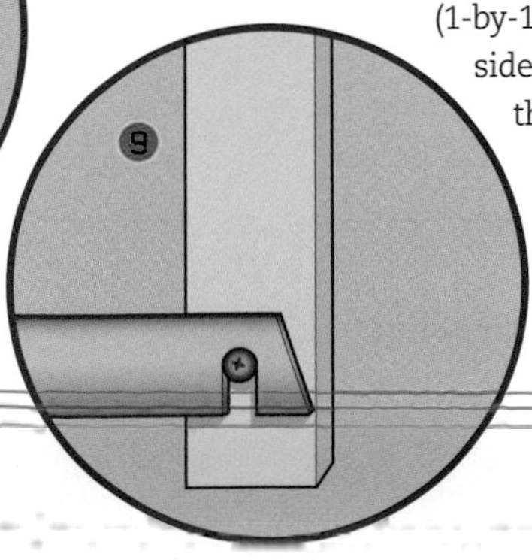

STEP TWELVE Secure one end of a length of Velcro to the back upright. It should be long enough to wrap around the seat when folded up, securing it for transport. Test your chair out on a nice clear night with your telescope!

RESOURCES

ORGANIZATIONS

NATIONAL AERONAUTICS AND SPACE ADMINISTRATION (NASA) It's NASA! Dive into its official website for the latest news on NASA science; missions past, present, and future; amazing images; and much, much more.
www.nasa.gov

EUROPEAN SPACE AGENCY (ESA) The ESA is a leader in space technologies. This organization's website rivals NASA as a source for incredible astronomical information.
www.esa.int

JAPANESE AEROSPACE EXPLORATION AGENCY (JAXA) The Japanese Space Agency's official website is full of excellent information on its past, present, and future missions, as well as tons of excellent science and photos.
global.jaxa.jp

INDIAN SPACE RESEARCH ORGANIZATION Dedicated to harnessing space technology and pursuing space science research and planetary exploration, these folks are doing great research for a fraction of what other countries are spending on space.
www.isro.gov.in

ASTRONOMICAL SOCIETY OF THE PACIFIC (ASP) Uniting lovers of astronomy on all levels, the ASP has pushed fervently toward its goal of science literacy through education for more than 126 years. Many articles on astronomical research, astronomy education and outreach, workshops, materials, and more can be found on its website (and member magazines: *Mercury, Astronomy Beat,* and *Universe in the Classroom*).
www.astrosociety.org

THE ASTRONOMICAL LEAGUE Dedicated to supporting amateur astronomers, the Astronomical League sponsors conferences as well as its famous Observing Programs, giving prized pins to members who complete their observing challenges.
www.astroleague.org

ROYAL ASTRONOMICAL SOCIETY OF CANADA (RASC) The RASC is a huge supporter of astronomy in Canada and elsewhere. Beyond supporting astronomy in dozens of centers across the uppermost region of the northern hemisphere, it publishes excellent and detailed annual observing guides, as well as a pretty fantastic observing calendar.
www.rasc.ca

SEARCH FOR EXTRA-TERRESTRIAL INTELLIGENCE (SETI) Is there life on other worlds? Can we detect that life? Is it intelligent? SETI has dedicated itself to the search for life beyond Earth with the goal of one day finding the answer to the ultimate question: Are we alone?
www.seti.org

THE ROYAL ASTRONOMICAL SOCIETY (RAS) The famed RAS has brought together astronomers for hundreds of years in its pursuit of knowledge about the night sky.
www.ras.org.uk

ASTRONOMERS WITHOUT BORDERS (AWB) The AWB is dedicated to forging relationships across international and cultural boundaries through a shared love of stargazing. The AWB sponsors outreach programs across the globe, spreading the knowledge of the night sky to all people.
www.astronomerswithoutborders.org

LOCAL ASTRONOMY CLUBS Find them via the Night Sky Network, Astronomical League, Astronomers Without Borders, and other resources.

MAGAZINES

SKY AND TELESCOPE A venerable magazine focused on the U.S.-based amateur astronomer, *Sky and Telescope* has tons of useful info in its monthly observing guides.

MERCURY (ASTRONOMICAL SOCIETY OF THE PACIFIC) With the latest news and research from the world of astronomy, *Mercury* includes contributions by some of the leading lights in the field.

ASTRONOMY *Astronomy* prints news and amateur observing guides for observers in the United States and the northern hemisphere.

ASTRONOMY NOW! (BRITISH) A giant monthly guide to the stars, *Astronomy Now!* is focused on observers in the United Kingdom and northern Europe.

L'ASTRONOMIE (FRENCH) A wonderful and beautiful French-language publication, *L'Astronomie* covers science as well as articles on the latest in amateur equipment and observing.

THE REFLECTOR (ASTRONOMICAL LEAGUE) The Astronomical League's monthly magazine, *The Reflector,* contains articles aimed at the more hard-core amateur astronomer, with a special focus on achieving goals for observing programs and distributing news on upcoming conferences and star parties.

WEBSITES

NIGHT SKY NETWORK NSN is your one-stop shop for stargazing tips, astronomy demos, and links to your local astronomy club and local stargazing events. Find a star party and astronomy club near you and check out the night sky!
www.nightskynetwork.org

THE EVENING SKY MAP A PDF-printable sky map for each month, the Evening Sky Map includes a sky map and a list of events with dates to watch out for that month. They have three editions each month—one each for northern hemisphere-, equatorial-, and southern hemisphere-based observers.
skymaps.com

ZOONIVERSE Starting with the Galaxy Zoo project, the Zooniverse team has assisted major scientific research in many scientific disciplines—in addition to astronomy—with the help of dedicated volunteer citizen scientists. Perhaps you can help make a major discovery?
www.zooniverse.org

UNIVERSE TODAY Universe Today is a long-running blog with astronomy information and many great news items.
www.universetoday.com

SPACE.COM One of the oldest astronomy news sites on the web, space.com has a virtual treasure trove of articles.
www.space.com

EARTHSKY An institution on the web since the 1990s, EarthSky offers daily tips on what to look for in the night sky, along with other astronomy- and earth science-related news.
www.earthsky.org

GLOBE AT NIGHT A global campaign to monitor light pollution, Globe at Night lets you contribute with your smartphone or your own eyes!
www.globeatnight.org

NASA'S SPACE PLACE News and information for space explorers of all ages are here.
www.spaceplace.nasa.gov

SPACEWEATHER The latest news and information on solar storms, sunspots, auroras, and meteor showers are just some of the things Spaceweather covers.
www.spaceweather.com

SPOT THE STATION Find out where the International Space Station is orbiting and sign up for alerts for when it will be above your head!
www.spotthestation.nasa.gov

APPS

GOOGLE SKY (ANDROID APP/WEB APP FOR OTHER DEVICES) One of the first point-at-the-sky-and-see-the-names-of-the-objects-above-you apps, Google has added tremendous amounts of data in its attempt to make Google Sky the "Google Maps for Space."
www.google.com/sky

STELLARIUM (ALL PLATFORMS, MOBILE/PC/MAC) This long-running planetarium program is open source and filled with many, many plug-ins. It is also very customizable to fit your needs, whether you're running your own planetarium, planning your next observing sessions, or trying to find that hard-to-spot new asteroid or comet!
www.stellarium.org

STARRY NIGHT (ALL PLATFORMS) Long-running planetarium software Starry Night has many versions tailored to different levels of use.
www.starrynight.com

DISTANT SUNS (IOS/ANDROID/KINDLE) Venerable star-finding software, Distant Suns started on the Amiga in the 1980s and is available on numerous mobile platforms today.
www.distantsuns.com

MICROSOFT WORLDWIDE TELESCOPE (PC) Microsoft WorldWide Telescope is a very robust virtual telescope software with many different modes of use, including a 3-D mode compatible with headsets such as the Oculus Rift and videogame controllers. Explore the universe as a virtual astronaut or plot your next star party. You can even import your own research data and use it for presentations.
www.worldwidetelescope.org

IEPHEMERIS (IOS) Find out ephemera for the Sun and Moon, including sunrise and sunset times, moonrise and moonset times, the phases of the Moon, and more.

METEOR COUNTER (IOS) This app from NASA allows iPhone users to receive alerts when the next meteor shower is going to take place. It also allows meteor watchers to record the amount and brightness of any meteors they may happen to see. The results are automatically shared with NASA researchers.

DARK SKY METER (IOS AND ANDROID) This app takes a measurement of the brightness of the night sky and can be used in conjunction with the Globe at Night campaign to measure light pollution across the world's night skies.
www.darkskymeter.com

GLOSSARY

A

AFTERGLOW A rosy or orange color high in the sky after sunset that is caused by atmospheric dust refracting light from the Sun.

ALTAZIMUTH MOUNT A simple, two-axis telescope mount that allows you to track celestial objects by adjusting the vertical and horizontal controls.

ANALEMMA A figure eight-shaped documentation of the Sun's path during a year.

ANDROMEDA GALAXY A spiral galaxy in the constellation of Andromeda.

ANGULAR SEPARATION The angle of distance between celestial objects as we perceive them on Earth, measured in degrees, arc minutes, and arc seconds.

APOCHROMAT A high-quality lens that lacks the chromatic and spherical aberration of less-refined lenses.

APOGEE The point in the Moon's orbit in which it is farthest from Earth.

AQUARIUS THE WATER BEARER The 11th constellation of the zodiac.

AQUILA THE EAGLE A constellation viewable in the northern sky that is a few degrees north of the celestial equator.

ARC MINUTE A unit of angular measurement equal to 1/60 of 1 degree.

ARC SECOND A unit of angular measurement equal to 1/60 of 1 arc minute.

ARCTURUS The fourth-brightest star and the brightest in the constellation Boötes.

ARIES THE RAM A small constellation between Pisces and Taurus.

ASTERISM Any group of stars that form a pattern recognized in Earth's night sky, though not as popular or prominent as a constellation.

ASTEROID Also called a minor planet. A rocky object orbiting the Sun that is less than 621 miles (999.4 km) in diameter.

ASTEROID BELT A region between the orbits of Mars and Jupiter that contains most of our solar system's asteroids.

ASTRONOMICAL UNIT (AU) The average distance between the Earth and the Sun, 93 million miles (150 million km).

ATMOSPHERE The various layers of gases enveloping a celestial object. Earth's atmosphere includes the troposphere, stratosphere, mesosphere, thermosphere, and exosphere.

AURORA Colorful, glowing lights in the sky visible over high or low latitudes and caused by particles from the Sun hitting gases in Earth's atmosphere. They occur in the northern hemisphere (called the aurora borealis) and the southern hemisphere (the aurora australis).

AVERTED VISION The technique of using peripheral vision to see objects better at night.

AXIS The imaginary line through the center of a planet, star, or galaxy around which it rotates.

B

BAILEY'S BEADS A feature of total solar eclipses wherein sunlight shines through valleys in the Moon's surface, creating dots of light.

BETELGEUSE The ninth-brightest star in the night sky and the second-brightest in the constellation Orion.

BIG BANG Currently the best theory for the origin of the universe. Proposes that it began as an explosion of a tiny, superhot bundle of matter 13.8 billion years ago.

BIG DIPPER An asterism of seven bright stars in the constellation Ursa Major that points to the North Star.

BINARY STAR Either of two stars linked by mutual gravity and revolving around a common center of mass, or the alignment of two stars as seen from Earth. Also called a double star.

BINOCULARS A portable pair of identical telescopes mounted side by side that allow users to view with their binocular vision.

BLACK HOLE A massive space object with gravity so strong that no light or other radiation can escape from it. Forms after a huge star collapses, or supernovas.

BOÖTES A constellation visible in the northern hemisphere that contains the sky's fourth-brightest star, Arcturus.

C

CANOPUS The brightest star in the southern constellation Carina and the second-brightest star in the night sky after Sirius.

CELESTIAL EQUATOR A projection of the Earth's equator into space.

CHROMATIC ABERRATION The "fringing" that occurs when a lens is not able to bring all wavelengths of light to the same focal point.

CHROMOSPHERE The second of three layers in the Sun's atmosphere. It appears pink when viewed during a total solar eclipse.

CIRCUMPOLAR Surrounding or found in the vicinity of a terrestrial pole.

COLLIMATION Aligning the optics of a telescope for optimum viewing.

COMET A small body of ice and dust that orbits the Sun on an elongated path.

CONE A light-sensitive cell in the retina that detects color.

CONJUNCTION The moment when two celestial objects lie closest together in the sky, technically within 1 degree.

CONSTELLATION Various groups of stars with a known name and pattern. One of the 88 official patterns of stars into which the night sky is divided.

CONVECTION LAYER The section of a star where hot, turbulent plasma and gas boil and energy is transferred to the star's surface.

CORE The innermost section of a star where nuclear fusion occurs by fusing hydrogen and helium, thus creating huge amounts of energy.

CORONA The high-temperature outermost atmosphere of the Sun, visible from Earth only during a total solar eclipse.

CORONAL MASS EJECTION A massive burst of electrons and atomic nuclei erupting through the Sun's corona into space.

CRATER A depression on the surface of a celestial body with a raised rim. Usually marks the opening of a volcano, or formed as a result of a meteorite impact.

CROWDSOURCED SCIENCE The delegating of tasks in a scientific project to the public. Also called citizen science.

D

DARK ENERGY A force that scientists are exploring as the root cause of our increasingly accelerating galaxy.

DARK MATTER A mysterious, yet-to-be-detected force that scientists propose holds galaxies together.

DECLINATION (dec) The angular distance of a point north or south of the celestial equator.

DISABILITY GLARE A reduction in visibility caused by excessive brightness.

DOBSONIAN Invented by John Dobson, this lightweight, simple Newtonian reflector telescope has a large aperture that collects tons of light. Beloved by amateurs.

DOG A relatively rare atmospheric phenomenon, a Sun or Moon dog consists of bright spots on either side of the celestial object, usually as part of an airy white circle.

DRAKE'S EQUATION Invented by astrobiologist Frank Drake, this equation isolates the factors that may affect the number of intelligent, advanced life forms in our universe, if any.

E

EARTH The third planet from the Sun and our home in the solar system.

ECLIPSE When one celestial body passes in front of another, dimming or obscuring its light. Can be lunar (in which the Moon passes into Earth's shadow) or solar (in which the Moon passes between Earth and the Sun).

ECLIPTIC The apparent path of the Sun and Moon against the background of stars.

EQUATORIAL MOUNT A motorized telescope mount that follows the rotation of the sky, allowing you to track celestial objects with little effort.

EQUINOX Either of the two times each year (around March 21 and September 21) when the Sun crosses the equator and day and night are of equal length everywhere on Earth.

ESCARPMENT Long, precipitous ridge of land or rock. There are many on the Moon.

EXOPLANET A planet that orbits a star other than our Sun and that might potentially host other life forms.

EYEPIECE The part of a telescope that you look through and that determines the magnification and the field of view.

F

FIELD OF VIEW The amount of the sky you're able to see through an optical instrument.

FILTER A telescope accessory that transmits light of different wavelengths, usually due to special dyes or coatings. There are specific types for different purposes: solar for safe viewing of the Sun, broadband for blocking light pollution, and so on.

FINDER SCOPE A small, low-power telescope attached to and aligned with a larger one. Its wider field of view makes it useful for locating celestial objects before getting in close with a larger telescope.

G

GALAXY A gravitationally bound system containing stars, stellar remnants, interstellar gas, and dark matter.

GANYMEDE Jupiter's moon and the largest moon in our solar system.

GLOBULAR CLUSTER A large, compact, and spherical collection of older stars that orbit a core.

GLORY An optical phenomenon in which light from behind the viewer is bent as it passes through water droplets in front of the viewer, creating concentric rings of color like a rainbow.

GNOMON An object that by the position or length of its shadow serves as an indicator of the hour of day; also, the raised part of a sundial that creates the shadow.

GO-TO TELESCOPE A computerized telescope that automatically points itself at a space object selected by the user.

GLOSSARY

GRAVITY One of the four fundamental forces of nature by which all physical bodies attract each other and objects fall toward the Earth.

GREAT RED SPOT A persistent storm that has raged on Jupiter for more than 300 years.

GREEN FLASH An optical phenomenon in which a green spot is visible above the upper rim of the Sun's disk as it sets. It results from Earth's atmosphere bending and separating sunlight into distinct colors, much like a prism.

H

HELIUM An inert, gaseous element and the second most-abundant element in the universe.

HERCULES The fifth-largest of the modern constellations.

HUBBLE SPACE TELESCOPE The world's first space-based optical telescope, named after American astronomer Edwin P. Hubble.

HYDROGEN The most abundant substance in the universe.

HYDROGEN-ALPHA FILTERS A telescope filter designed to transmit a narrow bandwidth of light on the hydrogen-alpha wavelength, which reduces the effects of light pollution.

I

INTERNATIONAL SPACE STATION (ISS) A habitable artificial satellite in low Earth orbit.

J

JAMES WEBB SPACE TELESCOPE A space observatory scheduled to launch in October 2018 that will replace the Hubble, the James Webb will gather data from 930,000 miles (1.5 million km) away.

JUPITER The fifth planet from the Sun and the largest planet in the solar system.

K

KEPLER A space observatory launched by NASA to discover Earth-like planets orbiting other stars.

KUIPER BELT A region of the solar system beyond the planets that contains trillions of space objects, such as Pluto. Extends from Neptune's orbit to about 50 au from the Sun.

L

LATE HEAVY BOMBARDMENT A large number of asteroids that collided with the terrestrial planets in the inner solar system about 4 billion years ago.

LATITUDE A geographical coordinate representing the north-south position of a point on the surface of Earth, expressed in degrees.

LEONIDS A meteor shower that appears to radiate from the constellation Leo. Typically occurs annually around November 16–17.

LIGHT POLLUTION Excessive or obtrusive artificial light, which has adverse effects on our health, environment, and viewing conditions.

LIGHT YEAR The distance that light travels in one year: 6 trillion miles (9.5 trillion km).

LONGITUDE A geographical coordinate representing the east-west position of a point on the surface of Earth, expressed in degrees.

LUNAR ECLIPSE Occurs when the Moon passes directly behind Earth into its shadow.

LUNAR MARE Dark basaltic plains on our Moon that formed when volcanoes erupted. Early astronomers named them "mare" (Latin for "sea"); they believed they were oceans.

M

MAGELLANIC CLOUDS Named after voyager Ferdinand Magellan, these galaxies are located near the Milky Way Galaxy and are stunning to view from the southern hemisphere.

MAGNETIC FIELD A region around a magnetic material or a moving electric charge within which the force of magnetism acts.

MAGNIFICATION The magnifying power of an instrument, or the amount by which an object will appear larger across a distance.

MAGNITUDE A star's brightness, which depends on its size, temperature, and distance from us. There are two types: intrinsic (how bright stars would appear if they were all the same distance from us) and apparent (how bright they look to us at their actual distance from Earth).

MAIN-SEQUENCE STAR A star that has a core hot enough for nuclear fusion.

MARS The fourth planet from the Sun and the second-smallest, after Mercury.

MERCURY The smallest and closest planet to the Sun in our solar system.

METEOR The bright, brief streak of light produced by a piece of space debris burning up as it enters the atmosphere at a high speed.

METEORITE Any piece of interplanetary debris that reaches Earth's surface intact.

METEOROID A small piece of rock or debris that orbits millions of miles through space.

METEOR SHOWER An event in which meteors can be seen to radiate, or originate, from one point in the night sky, caused by streams of cosmic debris entering Earth's atmosphere at high speeds.

MOON Earth's only natural satellite, which scientists propose was created when a massive object hit Earth and a big chunk was flung into space, becoming our Moon.

MOON ILLUSION An optical illusion that causes the Moon to appear larger near the horizon than when it is higher up in the sky.

MORNING STAR The nickname given to the planet Venus when it appears in the east before sunrise.

N

NATIONAL AERONAUTICS AND SPACE ADMINISTRATION (NASA) A U.S. government agency spearheading the civilian space program, aeronautics, and aerospace research.

NEBULA A cloud of gas or dust in space; may be luminous or dark, and it often serves as a birth place for stars.

NEPTUNE The eighth planet from the Sun.

NEW MOON The first phase of the monthly lunar cycle in which the Moon and Sun are in conjunction, making the Moon invisible. Great for stargazing, as the night sky will be dark.

NORTH CELESTIAL POLE As seen from the northern hemisphere, it is the point in the sky about which all visible stars rotate.

NORTHERN HEMISPHERE Region of the sky that lies north of the equator.

NORTH STAR Also called Polaris, the star in the northern hemisphere appearing still in the night sky, located nearly at the north pole, around which the entire northern sky turns. Given Earth's precession, Polaris will not always be the North Star.

NUCLEAR FUSION A thermonuclear reaction in which nuclei of light atoms join to form nuclei of heavier atoms. This happens in the cores of stars and creates energy.

OBJECTIVE LENS In an optical instrument, the lens that gathers light from an observed object to produce a real image.

OBSERVATORY A room or building housing an astronomical telescope or other scientific equipment to use for research purposes.

OCCULTATION The covering up of one celestial object by another, such as the Moon passing in front of a star or planet, as we see it from Earth.

OORT CLOUD A vast group of icy objects between 5,000 and 100,000 au from the Sun. Scientists believe that many comets originate from the Oort Cloud.

OPEN CLUSTER A group of thousands of young stars that came to be at the same time in the same molecular cloud and that are bound loosely by weak gravitational forces but that will eventually drift apart.

OPHIUCHUS THE SERPENT BEARER A constellation located around the celestial equator between Libra and Aquila.

OPPOSITION When two celestial bodies are on opposite sides of the sky, as viewed from Earth.

ORBIT A curved, usually elliptical path, taken by a celestial body around another celestial body, such as the Sun.

ORIONIDS A meteor shower that originates in the constellation Orion and is visible annually in late October.

ORION NEBULA A diffuse nebula in the constellation Orion, located in the center of Orion's sword.

P

PARSEC A unit of distance equal to 3.26 light years that astronomers use to measure a space object's apparent distance from Earth. It is the distance at which a star would have a parallax of 1 second of arc.

PEGASUS THE WINGED HORSE A northern constellation between Cygnus and Aquarius.

PENUMBRA The region in which only a portion of a lightsource is obscured by an occluding body. In a penumbral eclipse, this results in a subtle shadow over the Moon.

PERSEIDS A group of meteors that appear in August and radiate from the constellation Perseus.

PERSEUS A large northern constellation that contains several clusters and the variable star Algol.

PHOBOS The inner and larger of two satellites orbiting Mars, discovered in 1877.

PHOTOSPHERE The luminous visible surface of a star from which light radiates.

PISCES THE FISH A zodiacal constellation between Aries and Aquarius.

PLANET A celestial body orbiting a star that is big enough to be rounded by its own gravity.

PLANISPHERE An adjustable star chart that consists of two disks that rotate and display the visible stars for any time, date, and latitude.

PLASMA One of the four states of matter, plasma is the most abundant matter in the universe. It is an electrically neutral cloud of positive ions and free electrons.

PLEIADES An open star cluster in the constellation Taurus. Also called the Seven Sisters.

GLOSSARY

PLUTO A dwarf planet in the Kuiper Belt, classified in the past as the ninth planet from the Sun.

POLARIS Also called the North Star. The brightest star in the constellation Ursa Minor and the outermost star in Little Dipper's handle.

PRECESSION A slow change in the orientation of a rotating body's axis.

PRIME MERIDIAN Earth's zero of longitude, which passes through Greenwich, England.

PROCYON The eighth-brightest star in the sky and the brightest in the constellation Canis Minor.

Q

QUADRANTIDS A meteor shower that comes into view annually in early January, appearing to radiate from the constellation Draco.

R

RADIATIVE ZONE The interior section of a star in which energy from the core passes to the star's outer shell as photons.

RED DWARF A relatively small, cool, and dim main-sequence star. The most common and long-living stars in our galaxy.

RED LIGHT Low-intensity light that doesn't adversely affect night vision when stargazing.

REFRACTION The bending of an electromagnetic wave when it enters a medium in which it travels at a different speed, such as light through the atmosphere.

REGULUS The brightest star in the constellation Leo and one of the brightest stars in the night sky.

RETROGRADE Orbital motion of a celestial body in a direction opposite that of other celestial bodies in the same system.

RIGEL KENTARUS (ALPHA CENTAURI) A triple-star system that is the brightest celestial object in the constellation Centaurus and the closest star system to Earth other than the Sun.

RIGHT ASCENSION Analogous to longitude on Earth, right ascension measures a celestial object's rise in the eastern sky in hours, minutes, and seconds.

ROD The photoreceptor cells in the retina that are capable of functioning in low light and are used in peripheral vision.

S

SAGITTARIUS A southern zodiacal constellation represented by a centaur shooting an arrow, containing a point in the sky where the center of the Milky Way is located.

SATELLITE An object—natural or artificial—that orbits another body, such as our Moon.

SATURN The sixth planet from the Sun and the second-largest planet in the solar system.

SCORPIUS A southern zodiacal constellation between Libra and Sagittarius.

SEEING In the astronomical sense, the blurring and twinkling of objects in the sky caused by turbulence and other factors in Earth's atmosphere.

SHADOW BANDS Waves of light and dark that shift across surfaces on Earth before and after totality in a solar eclipse, thought to originate from irregular atmospheric refraction.

SIRIUS The brightest star in the Earth's night sky. A binary star in the constellation Canis Major.

SOFIA The Stratospheric Observatory for Infrared Astronomy, which travels on a modified Boeing airplane and collects infrared imagery.

SOLAR FLARE A flash of brightness that emanates from the Sun's surface and releases a tremendous amount of energy.

SOLARGRAPH Long-exposure photography that involves using a long shutter speed to capture the Sun's path as it moves across the sky during the course of six months to a year.

SOLAR SYSTEM The Sun, the planets, and all the celestial bodies that revolve around it.

SOLAR WIND A stream of plasma containing charged electrons and protons emanating from the upper atmosphere of the Sun.

SOLSTICE Either of the two times each year (around June 21 and December 21) when days are either longest or shortest, depending on the hemisphere.

SOUTH CELESTIAL POLE The region in the sky directly above Earth's south pole.

SOUTHERN CROSS A southern constellation between Centaurus and Musca used to navigate the southern hemisphere.

SOUTHERN HEMISPHERE The region in the sky between the south pole and the equator.

SPHERICAL ABERRATION A defect in image quality caused by light waves hitting a curved mirror in an optical instrument.

SPICA The brightest star in the constellation Virgo and the 15th-brightest star in the sky.

STAR A luminous sphere of plasma held together by its own gravity.

STAR CHART A map of the night sky showing the relative positions of the stars as seen from Earth and in a particular area of the sky.

STAR TRAIL A photographic technique using long exposures to capture the motion of stars in the sky. Shows individual stars as streaks across the image.

SUMMER TRIANGLE An imaginary triangle drawn on the northern hemisphere's celestial sphere, with vertices at Altair, Deneb, and Vega.

SUN The star, around which Earth and other planets orbit, that gives Earth heat and light.

SUNSPOT A dark spot sometimes appearing on the surface of the Sun as a result of fluctuations in the magnetic fields below the surface.

SUPERGIANT Among the most massive and luminous stars, supergiants have a mass around 10 times that of the Sun.

SUPERNOVA The explosion of a star caused by gravitational collapse, sometimes leaving behind an extremely dense core.

SYZYGY An alignment of three celestial objects, such as Earth, the Sun, and the Moon.

T

TAURUS THE BULL A zodiacal constellation between Gemini and Aries, containing the star Aldebaran and the Crab Nebula.

TELESCOPE A device shaped like a long tube that you look through in order to see things that are far away. Can be classified as a reflector, which uses a single or combination of curved mirrors to reflect light and form an image, or as a refractor, which has a convex objective lens at one end that gathers light and an eyepiece at the other that magnifies the image formed by the lens.

TIDE The regular upward and downward movement of the ocean, which is caused by the pull of the Sun and the Moon on Earth.

TRANSIT The passage of a smaller body (such as Venus) across the disk of a larger (such as the Sun).

TRIPOD A stool, table, or pedestal with three legs. Specific models are produced for telescope use to ensure steady, clear views of night-sky objects.

U

UMBRA The dark inner part of a shadow.

UNIVERSE All of the known or supposed objects throughout space.

URANUS The seventh planet from our Sun.

URSA MAJOR (THE GREAT BEAR) The most prominent northern constellation; contains the asterism the Big Dipper.

URSIDS A meteor shower originating in the constellation Ursa Minor around December 22.

V

VARIABLE STAR A star whose apparent magnitude (brightness as seen from Earth) fluctuates.

VEGA The fifth-brightest star in the sky and the brightest in the constellation Lyra, observable to viewers in the northern hemisphere.

VELA THE SAIL A southern constellation and one of the constellations into which Argo is divided.

VENUS The second planet from the Sun in our solar system and the second-brightest object in the night sky after the Moon.

VIRGO THE VIRGIN The second-largest constellation. Contains several bright stars, including Spica, and a dense cluster of galaxies.

VISIBLE LIGHT The portion of the electromagnetic spectrum that is visible to the human eye.

W

WHITE DWARF The small and hot, but intrinsically faint, stellar remnant left when a star loses its outer layers as a planetary nebula.

Z

ZENITH The highest point reached by a celestial body in the observed sky.

ZODIAC A circle of 12, 30-degree divisions of celestial longitude that are centered on the ecliptic.

ZODIACAL LIGHT A faint, diffuse glow seen in the sky that rises from near the Sun along the ecliptic or zodiac.

INDEX

A

B

C

D

INDEX

INDEX

N

O

P

Q

R

S

INDEX

T

U

V

W

Y

Z

PUBLISHER ACKNOWLEDGMENTS

Weldon Owen would like to thank Amy Bauman, Kevin Broccoli, Jan Hughes, Lisa Marietta, and Marisa Solís for editorial help and Kevin Gan Yuen for storyboarding and image-licensing assistance. We would also like to thank Margaret Berendsen and Sarah Scoles for their expert reviews of the content.

Further thanks to Frank Espinack, NASA's GSFC, for his eclipse predictions (#225); Charles P. Carlson and Dave Trott of the Denver Astronomical Society for their observation chair tutorial (#298); and Berislav Bracun for his DIY Dobsonian mount (#289).

AUTHOR ACKNOWLEDGMENTS

The Astronomical Society of the Pacific has been supporting astronomy enthusiasts for more than 125 years. This book includes the ideas and wisdom of literally hundreds of people who have been associated with the ASP during that time—too many to individually name. Yet the authors wish to gratefully acknowledge several people whose wisdom, time, talent, and leadership contributed to the activities in this book: Michael Bennett, Margaret Berendsen, Andrew Fraknoi, James Manning, Dennis Schatz, and Dan Zevin. The authors additionally wish to thank current ASP staff whose work is represented in these pages (Suzy Gurton, Anna Hurst, and Brian Kruse) and whose support made it possible for us to devote our time to writing this book (Noel Encarnacion, Kathryn Harper, Mannan Latif, Pablo Nelson, Leslie Proudfit, Greg Schultz, Michael Sowle, and Perry Tankeh). We are also deeply grateful for the leadership provided by the ASP Board of Directors and especially our past and current board presidents (Gordon Myers and Constance Walker) for their vision and leadership. Last, the ASP wishes to thank its many members and generous donors for their steadfast dedication to the society.

ILLUSTRATION CREDITS

Conor Buckley: 01–02, 18, 21, 25, 34–36, 38, 49, 61–64, 69–71, 73, 75–77, 87, 89, 91, 96, 98, 107, 125–126, 142–144, 150, 160, 163, 226, 230, 247, 289, 291, 298 Hayden Foell: 148–149, 172, 240, 249 Aaron John Gregory: 12–13, 29–32, 40–47, 52–59, 79–86, 112–121, 130–132, 137–140, 154–157, 186–189, 193–196, 203–206, 210–217, 225, 235–238, 261 Vic Kulihin: 179, 262, 294, 296 Liberum Donum: 17, 68, 78, 93, 103, 123, 185, 207, 229 Christine Meighan: 09–10, 22, 28, 101, 110, 151, 158–159, 169, 171, 174, 180, 221–222, 232, 248, 251, 263 Robert L. Prince: 127, 141, 147, 266, 271, 293 Bryon Thompson: 27 Lauren Towner: 08, 23–24, 128, 161, 165–166, 219, 223, 239, 265, 292

PHOTOGRAPHY CREDITS

All images courtesy of Shutterstock unless otherwise noted. Aeronet: 271 Miguel Araoz: 63 Aristarchus of Samos: 122 AstroZap: 181 Astronomical Society of the Pacific: 152 Rogelio Bernal Andreo: 200 Bob Atkins: 182 Jeff Barton: 199 Yuri Beletsky: p. 4, 14 Giuseppe Bertini: 122 Andrew Borde: 122 Bushnell: 146 Celestron: 146 Kevin Cho: 65 Corbis: 124 ESA/Hubble: 208 Explore Scientific USA: 144, 174 Getty Images: 100 Graham Green: 287 David Hayworth: 153 The Hubble Heritage Team (STScI/AURA): 109 Brocken Inaglory: p. 5, 67 Jim's Mobile, Inc. D/B/A JMI Telescopes: 178 Steven R. Majewski: 137 Shevill Mathers: 108 Meade: 183 Terry McKenna: 177 Scott D. McLeod: 88 Valetine Naboth: 122 Gautham Narayan: p. 7, 167 NASA: 04, 07, 11, 37, 95, 97, 133, 136, 264–265, 268 NASA, ESA and AURA/Caltech: 202 NASA/ESA, M. Robberto (Space Telescope Science Institute/ESA) and the Hubble Space Telescope Orion Treasury Project Team: 263 NASA Scientific Visualization Studio: 271 Barnaby Norris: 61–62, 64 Orion: 296 Wally Pacholka: 37 Matt Payne: 89 Planetary Society: 271 Lorraine Rath: 256 Gote Reber: 253 Joe Roberts: 162 Rolls Press: 274 Mike Salway: 39 SciencePhotoLibrary: 104, 105, 134, 275, p. 272 Science Source: pp. 2–3, pp. 8–9, 04, 15, 20, 33, 48, 102, 164, 184, 191–192, 197, 208–209, 227–228, 231, 241, 269–271, 277, 283, 295 Joshua Stevens: 271 Jimmy Thomas: 224 Mike Trenchard: 271 Chase Turner: 99 White Sand Missile Range/Applied Physics Laboratory: 274 Wikicommons: 122, 253, 274 Wikipedia: 208, 253 Hurst Wilson: 22 Bartolomeu Velho: 122 Frank Zulio: 21

weldon**owen**

PRESIDENT & PUBLISHER Roger Shaw
SVP, SALES & MARKETING Amy Kaneko
FINANCE MANAGER Phil Paulick

SENIOR EDITOR Lucie Parker
PROJECT EDITOR Nic Albert
EDITORIAL ASSISTANT Jaime Alfaro

CREATIVE DIRECTOR Kelly Booth
ART DIRECTOR Lorraine Rath
DESIGNER Allister Fein
ILLUSTRATION COORDINATOR Conor Buckley
SENIOR PRODUCTION DESIGNER Rachel Lopez Metzger

PRODUCTION DIRECTOR Chris Hemesath
ASSOCIATE PRODUCTION DIRECTOR Michelle Duggan
DIRECTOR OF ENTERPRISE SYSTEMS Shawn Macey
IMAGING MANAGER Don Hill

1045 Sansome Street, Suite 100
San Francisco, CA 94111
www.weldonowen.com

Weldon Owen is a division of

BONNIER

Library of Congress Cataloging-in-Publication data is available.

ISBN 13: 978-1-61628-871-6
ISBN 10: 1-61628-871-X

10 9 8 7 6 5 4 3 2 1
2015 2016 2017 2018 2019

Printed in China by 1010 Printing.

ABOUT THE ASTRONOMICAL SOCIETY OF THE PACIFIC

On a chilly February evening in 1889 in San Francisco, astronomers from the Lick Observatory and members of the Pacific Coast Amateur Photographic Association—fresh from viewing the New Year's Day total solar eclipse north of the city—met to share pictures and experiences. Edward Holden, Lick Observatory's first director, complimented the amateurs on their service to science and proposed to continue the good fellowship through the founding of a society to advance the science of astronomy. The Astronomical Society of the Pacific (ASP) was born.

Through more than a century of operation, as human understanding of the universe has advanced, so has the ASP—connecting scientists, educators, amateur astronomers, and the public to share astronomical research, conduct professional development in science education, and provide resources that engage students and adults alike in the adventure of scientific discovery.

As a nonprofit membership organization, international in scope, the ASP's mission is to advance science and science literacy through the excitement of astronomy. We do this by directly engaging astronomy enthusiasts of every kind and connecting them to resources, tools, and programs. We invite you to check out our website (www.astrosociety.org) to learn more about us, get involved by becoming an ASP member, and consider donating to our mission.